Textile-Based Advanced Materials

Textile-Based Advanced Materials

Construction, Properties and Applications

Editors

Thomas Bechtold
Avinash P. Manian

MDPI • Basel • Beijing • Wuhan • Barcelona • Belgrade • Manchester • Tokyo • Cluj • Tianjin

Editors
Thomas Bechtold
Universität Innsbruck,
Research Institute for Textile
Chemistry and Textile Physics
Austria

Avinash P. Manian
Universität Innsbruck,
Research Institute for Textile
Chemistry and Textile Physics
Austria

Editorial Office
MDPI
St. Alban-Anlage 66
4052 Basel, Switzerland

This is a reprint of articles from the Special Issue published online in the open access journal *Materials* (ISSN 1996-1944) (available at: https://www.mdpi.com/journal/materials/special_issues/Textile-Based_Materials).

For citation purposes, cite each article independently as indicated on the article page online and as indicated below:

LastName, A.A.; LastName, B.B.; LastName, C.C. Article Title. *Journal Name* **Year**, *Volume Number*, Page Range.

ISBN 978-3-0365-0420-9 (Hbk)
ISBN 978-3-0365-0421-6 (PDF)

© 2021 by the authors. Articles in this book are Open Access and distributed under the Creative Commons Attribution (CC BY) license, which allows users to download, copy and build upon published articles, as long as the author and publisher are properly credited, which ensures maximum dissemination and a wider impact of our publications.

The book as a whole is distributed by MDPI under the terms and conditions of the Creative Commons license CC BY-NC-ND.

Contents

About the Editors vii

Avinash P. Manian and Thomas Bechtold
Special Issue "Textile-Based Advanced Materials: Construction, Properties and Applications"
Reprinted from: *Materials* **2020**, *13*, 5766, doi:10.3390/ma13245766 1

Haleema Saleem and Syed Javaid Zaidi
Sustainable Use of Nanomaterials in Textiles and Their Environmental Impact
Reprinted from: *Materials* **2020**, *13*, 5134, doi:10.3390/ma13225134 5

Waleri Root, Thomas Bechtold and Tung Pham
Textile-Integrated Thermocouples for Temperature Measurement
Reprinted from: *Materials* **2020**, *13*, 626, doi:10.3390/ma13030626 33

Christine Ann Spencer, Leon van Paassen and Henrik Sass
Effect of Jute Fibres on the Process of MICP and Properties of Biocemented Sand
Reprinted from: *Materials* **2020**, *13*, 5429, doi:10.3390/ma13235429 55

Dayong Hu, Linwei Dang, Chong Zhang and Zhiqiang Zhang
Mechanical Behaviors of Flax Fiber-Reinforced Composites at Different Strain Rates and Rate-Dependent Constitutive Model
Reprinted from: *Materials* **2019**, *12*, 854, doi:10.3390/ma12060854 79

Fen Zhou, Huanhui Liu, Yunxing Du, Lingling Liu, Deju Zhu and Wei Pan
Uniaxial Tensile Behavior of Carbon Textile Reinforced Mortar
Reprinted from: *Materials* **2019**, *12*, 374, doi:10.3390/ma12030374 97

Stefan Rothe, Ellen Wendt, Sybille Krzywinski, Marianna Halász, Peter Bakonyi, Peter Tamás and Attila Bojtos
Investigation of Shear-Induced Deformation of Reinforcing Textiles by Optical Measurement Devices
Reprinted from: *Materials* **2019**, *12*, 1029, doi:10.3390/ma12071029 117

Julia Ullrich, Martin Eisenreich, Yvonne Zimmermann, Dominik Mayer, Nina Koehne, Jacqueline F. Tschannett, Amalid Mahmud-Ali and Thomas Bechtold
Piezo-Sensitive Fabrics from Carbon Black Containing Conductive Cellulose Fibres for Flexible Pressure Sensors
Reprinted from: *Materials* **2020**, *13*, 5150, doi:10.3390/ma13225150 135

Isidoro Ibanez-Labiano and Akram Alomainy
Dielectric Characterization of Non-Conductive Fabrics for Temperature Sensing through Resonating Antenna Structures
Reprinted from: *Materials* **2020**, *13*, 1271, doi:10.3390/ma13061271 149

Jae Seon Kim and Chung Kun Song
Textile Display with AMOLED Using a Stacked-Pixel Structure on a Polyethylene Terephthalate Fabric Substrate
Reprinted from: *Materials* **2019**, *12*, 2000, doi:10.3390/ma12122000 163

Kathleen A. Montoya-Villegas, Alejandro Ramírez-Jiménez, Ángel Licea-Claverie, Sergio Pérez-Sicairos, Emilio Bucio, Johanna Bernáldez-Sarabia and Alexei F. Licea-Navarro
Surface Modification of Polyester-Fabric with Hydrogels and Silver Nanoparticles: Photochemical Versus Gamma Irradiation Methods
Reprinted from: *Materials* **2019**, *12*, 3284, doi:10.3390/ma12203284 175

Mari Pekkanen-Mattila, Martta Häkli, Risto-Pekka Pölönen, Tuomas Mansikkala, Anni Junnila, Elina Talvitie, Janne T Koivisto, Minna Kellomäki and Katriina Aalto-Setälä
Polyethylene Terephthalate Textiles Enhance the Structural Maturation of Human Induced Pluripotent Stem Cell-Derived Cardiomyocytes
Reprinted from: *Materials* **2019**, *12*, 1805, doi:10.3390/ma12111805 193

Iwona Karbownik, Olga Rac-Rumijowska, Marta Fiedot-Toboła, Tomasz Rybicki and Helena Teterycz
The Preparation and Characterization of Polyacrylonitrile-Polyaniline (PAN/PANI) Fibers
Reprinted from: *Materials* **2019**, *12*, 664, doi:10.3390/ma12040664 209

Dongyan Shao, Changhai Xu, Hongbo Wang and Jinmei Du
Enhancing the Dyeability of Polyimide Fibers with the Assistance of Swelling Agents
Reprinted from: *Materials* **2019**, *12*, 347, doi:10.3390/ma12030347 229

About the Editors

Thomas Bechtold is the former head of the Research Institute of Textile Chemistry and Textile Physics. His research career spans over 30 years, and his interests include the application of electrochemistry in textile processing, textile dyeing with natural colorants, cellulose fiber structure and chemistry, and the technical application of textiles. He has published over 230 journal articles, book chapters, and patents, and has edited one book.

Avinash P. Manian obtained his Bachelor's and Master's degrees in Textile Chemistry from the Institute of Chemical Technology (Mumbai, India) and his PhD from the University of Georgia (USA), before joining the Research Institute of Textile Chemistry and Textile Physics in 2003. His research interests include textile coloration, cellulose fiber structure and chemistry, and the technical application of textiles. He has published over 45 journal articles, patents, and book chapters.

Editorial

Special Issue "Textile-Based Advanced Materials: Construction, Properties and Applications"

Avinash P. Manian * and Thomas Bechtold *

Research Institute of Textile Chemistry/Physics, University of Innsbruck, Hoechsterstrasse 73, 6850 Dornbirn, Austria
* Correspondence: avinash.manian@uibk.ac.at (A.P.M.); thomas.bechtold@uibk.ac.at (T.B.); Tel.: +43-5572-28533 (A.P.M. & T.B.)

Received: 16 December 2020; Accepted: 16 December 2020; Published: 17 December 2020

Developments in the science and technology of textiles is not only limited to apparel and fashion. Certainly, there are research efforts on improving the construction and processing of textiles for clothing–such as on cleaner production to reduce environmental impact, increasing the utilization of fibers and process chemicals from renewable resources, and on the recycling of materials from post-consumer waste apparel back into the manufacture of new clothing articles. In addition, technological concepts developed for the creation of clothing over centuries, are now being investigated for use in a diverse array of fields—such as in the manufacture of engineering composites, personal protective equipment, and medicine. Further, developments in other fields–such as electronics, nanotechnology, and information and communication technologies–are being investigated for their incorporation into apparel and clothing to create "smart textiles". The aim of this Special Issue was to put together a collection of scientific reports on such efforts, to highlight the range of scientific and technological questions that are being targeted, and the ingenuity of the methodologies employed to find answers. It is hoped that readers of this issue will come away with an appreciation of the research being conducted in this area, and perhaps gain inspiration for their own scientific endeavors.

The issue contains eleven research articles, on composites, geotechnical engineering, medical applications, high-performance polymers, and the development and integration of sensors in textile articles. Furthermore, there are two review articles–one on the state of technology in incorporation of nanoparticle into apparel and clothing textiles including potential safety and toxicological concerns, and the other on the integration of temperature sensors in apparel and clothing.

Spencer et al. [1] studied microbially induced calcium carbonate precipitation (MICP) in silica sand, and what effect the incorporation of jute fibers into the sand may have on the process. MICP or bio-cementation is a process where non-pathogenic bacteria are employed to catalyze the precipitation of calcium from ground water as its carbonate salt between sand particles, so that the precipitates bind the sand and increase its strength. They found that the efficiency of calcium conversion to $CaCO_3$ increased in the presence of jute, which they attributed to a possible positive influence of the fibers on fixation and viability of the microbes. The uniaxial (or unconfined) compressive strength of the jute-sand mixtures were greater than those of sand alone, but it was difficult to ascertain the contributions arising from the fibers alone compared to the greater contents of precipitated $CaCO_3$. A further goal was to investigate whether the fibers would retain viable bacterial cultures and thereby exert a 'self-healing' effect through the facilitation of further $CaCO_3$ precipitation to fill in cracks that developed over time. However, the results proved inconclusive.

In a similar vein, Zhou et al. [2] investigated woven carbon mesh as a textile reinforcement for constructions of mortar on their uniaxial tensile behavior, and also studied the effect of dispersing 12–15 mm long copper-coated steel fibers in the mix as additional reinforcement. The tensile strength increased with the volume fraction of the textile reinforcement, but at high reinforcement levels, the cracking propensity of constructions increased, which indicated a heightened risk of delamination between the mortar and the reinforcement. Applying a pre-stress to the carbon mesh on integration into the mortar mix increased the load at which the crack first appeared in tests, but the dispersion of short length steel fibers both reduced the length and width of cracks and increased the tensile strength significantly. That suggests that the short-length fibers acted to improve reinforcement-matrix bonding and reduced the delamination risks.

Two papers in the issue deal with modeling. One is on the in-plane shear deformation behavior of textile fabric reinforcements, which if predicted with high reliability, can significantly improve the efficiency in the design and construction of composites. The aspect dealt with by Rothe et al. [3] was the critical point at which shear-induced folds occur in extension tests. In normal course, that point is estimated from inflections in load-extension profiles but that is not always valid, as visual observations have shown folds to form before an inflection occurs. The authors report on the development of a test apparatus and laser measurement system that can detect and quantify the appearance of folds in extension tests, and the measured parameters can be fed into software, also developed by the authors, to more accurately characterize the load-bearing and deformation characteristics of the reinforcements. The other is on the compression behavior of vacuum-assisted resin-impregnated composites of multi-layered woven flax fiber composites, by Hu et al. [4]. Flax shows promise as a bio-based alternative to synthetic materials as reinforcement in lightweight composites, and work such as theirs on understanding the dynamic mechanical behavior of flax-reinforced composites will contribute to an increase in confidence on their use in engineering applications.

A way to improve the biocompatibility of polyethylene terephthalate (PET) for medical applications was investigated by Montoya-Villegas et al. [5] who studied the grafting of PET fabrics with 2-hydroxyethyl methacrylate (HEMA) and co-polymer of HEMA and polyethylene glycol methacrylate under irradiation with UV or gamma-ray. The resulting substrates were then loaded with silver nanoparticles and then tested for their antimicrobial activity against gram positive and gram negative bacteria. The grafting under UV irradiation was found to yield a homogeneous, thin coating of the graft polymer whereas far thicker coatings were obtained under gamma-ray. Silver nanoparticles could be synthesized in situ on the grafted substrates from both treatments, and the resulting materials exhibited significant activity against both microorganisms. In tandem, one may be interested to read the review by Saleem et al. [6] on the application of nanomaterials in the textiles sector for end-uses ranging from improving comfort and aesthetics to UV protection and anti-static finishing. The authors also summarize the state of knowledge on the potential risks of nanomaterials to both humans and the environment.

A second investigation on application of PET fabrics in medicine is reported by Pekkanen-Mattila et al. [7], who studied their use as scaffolds for the growth of cells of the human heart muscle (cardiomyocytes). The hypothesis was that the three-dimensional structure of the woven substrates could help in the proliferation, differentiation and maturation levels of the cells. A variety of weave patterns were investigated, and all were found to support cell growth, but no one pattern appeared to present an advantage over another. The culturing of these cells on fabrics improved their structural properties, and it is hoped that the results will be beneficial in the development of materials to obtain reliable cell cultures for toxicology, drug screening and disease modeling investigations.

Ullrich et al. [8] report on investigations of conductive viscose fibers produced by the incorporation of carbon black, and their use as pressure sensors when converted to needle-punched webs together with polyester fibers. The conductivity of fibers increased with the addition of carbon black, and when these fibers were assembled into webs together with virgin polyesters, they exhibited a variation in resistance as a function of pressure with a relationship that did not change significantly over multiple pressure cycles. The color of the fibers, black, will limit their application in visible areas of apparel and additional work is required to develop means of protecting these fibers against mechanical damage and atmospheric moisture. However, these results are a source of optimism for the authors that such constructions could be employed in applications such as pressure pads inside bandages. Another investigation on imparting conductivity to textile fibers was reported by Karbownik et al. [9]

who investigated the effects of incorporating polyaniline in polyacrylonitrile. Two modes were attempted: the addition of preformed polyaniline to the polyacrylonitrile spin dope with, and in situ synthesis of polyaniline in the polyacrylonitrile spin dope. The simple mixing of preformed polyaniline produced fibers exhibiting good mechanical strength, but the fibers behaved as dielectrics, and significant levels of polyaniline could not be detected. The in situ synthesis led to fibers exhibiting poor mechanical strength but high electrical conductivity, and high levels of polyaniline could be detected in the fibers. The authors recognize that further work is required to improve the process but hope that ultimately, such fibers may be used to integrate conductive lines in textiles.

Ibanez-Labiano et al. [10] report on their investigations into measurements of the dielectric properties of cotton textiles as a function of temperature, which according to the authors has not been reported previously. They employed two methods, a resonant perturbation and resonator method, and found the dielectric constant to change linearly with the temperature up to 50 °C. An innovative experimental design created for their experiments, a microstrip patch antenna, exhibits potential for use as a passive temperature sensor that can be integrated into textile structures.

Kim et al. [11] report on their work to create light emitting diode-based textile displays with polyethylene terephthalate substrates. They describe their efforts to enhance the aperture ratio, i.e., the area ratio of a light emitting diode to the total pixel area. That was achieved through a combination of reducing roughness of the substrate by depositing alternating layers of polyurethane and acryl polymers, creating a protective layer with a combination of polyvinyl alcohol (neat and derivatized), and improving conductivity through the use of carbon nanotubes.

High performance fibers often are reported for their poor dyeability that may act to limit their application, and Shao et al. [12] report on their work to improve the dyeability of polyimide fibers with the assistance of swelling agents. The addition of acetophenone, N-methyl formamide or phenoxyisopropanol to dyebaths significantly improved the dye uptake without significant evidence of volume swelling in the fibers, and the fibers exhibited good fastness to washing.

Root et al. [13] present a review of the literature on the integration of thermocouples into textile substrates as temperature sensors, and compare methods, such as the screen printing of conductive polymers, soldering and sputter deposition of metals, the use of conductive glues and interweaving of metal wires. They also highlight the work with electroless metal deposition to achieve a conductive coating on fabric surfaces. They compare different sensing principles, such as resistance temperature detector, fiber Bragg grating, etc. Finally, they discuss the challenges that exist in such technologies, related primarily to the robustness of the integrated materials to the environment (pH, light and mechanical forces) as well as the need to maintain maximum comfort to the wearer.

The guest editors thank all authors for their valuable contributions and are grateful for the efforts they undertook to submit their contributions. Thanks also go to the reviewers who contributed with insightful and constructive comments that helped improve the overall quality of the presented work, and to the editorial staff at the publishers who ably supported the authors, reviewers and us through the whole process.

Funding: This research received no external funding.

Conflicts of Interest: The authors declare no conflict of interest.

References

1. Spencer, C.A.; Van Paassen, L.; Sass, H. Effect of Jute Fibres on the Process of MICP and Properties of Biocemented Sand. *Materials* **2020**, *13*, 5429. [CrossRef] [PubMed]
2. Zhou, F.; Liu, H.; Du, Y.; Liu, L.; Zhu, D.; Pan, W. Uniaxial Tensile Behavior of Carbon Textile Reinforced Mortar. *Materials* **2019**, *12*, 374. [CrossRef] [PubMed]
3. Rothe, S.; Wendt, E.; Krzywinski, S.; Halász, M.; Bakonyi, P.; Tamás, P.; Bojtos, A. Investigation of Shear-Induced Deformation of Reinforcing Textiles by Optical Measurement Devices. *Materials* **2019**, *12*, 1029. [CrossRef] [PubMed]

4. Hu, D.; Dang, L.; Zhang, C.; Zhang, Z. Mechanical Behaviors of Flax Fiber-Reinforced Composites at Different Strain Rates and Rate-Dependent Constitutive Model. *Materials* **2019**, *12*, 854. [CrossRef] [PubMed]
5. Montoya-Villegas, K.A.; Ramírez-Jiménez, A.; Licea-Claverie, A.; Sicairos, S.P.; Bucio, E.; Bernáldez-Sarabia, J.; Licea-Navarro, A. Surface Modification of Polyester-Fabric with Hydrogels and Silver Nanoparticles: Photochemical Versus Gamma Irradiation Methods. *Materials* **2019**, *12*, 3284. [CrossRef]
6. Saleem, H.; Zaidi, S.J. Sustainable Use of Nanomaterials in Textiles and Their Environmental Impact. *Materials* **2020**, *13*, 5134. [CrossRef]
7. Pekkanen-Mattila, M.; Häkli, M.; Pölönen, R.P.; Mansikkala, T.; Junnila, A.; Talvitie, E.; Koivisto, J.T.; Kellomäki, M.; Aalto-Setälä, K. Polyethylene Terephthalate Textiles Enhance the Structural Maturation of Human Induced Pluripotent Stem Cell-Derived Cardiomyocytes. *Materials* **2019**, *12*, 1805. [CrossRef] [PubMed]
8. Ullrich, J.; Eisenreich, M.; Zimmermann, Y.; Mayer, D.; Koehne, N.; Tschannett, J.F.; Mahmud-Ali, A.; Bechtold, T. Piezo-Sensitive Fabrics from Carbon Black Containing Conductive Cellulose Fibres for Flexible Pressure Sensors. *Materials* **2020**, *13*, 5150. [CrossRef]
9. Karbownik, I.; Rac-Rumijowska, O.; Fiedot-Toboła, M.; Rybicki, T.; Teterycz, H. The Preparation and Characterization of Polyacrylonitrile-Polyaniline (PAN/PANI) Fibers. *Materials* **2019**, *12*, 664. [CrossRef] [PubMed]
10. Ibanez-Labiano, I.; Alomainy, A. Dielectric Characterization of Non-Conductive Fabrics for Temperature Sensing through Resonating Antenna Structures. *Materials* **2020**, *13*, 1271. [CrossRef] [PubMed]
11. Kim, J.S.; Song, C.-K. Textile Display with AMOLED Using a Stacked-Pixel Structure on a Polyethylene Terephthalate Fabric Substrate. *Materials* **2019**, *12*, 2000. [CrossRef] [PubMed]
12. Shao, D.; Xu, C.; Wang, H.; Du, J. Enhancing the Dyeability of Polyimide Fibers with the Assistance of Swelling Agents. *Materials* **2019**, *12*, 347. [CrossRef] [PubMed]
13. Root, W.; Bechtold, T.; Pham, T. Textile-Integrated Thermocouples for Temperature Measurement. *Materials* **2020**, *13*, 626. [CrossRef] [PubMed]

Publisher's Note: MDPI stays neutral with regard to jurisdictional claims in published maps and institutional affiliations.

© 2020 by the authors. Licensee MDPI, Basel, Switzerland. This article is an open access article distributed under the terms and conditions of the Creative Commons Attribution (CC BY) license (http://creativecommons.org/licenses/by/4.0/).

Review

Sustainable Use of Nanomaterials in Textiles and Their Environmental Impact

Haleema Saleem and Syed Javaid Zaidi *

Center for Advanced Materials (CAM), Qatar University, Doha 2713, Qatar; haleema.saleem@qu.edu.qa
* Correspondence: szaidi@qu.edu.qa; Tel.: +974-4403-7723

Received: 15 October 2020; Accepted: 9 November 2020; Published: 13 November 2020

Abstract: At present, nanotechnology is a priority in research in several nations due to its massive capability and financial impact. However, due to the uncertainties and abnormalities in shape, size, and chemical compositions, the existence of certain nanomaterials may lead to dangerous effects on the human health and environment. The present review includes the different advanced applications of nanomaterials in textiles industries, as well as their associated environmental and health risks. The four main textile industry fields using nanomaterials, nanofinishing, nanocoatings, nanofibers, and nanocomposites, are analyzed. Different functional textiles with nanomaterials are also briefly reviewed. Most textile materials are in direct and prolonged contact with our skin. Hence, the influence of carcinogenic and toxic substances that are available in textiles must be comprehensively examined. Proper recognition of the conceivable benefits and accidental hazards of nanomaterials to our surroundings is significant for pursuing its development in the forthcoming years. The conclusions of the current paper are anticipated to increase awareness on the possible influence of nanomaterial-containing textile wastes and the significance of better regulations in regards to the ultimate disposal of these wastes.

Keywords: nanomaterials; environmental impacts; textiles; toxicity; health and safety

1. Introduction

The idea of nanotechnology was initially presented by Richard Feynman in the year 1959, through his speech "There's Plenty of Room at the Bottom", which was delivered at an American Physical Society conference, at the California Institute of Technology (Feynman, 1959, [1]). The concepts introduced by Feynman were unobserved until 1974, when Norio Taniguchi presented the term "nanotechnology" (Taniguchi, 1974, [2]). The term "nano" means one-billionth or 10^{-9} henceforth, one nanometer is referred to as one billionth of a meter. Currently, an extensive range of fabrication systems are present that are capable of controlling and producing nanostructures to the preferred composition, size, morphology, shape, and crystalline structure. The two typical fabrication methods utilized are "top-down" and "bottom-up". These days, nanotechnology contributes to the prospective opportunities in developing improved materials with advanced properties for utilization in different application fields. The atoms inside nanoparticles are perfectly ordered and consequently, while the material dimensions change from macro-size to nano-size, extensive variations happen in the material properties (Yang et al., 2019, [3]).

At the present time, researchers can develop various nano-sized materials such as nanoclays, carbon nanotubes, nanofibers, and graphene with lighter, stronger, increased chemical reactivity, and more prolonged control on the light spectrum (Khan et al., 2019, [4]). An improved understanding of the properties of nanomaterials provides a way for developing progressive materials in the upcoming years with the probability for improving the life quality. Nanomaterials are gradually turning out to be commercialized, starting to progress as commodities, and used in numerous advanced technological applications and products, including a wide variety of consumer products. The design as well as preparation of nanomaterials with a unique combination of textile material is anticipated to expand the demanding scope in the future (Verma et al., 2020, [5]).

At present, the engineered nanomaterials are being examined extensively by research institutes as well as industries for improving prevailing functions in products together with implementing new ones. Regardless of such developments in nanomaterial technology, data regarding the probable effects of nanomaterials on human health and the environment has been inadequate until now (Kumar et al., 2018, [6]) (Mishra et al., 2018, [7]), (Kumar et al., 2018, [8]). Nanosafety is an increasing concern as exposure to engineered nanomaterials has been related to several health effects inclusive of carcinogenicity, genotoxicity, pulmonary inflammation, and circulatory effects (Leong, 2017, [9]), (Johnston et al., 2020, [10]), (Mirshafiee et al., 2018, [11]), (Karim et al., 2020, [12]), (Dobrovolskaia et al., 2013, [13]). Due to the fact that the nanomaterials may not be recognizable subsequent to its discharge into the surroundings, these materials could cause various kinds of ecological problems as long as the remediation scheme is unsafe. The data acquisitions on emission as well as ecological concentrations of nanomaterials of the nanotextiles is extremely important. Subsequently, additional study is crucial for scientifically describing the structure-function relation of nanomaterials with respect to the fundamental chemistry (as an illustration, functionality and toxicity). Additionally, comprehensive risk assessments should be performed on nanomaterials that present an actual exposure danger throughout its fabrication or usage (Ahmad et al., 2020, [14]), (Kawai et al., 2019, [15]), (Auffan et al., 2019, [16]), (Oomen et al., 2018, [17]), (Schulte et al., 2018, [18]). Henceforward, green nanoscience has been suggested for reducing the probable environmental threats and human health risks from the fabrication and usage of nanomaterials and to develop the substitution of prevalent items with progressive nanomaterials that are more eco-friendly (Iavicoli et al., 2014, [19]), (McKenzie et al., 2004, [20]), (Hutchison et al., 2008, [21]), (Bamoharram et al., 2011, [22]). In this review paper, we discuss the application of nanomaterials in the textile industries. These sorts of studies may be advantageous for the suitable advancement of applications and research interest towards the further development of nanomaterials. To the best of our knowledge, there are not many works about the state-of-the-art progress in nanotechnology for application in textile industries. This review paper highlights the sustainable use of nanomaterials in textiles, their release from textiles, and the different approaches for examining nanomaterial toxicity. In addition, we have comprehensively studied the hazardous effects of textile field nanomaterials on human health and the environment.

2. Application of Nanomaterials in Textile Industry

Presently, the textile industry is a significant user of nanotechnology and there are a remarkable number of nanotextiles present in the market, inclusive of several consumer goods, which includes nanomaterials (Karst, D et al., 2006, [23]), (Jatoi et al., 2021, [24]), (Darwesh et al., 2021, [25]), (Schoden et al., 2021, [26]), (Yilmaz, 2018, [27]), (Ehrman et al., 2020, [28]), (Abdullaeva, 2017, [29]), (Riaz et al., 2019, [30]). Nanotextiles are regarded as conventional textiles with the inclusion of nanomaterials. These advanced textiles offer different functionalities like flame retardancy, self-cleaning, dirt repellency, water repellency, ultraviolet radiation protection, or antibacterial property (Almeida et al., 2017, [31]), (Brown et al., 2007, [32]), (Radetic et al., 2013, [33]), (Ibrahim et al., 2015, [34]), (Sundarrajan et al., 2010, [35]), (Afzali, A. et al., 2016, [36]), (Sharon et al., 2019, [37]), (El-Naggar et al., 2018, [38]), (Xue et al., 2020, [39]), (Gadkari et al., 2020, [40]), (Elsayed et al., 2020, [41]), (Mejia et al., 2017, [42]). Nanocoatings and nanofinishings are enhancing the possible utilizations of textile materials in different fields (Banerjee et al., 2019, [43]), (Jadoun et al., 2020, [44]), (Gokarneshan et al., 2017, [45]), (Perera et al., 2013, [46]), (Ferraris et al., 2014, [47]). The usage of nanofibers and nanocomposite based coatings/finishings have demonstrated a huge possibility in emerging functional and high-performance textiles (Bashari et al., 2018, [48]), (Riaz et al., 2018, [49]), (Haque et al., 2019, [50]), (Lund et al., 2018, [51]), (Silva et al., 2019, [52]), (Shabbir et al., 2020, [53]), (Ul-Islam et al., 2018, [54]). The study by (Singh et al., 2020 [55]) reviewed the latest studies involving the modification and characterization of textile, highlighting plasma and nano-pretreatment. Figure 1 diagrammatically presents various nanotechnology-enhanced textiles. Due to its higher surface area to volume ratio and nanoscale dimensions, the nanomaterials have increased potential for providing different functionalities in the textiles. Various nanomaterials that are utilized for textile utilization are mostly: (1) Carbon-based nanomaterials such as graphene, carbon nanofibers, and carbon nanotubes; (2) inorganic nanoparticles such as metal oxide, metal, and nanoclay; (3) core-shell nanoparticles; (4) composite nanomaterials; (5) hybrid nanomaterials; and (6) polymeric nanomaterials. Table 1 provides information on the different nanomaterials most frequently utilized for functionalization in textiles.

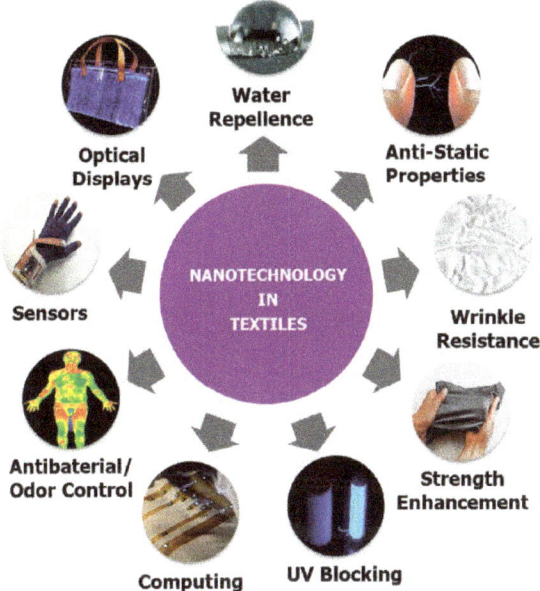

Figure 1. A diagrammatic representation of various utilizations of nanotechnology-based textiles. Reproduced from reference (Yetisen et al., 2016, [56]).

2.1. Innovations in Nanotechnology-Based Textile Industry Applications

The four main fields through which nanomaterials and nanotechnology find utilizations in the textile industry are analyzed below.

2.1.1. Nanofinishing

Nanofinishing is the process in which the colloidal solution or ultrafine dispersion of nanomaterials is applied to a textile material for enhancing some of the functionalities (Joshi et al., 2018, [57]), (Haji et al., 2016, [58]), (Ghosh et al., 2018, [59]), (Radetic et al., 2019, [60]). It has certain benefits over traditional finishing, which are principally: (i) In general, in the case of nanofinishing, it only requires a lesser quantity of nanomaterials relative to bulk materials utilized in traditional finishing, for obtaining a similar effect. (ii) They do not influence the aesthetic feel of the textile materials. (iii) These nanofinishings are more durable due to the increased surface area-to-volume ratio of nanomaterials along with its homogenous distribution in textile material. (iv) Certain functionalities that are hard to attain by traditional finishes could possibly be developed by nanofinishing (Gokarneshan et al., 2018, [61]), (Ghosh et al., 2020, [62]). For almost two decades, the nano-finishing of cellulose textile material while using copper and copper oxide nanoparticles has been in the focus of science and textile industries (Radetic et al., 2019, [63]).

2.1.2. Nanocoating

In a nanocoating process, a thin layer approximately less than 100 nm thickness is deposited on the substrate in order to improve certain properties or for contributing advanced functionality. Traditional coatings possess certain disadvantages such as (i) less durability, (ii) poor abrasion resistance, (iii) strength loss, (iv) less flexibility, and (v) improper adhesion between the substrate and coating layer (Nguyen et al., 2018, [64]). The aforementioned problems of traditional coatings could be solved by the utilization of nanocoatings (Joshi et al., 2011, [65]), (Smole et al., 2006, [66]), (Peng et al., 2019, [67]). The nanomaterial coating on fabrics will not influence their breathability or hand-feel (Temesgen et al., 2018, [68]).

2.1.3. Nanofibers

For the fabrication of nanofibers, different techniques are used, like electrospinning (Figure 2), self-assembly, force spinning, melt blowing, and island-in-sea (bicomponent nanofiber) (Almetwally et al., 2017, [69]), (Nayak et al., 2019, [70]), (Naeem et al., 2019, [71]). Out of these techniques, the electrospinning is considered the most convenient one due to its low cost, higher rate of production, higher porosity, and ability to control nanofiber morphology and diameter. A stretchable piezo-resistive carbon nanotube-incorporated nanofiber sensing yarn (Figure 3) was first designed as well as prepared by a facile electrospinning technique by (Qi et al., 2020, [72]). Moreover, the nanofibers are unique with high capability as active layers in face masks, to protect people against diseases such as coronavirus (Tebyetekerwa et al., 2020, [73]).

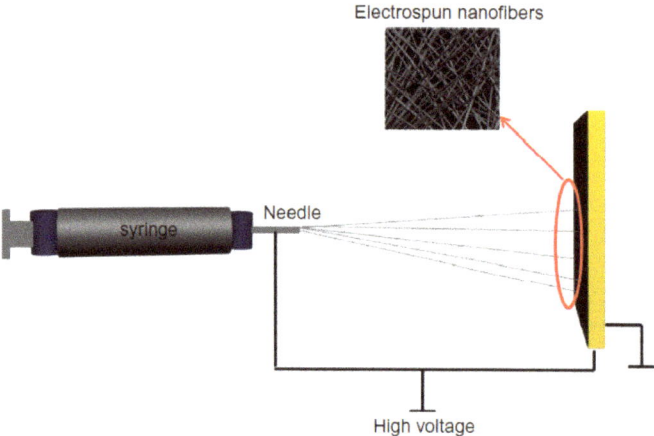

Figure 2. Electrospinning design. Reproduced from (Montazer et al., 2018, [74]).

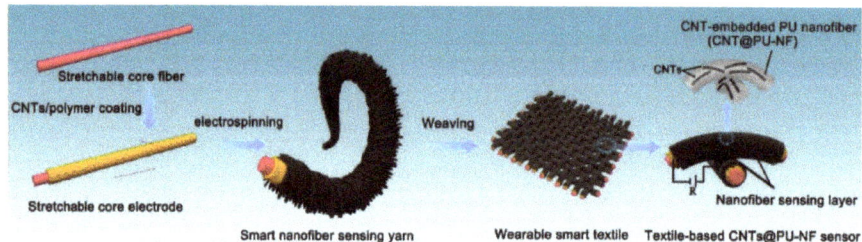

Figure 3. Schematic of the fabrication processes of carbon nanotube-incorporated nanofiber sensing yarn. Reproduced from (Qi et al., 2020, [72]).

2.1.4. Nanocomposites

Nanocomposite is a multiphase solid material in which the minimum one dimension of the reinforcing phase is in nano-level (Malhotra et al., 2017, [75]). In the case of polymer-based nanocomposites, nanomaterials are dispersed in polymer matrices. The polymer nanocomposite-based coatings and fibers have enormous possibility in the production of functional as well as superior-performance textiles. Novel nanocomposite materials have been developed using facile one pot method. In a study by (Attia et al., 2016 [76]), the team developed nanocomposites based on silver nanoparticles as well as diphosphate malonate (DPHM) as organic phosphates (Figure 4). The mass ratios of both organic phosphate and silver nanoparticles in the nanocomposites were varied as well as optimized. Various textile fabrics have been treated with the nanocomposites developed to improve their fire retardancy and antibacterial properties. It was found that the flame retardancy of the treated textiles increased significantly.

Table 1. Utilization of nanomaterials in textile functionalization.

Sl. No.	Nanomaterial	Function	Reference
1	Nanoclays	Active ingredient support, flame retardance, abrasion resistance	(Gocek et al., 2019, [77])
2	Aluminium oxide	flame retardance, abrasion resistance	(Korkmaz et al., 2016, [78])
3	Silicon dioxide	Reinforcement enhanced the dyeability, abrasion resistance, water repellence, dirt repellence	(Dogan et al., 2017, [79])
4	Zinc oxide	Stiffness, abrasion resistance, self-cleaning, antibacterial property and UV protection	(Verbic et al., 2019, [80])
5	Titanium dioxide	Water repellence, dirt repellence, self-cleaning, UV protection	(Abbas et al., 2018, [81])
6	Silver	Electrically conductive, antibacterial property	(Xu et al., 2017, [82])

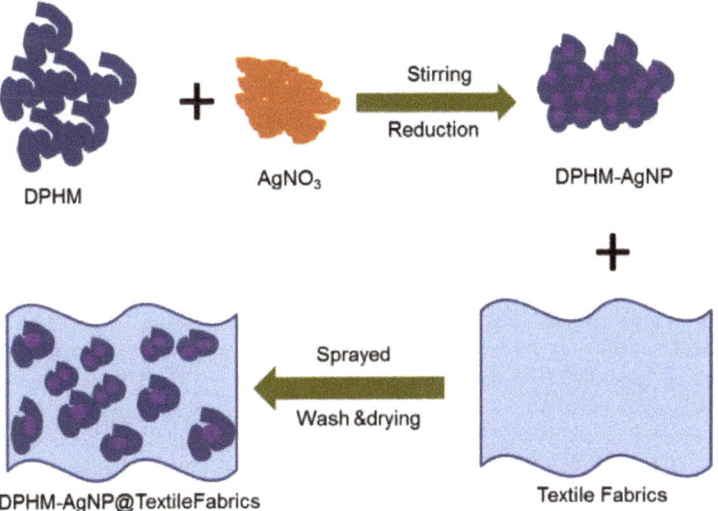

Figure 4. Schematic diagram showing the synthesis of diphosphate malonate-silver nanoparticle (DPHM-AgNP) nanocomposites and their treatment on textile fabrics. DPHM: Diphosphate malonate, $AgNO_3$-Silver nitrate. Reproduced from (Attia et al., 2016, [76]).

2.2. Functional Textiles with Nanomaterials

In the following section, we discuss nanomaterial included different functional textiles, such as flame-retardant textiles, UV Protective textiles, antimicrobial and antibacterial textiles, water and oil repellent textiles, anti-odor textiles, wrinkle resistance textiles, and antistatic textiles. The study by (Vigneshwaran et al., 2020 [83]) demonstrated the fundamental mechanisms involved in the usage of nanomaterials for contributing the functional properties in cotton textiles, applicable issues as well as future scope for commercial utilization.

2.2.1. Flame-Retardant Textiles

Material scientists are planning to replace traditional flame-retardant systems with eco-friendly substitutes like the utilization of nanotechnology or flame retardant nano-additives (Attia et al., 2016, [76]), (Erdem et al., 2009, [84]), (Attia et al., 2017, [85]), (Shariatinia et al., 2015, [86]), (Kundu et al., 2020, [87]), (Rivero et al., 2015, [88]), (Ortelli et al., 2019, [89]), (Ortelli et al., 2018, [90]), (Carosio et al., 2012, [91]), (Saleemi et al., 2020, [92]), (Ali et al., 2020, [93]), (Butola et al., 2020, [94], (Li et al., 2018, [95], (Sharma et al., 2018, [96]). (Norouzi et al., 2015 [97]) studied the influence of nanoparticles along with standard flame retardants on the flame retardation of several textile polymers. The results demonstrated that a majority of the nanoparticles could enhance the flame retardant properties and thermal stability of the textile polymers. The foremost mechanisms involved are the development of a shielding barrier layer consolidated with char promotion as well as free radical trapping. However, the level of improvement depends on numerous parameters like the composition and morphology of the nanoparticle, migration speed of nanoparticles to the surface, dispersion of the nanoparticles in the polymer matrix, and compatibility between polymer and nanoparticle.

In the work by (Yazhini et al., 2015, [98]), the team developed crosslinked cotton coated with nanocomposites (Polypyrrole-zinc oxide and polypyrrole-zinc oxide-carbon nanotube) for ultraviolet-protection as well as flame retardant finishes. It was confirmed that the polypyrrole-Zinc oxide-carbon nanotube composite-coated cotton was noted to demonstrate improved properties relative to uncoated cotton. (Fanglong et al., 2016, [99]) developed flame retardant mixtures of traditional intumescent flame retardant and nanosilica, and these were applied onto cotton fabric for investigating the synergistic influence of nano-silica on the fire resistance as well as thermal stability of the intumescent flame retardant system. The test results showed that a suitable inclusion of nano-silica into the conventional intumescent flame retardant system could enhance the fire protection properties of cotton fabric- intumescent flame retardant system to a definite extent, however it led to a reduction in the thermal stability of the system.

2.2.2. UV Protective Textiles

Currently, because of the depletion of the ozone-layer in the atmosphere, ultra-violet radiation is entering the surface of the earth, which is noted to have a harmful effect on both clothes and on human skin. The increased exposure to ultraviolet radiation enhances the probabilities of having several toxic diseases such as skin cancer. Consequently, protection against ultraviolet radiation has turned out to be a necessary property for clothing and textiles (Dhineshbabu et al., 2019, [100]). Certain metal oxide nanomaterials like magnetite nanoparticles, titanium dioxide nanoparticles, zinc oxide nanoparticles, and nano-ceria successfully block ultraviolet radiations, guaranteeing a sustainable as well as better performance relative to the organic ultraviolet absorbers (Sedighi et al., 2018, [101]), (Kathirvelu et al., 2009, [102]), (Becheri et al., 2008, [103]), (Fouda et al., 2018, [104]), (Tsuzuki et al., 2010, [105]), (Cakir et al., 2012, [106]), (Farouk et al., 2010, [107]), (Radetic et al., 2013, [108]), (Attia et al., 2017, [109]), (Attia et al., 2017, [110]). In recent times, the aforementioned nano-inorganic-ultraviolet additives are commonly preferred instead of the organic ones due to their exceptional properties such as harmlessness and chemical stability under UV radiation as well as higher temperature exposure. Properties such as particle size, phase composition, surface properties, crystallinity, and crystal structure are different factors that influence the ultraviolet blocking property of nano-sized ultraviolet additives (Lee, 2009, [111]), (Dhineshbabu et al., 2018, [112]).

In the latest research work by (Noorian et al., 2020, [113]), zinc oxide nanoparticles were in situ prepared on the modified cotton fabric for developing the multifunctional fabrics. This zinc oxide-4-aminobenzoic acid ligand oxidized cotton fabrics demonstrated superior ultraviolet-protection and substantial antibacterial effectiveness subsequent to 100 abrasion cycles and 20 washing cycles, and hence this could be used in innovative protective textiles. (Dhineshbabu et al., 2019 [100]), designed ultraviolet-blocking as well as fire resistant cotton fabric by coating polyurethane-based MnO_2-$FeTiO_3$ nanocomposites. The MnO_2-$FeTiO_3$ coated cotton fabrics showed a durable ultra-violet blocking capability and presented better fire resistant properties evaluated utilizing the limited oxygen index. Additionally, the coated cotton fabric maintained its properties in spite of 10 water-laundering cycles thus contributing smart, sustainable, and durable fabric for protective clothing utilization.

2.2.3. Antibacterial and Antimicrobial Textiles

Textile fabrics, particularly ones made up of cellulose fibers like lyocell, viscose, linen, and cotton have a greater tendency to be harmed by microorganisms, for example, protozoa, algae, fungi, virus, and bacteria, in the course of their service life (Ahmed et al., 2017, [114]), (Bu et al., 2019, [115]), (Hebeish et al., 2011, [116]), (Xue et al., 2012, [117]), (Zhang et al., 2009, [118]), (Budama et al., 2013, [119]), (Liu et al., 2014, [120]), (Perelshtein et al., 2008, [121]), (Zhang et al., 2014, [122]), (Attia et al., 2017, [123]). Recently, because of the enhancement in awareness about hygiene and health, the antimicrobial feature has developed into an essential prerequisite for all clothes, medial textiles, and household products. In recent times, various metal oxide (like copper oxide, zinc oxide, and titanium dioxide) and metal (such as silica, titanium, gold, zinc, copper, and silver) nanoparticles are receiving considerable research attention as prospective antimicrobial agents. Nanomaterials with a higher surface area-to-volume ratio contribute a superior antimicrobial characteristic relative to traditional antimicrobial agents. Figure 5 demonstrates different mechanisms of antimicrobial activity of metal-oxide and metal nanoparticles. Table 2 presents the textiles modified using different nanoparticles for antimicrobial effects. The utilization of nanocomposites of antimicrobial agents in textiles showed a positive synergistic antimicrobial property relative to a single nanomaterial. Economical and ecofriendly antibacterial properties of cotton fibers loaded with silver nanoparticles prepared from natural Chinese Holly plant extracts were studied by (Ullah N et al., 2014 [124]). The generation of silver nanoparticles from Chinese Holly plant extracts were noted by UV–vis spectrophotometer and noted to be less than 100nm in size, as confirmed by electron microscopy analysis. The antimicrobial properties of these cotton fibers incorporated with silver nanoparticles were assessed against gram-negative *Escherichia coli* bacteria. The test results confirmed superior antibacterial properties by incorporating 1.5% to 4.5% of Chinese Holly leave extracts. The cotton fibers also illustrated fine antibacterial efficacy after numerous washings, making it appropriate for medical usages with an ease. The process for the preparation of multifunctional polyester fabric coated by graphene/silver nanoparticles is shown in Figure 6 (Ouadil et al., 2019, [125]).

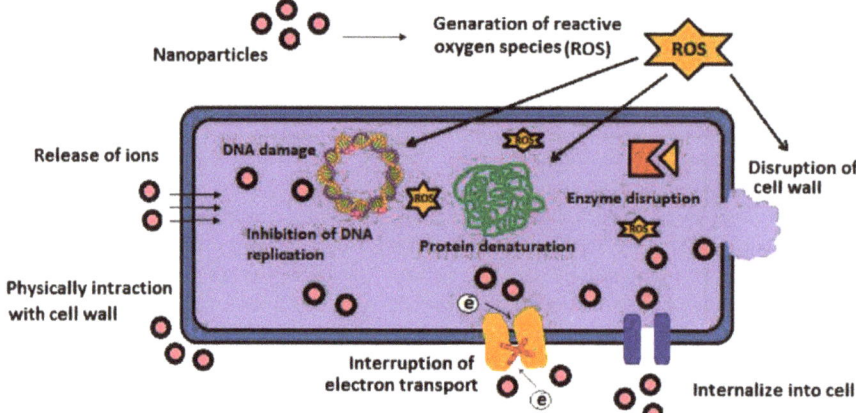

Figure 5. Different mechanisms of antimicrobial activity of metal-oxide and metal nanoparticles. Reproduced from (Dizaj et al., 2014, [126]).

Table 2. Textiles modified using different nanoparticles for antimicrobial effects.

Sl No.	Nanoparticles	Size	Fiber	Microorganisms	Results	Reference
1	Silver	2.3 nm	Cotton	E. coli and Staphylococcus aureus	99% reduction for Staphylococcus aureus and 92% reduction for E. coli	(Wu et al., 2019, [127])
2	Silver	60–100 nm	Cotton and rayon	E. faecalis, S. aureus, and E. coli	E. faecalis 97%, S. aureus-98%, and E. coli-100%	(Toh et al., 2017, [128])
3	Copper oxide	83 nm	Cotton	K. pneumoniae, E. coli, and S. aureus	Superior antimicrobial activity with antimicrobial durability of 93% subsequent to 50 washes.	(Vasantharaj et al., 2019, [129])
4	Titanium dioxide	50 nm	Cotton	E. coli and S. aureus	Greater than 95% decrease subsequent to 20 washes	(El-Naggar et al., 2016, [130])
5	Titanium dioxide -graphene oxide	Less than 100 nm	Cellulose acetate fibers	B. cereus and B. subtilis	Greater than 95% reduction	(Jia et al., 2019, [131])

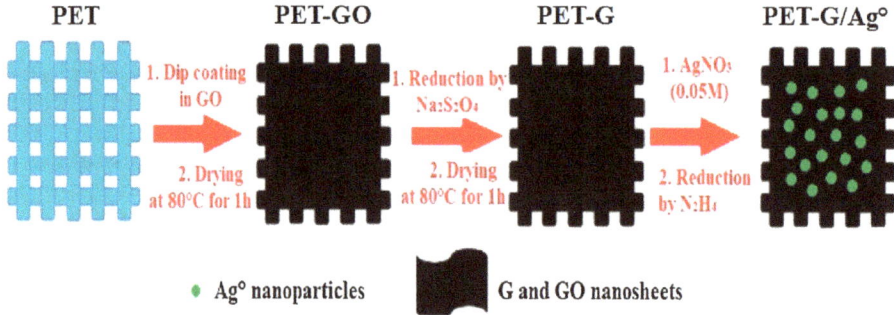

Figure 6. Process for the preparation of multifunctional polyester fabric coated by graphene/silver nanoparticles. Reproduced from (Ouadil et al., 2019, [125]).

2.2.4. Water and Oil-Repellent Textiles

Water and oil repellency has turned out to be a requirement for entire clothes and this has developed into one of the main targets for textile manufacturers and scientists for years (Asif et al., 2018, [132]). Presently, advanced nanocoatings or nanofinishings are satisfying a majority of similar market necessities with oil and water repellent or superhydrophobic textiles. Information on nanotechnology and textile mutually assist to progress an advanced conception of 'self-cleaning textiles', in which the textiles possess an ability to be cleaned with no laundry treatment (Katiyar et al., 2020, [133]), (Montazer et al., 2020, [134]). There exist two diverse methods prevalently utilized for the advancement of self-cleaning textiles, which are: (i) Photocatalytic activity and (ii) the lotus effect.

The lotus effect is produced by the surface modification of the textile fabric by nanocoating or nanofinishing, usually by utilizing surface modified carbon nanotubes, zinc oxide nanorods, nano-zirconia, or nano-silica (Das et al., 2015, [135]), (Joshi et al., 2012, [136]). In the case of the photo-catalytic activity method, zinc oxide or titanium dioxide nanoparticle-based coating or finish formulations are utilized to develop self-cleaning textiles. Titanium dioxide's photocatalytic activity is dependent on the crystal framework, and the anatase grade titanium dioxide demonstrated superior photo-catalytic activity against contaminants and colorants. (Wang et al., 2010 [137]) demonstrated that gold/titanium dioxide/silicon dioxide nanosol is a superior photocatalyst relative to titanium dioxide nanosol, displaying an improved self-cleaning property also in the existence of visible light. There are also several other works that demonstrate the cotton fabric's photocatalytic self-cleaning property by treatment with titanium dioxide nanowire and titanium dioxide nanowire doped Ag-PVP (Hebeish et al., 2013, [138]), graphene/TiO_2 nanocomposites (Karimi et al., 2014, [139]), etc. The work by (Landi et al., 2019 [140]) reported the preparation of photocatalytic nanocomposite materials based on nitrogen-doped nano-titanium dioxide, silicon dioxide, and various percentages of HY zeolite. It was noted that the fabrics coated with the photocatalysts, demonstrated similar RhB decolorization (almost 95%) after 5 h. Fluorine-free superhydrophobic cotton fabrics, having the self-cleaning photocatalytic ability, were fabricated by the combination of superhydrophobic SiO_2 and photoactive titanium dioxide (Xu et al., 2015, [141]).

2.2.5. Anti-Odor Textiles

Tourmaline nanomaterial-based nanofinishing on textiles contributes to an odor-resisting property and has the ability to separate till 75% sticky moisture, 99.99% of bacteria, and 90% of odor (Joshi et al., 2019, [142]). The inclusion of fragrant material (aroma) in antiodor/antimicrobial finishing by nanoencapsulation in synthetic fibers or by formulation might assist to release fragrance in the course of its utilization (Priyadarshinirajkumar et al., 2015, [143]).

2.2.6. Wrinkle Resistance Textiles

Cotton fabric is extremely vulnerable to developing creases at the time of usage. In traditional techniques, resin-based finishings are commonly utilized for imparting the wrinkle resistance feature to textile fabrics. The utilization of nanoparticles such as silicon dioxide and titanium dioxide have the ability to overcome certain restrictions to traditional crease resistant finish (Haque et al., 2019, [144]), (Tripathi et al., 2019, [145]). The study by (Hezavehi et al., 2015, [146]) evaluated the wrinkle behavior as well as wrinkle resistance ability of cotton fabrics dyed using Direct Blue 2B in the absence and presence of titanium dioxide nanoparticles. The test results confirmed that the wrinkle-resistance property of cross-linked fabrics was enhanced subsequent to direct dyeing. (Uğur et al., 2017 [147]) investigated the advanced flame retardant, wrinkle resistant, and durable finishing of linen by utilizing 1,2,3,4-Butanetetracarboxylic acid, nano-polyurethane for crosslinking process, and aluminum oxide nanoparticles for catalyst in the padding procedure. It was found that the flax fabrics applied with the aluminum oxide nanoparticles showed improved flame retardant and wrinkle resistance properties.

2.2.7. Antistatic Textiles

Due to increased moisture content, cellulose fibers like lyocell linen, viscose, cotton, etc. do not mount up the static charge. On the other hand, synthetic fibers like nylon are susceptible to the generation of static charge due to lower moisture regain. Consequently, several studies have been performed on the development of anti-static textiles by the incorporation of certain conducting nano-fillers (Memon et al., 2018, [148]). Titanium dioxide nanoparticle, nano-silver, zinc oxide nanoparticle, antimony-doped tin oxide, and silane nanosol can be used to provide anti-static ability in synthetic fibers (Zhang et al., 2009, [149]), (Hassan et al., 2019, [150]), (Hossain, M et al., 2013, [151]), (Yadav et al., 2006, [152]). Due to the fact that the aforementioned nanomaterials are conductive in nature, these materials dissipate static charges mount up on the surface of fiber.

3. Nanomaterials in Textile Industries—Environmental, Health, and Safety Concerns

Textile material is considered to be our second skin for an entire day. We use textiles for aesthetic, protective, decorative, and several other applications. Most textile materials are in direct and prolonged contact with our skin. Consequently, the influence of carcinogenic and toxic substances that are available in textiles must be comprehensively examined. One of the most modern groups of prospective dangerous substances belongs to the group of engineered nanomaterials. The significance of engineered nanomaterials has been acknowledged by the textile sector, due to the fact that these materials have the ability to modify the chemical and physical properties of textile materials and textile fibers—improve water and stain resistance, enhance the capability of materials for absorbing dyes, and alter the wettability depending on surface roughness and surface energy. The aforementioned properties are beneficial for several non-woven and woven textiles like automobile interior fabrics, sportswear, protective clothing, and rainwear (Harifi et al., 2017, [153]), (Shalaby et al., 2020, [154]), (Yu et al., 2018, [155]), (Kausar et al., 2018, [156]). Still there are no adequate studies, however some literature reviews and scientific papers have focused on safety issues related to nanotextiles. Some of the recently published studies include the works by (Rovira eta al., 2019, [157]) and a recent chapter by (Montazer et al., 2018, [158]) in a book entitled Nanofinishing of Textile Materials. Nanotextiles can definitely cause risks for human health, safety, the environment, and sustainability (Kohler et al., 2014, [159]), (Geranio et al., 2009, [160]), (Lorenz et al., 2012, [161]). On the basis of the location of the integration of nanomaterials in textiles, they could be less or more exposed to external influences (Rather et al., 2020, [162]). There have been different studies carried out for assessing the release of silver nanoparticle from antibacterial fabrics into artificial sweat (Wagener et al., 2016, [163]), (Kulthong et al., 2010, [164]), (Kim et al., 2017, [165]), (von Goetz et al., 2013, [166]), (Spielman et al., 2018, [167]), (Stefaniak et al., 2014, [168]), (Balakumaran et al., 2016, [169]), (Milosevic et al., 2014, [170]). (Wagener et al., 2016, [163]) studied the textile functionalization as well as its impacts on the discharge of silver nanoparticles into artificial sweat. Migration tests have been performed for four commercial textiles and for six lab-prepared textiles. Two of these laboratory-prepared textile signifies materials where silver nanoparticles were incorporated inside the textile fiber (composite), while the other laboratory-prepared textile consists of silver particles on the particular fiber surface (coating). The test results confirmed a lesser release of total silver from composites as compared to the surface-coated textile. The particulate portion found inside artificial sweat was noted to be less in the majority of textiles, confirming that the majority of the discharged silver is available as dissolved silver. Additionally, it can be noted that nanotextiles will not discharge more particulate silver, relative to the conventional silver textiles. Moreover, the results confirmed that the functionalization type is the significant parameter influencing the migration.

Despite the aforementioned reported works, attaining wide-ranging information on the risks associated with nanomaterials is still challenging. The aforementioned is because of the range of nanomaterials, manufacture techniques, nanoparticle shape, size, crystallinity, porosity, agglomeration and aggregation, and several other factors. Examination on the dangers associated with nanotextiles and nanofinishing techniques needs a complete understanding of the product's life-cycle (Figure 7). This take account of the fundamental properties of the nanomaterials, nanomaterial fabrication procedures, and different application techniques utilized for imparting nanocomposites and nanoparticles on the textile substrates or insertion of nanoparticles into fiber formation processing, nanofinishing durability on the surface, storage/transportation of the treated textiles, ultimate product usage, the conditions in which the final product will be exposed to, and product recycling/disposal. Depending on the utilization of the nanotextiles, the disposal of the product will influence soil, water, and commonly the environment, or straight skin contact of human causing threats on the health of a human with ultimate environmental effects. Nanomaterials could be accidentally discharged from the treated samples in the course of their lifetime or exposure of labors to the dangerous effects of nanomaterials might happen at the time of the fabrication process. The life cycle of the product and design of the product regulate the different environmental as well as health exposure situations. As an illustration, the engineered nanomaterials inadvertently discharged from geotextiles might possibly end up in soils, while the engineered nanomaterials involuntarily discharged from T-shirts might make direct contact with humans, finally entering wastewater. Therefore, recommending regulation based on the behavior and effects of nanotextiles needs complete product characterization based on particle stability, surface morphology, shape, porosity, size, chemical composition, as well as propensity to aggregation and agglomeration.

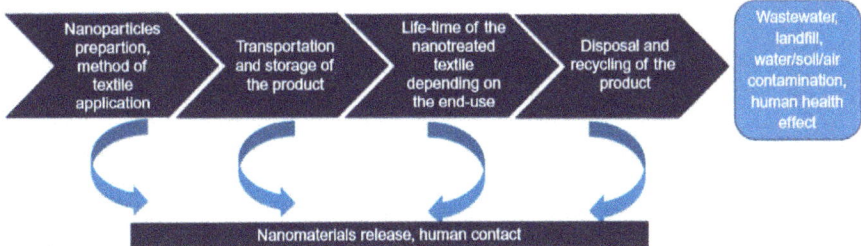

Figure 7. Life-cycle of nanotextiles. Reproduced from (Montazer et al., 2018, [158]).

3.1. Environmental Risks of Nanomaterials from Textile Industry

A moderately larger fraction of engineered nanomaterials in textiles are assumed to be discharged into wastewater (for example till 20% for silver nanoparticles) (Cucurachi et al., 2019, [171]), (Patnaik et al., 2019, [172]). Nevertheless, the quantity of silver nanoparticles and other engineered nanomaterials accidently discharged into the environment depends greatly on the textile design utilized, particularly how the nanomaterials are incorporated within the textile fiber. Straight discharge of nanomaterials into the air by means of abrasion looks to be of a lesser significance (just 5%). Other flows of nanomaterials from textiles, for example, into waste incineration, are greater from a bulk perspective though, they probably result in just minimal discharges to the surroundings. Therefore, the textiles might constitute a significant source of nanomaterials released into aqueous systems and into soils by the biosolid utilizations. The evaluation of the environmental dangers is extremely dependent on the corresponding product life cycles and on the quantities of engineered nanomaterials manufactured globally. The durability (stability) of the nanomaterials existing in the textile depends on its fabric binding, the influences on the fabric in the course of its life-cycle (manufacture, utilization, disposal/recycling), which could harm the textile material or the bonding between the fibers and nanomaterial, mechanical stress (like pressure, strains), abrasion, temperature changes, high temperatures (till 225 °C in textile finishing), detergents (either during laundry or in textile processing), solvents (during dry cleaning or textile processing), water (washing, rain), body fluids (urine, sweat, and saliva), and ultraviolet radiation. The nano-biocomposite application can be a good substitute to a majority of artificial antibacterial agents because of their biodegradability, increased environmental compatibility, as well as non-toxic nature. Considering a life cycle method for studying the possibility of the release of silver nanoparticles from functionalized textiles, it is possible to find the significance of various phases to silver discharge over time. In a study by (Mitrano et al., 2016 [173]), three distinct lab-prepared nanofabrics were exposed to one or 10 washing cycles under various laundering conditions. It was noted that the total entire metal discharged differed remarkably on the nanoparticle incorporation as well as the washing pattern variant. The test results confirmed that the active landfill environment will not mobilize nanoparticles from the surface of the fabric as easily subsequent to washing relative to the unwashed textile. Increased release of nanoparticles from textile have been noted at the time of the life cycle's utilization phase instead of the disposal phase.

Despite the fact that nanotextiles have been observed to be efficient in photocatalytic or the absorption degradation of pollutants from impure wastewaters, or the nanofibrous membranes could be employed as air/water filters, there have been several forewarnings in regards to the ecological dangers of nanomaterials. Due to the larger surface area of nanomaterials, these materials are extremely reactive to other materials while discharged during any stages of its life cycle. The aforementioned must be associated with both sides of danger and opportunity as the reaction of nanomaterials with other constituents might contribute to lesser toxicities, thus altering these materials from their unique properties. Conversely, these nanomaterials might carry other harmful materials, generating additional toxicity. If the aggregation/agglomeration of nanomaterials occurs, it is conceivable to separate the nanomaterials from the waste water effluent by filtration or sedimentation techniques. The consequence of sediments on the environment is not totally clear yet. The burning of sewage sludge was reported, even though the whole avoidance of the discharge of nanomaterials has not been demonstrated yet (Som et al., 2011, [174]). Certain nanomaterials could be dissolved in the surroundings producing no consequence, whereas other solutions consisting of metals are extremely harmful to the surroundings. Magnetic separation approaches based on the manufacture of magnetic nanomaterials and their utilization on textile substrates were also established as a capable inexpensive technique for the separation of the nanomaterials from waste water effluents (Harifi et al., 2014, [175]). Fascinatingly, certain engineered nanomaterials might influence the environment less harshly than they might influence human health, whereas the case for some others is vice versa. The aforementioned is specifically true for carbon nanotubes.

The engineered nanomaterials have several benefits depending on the properties, such as less textile laundering because if their antibacterial properties or quick wound healing (Temizel-Sekeryan et al., 2020, [176]). From the analysis of the study by (Lorenz et al., 2012, [161]), four out of the seven silver nanotextiles leached a noticeable amount of silver (Figure 8). Neither the rinsing nor the washing solutions of textiles 1 to 3 contained any silver. In a study by (Piontek et al., 2018 [177]), the team analyzed two Life Cycle Assessment studies on the textile value chain and on the product-service systems. It is good review paper, where the authors presented their findings on the dependence of Life Cycle Assessment studies, lack of accessible information on chemicals, advantages because of the technological development in the area, the significance of the use phase of clothes, and suggest additional routes of research in regards to user behavior as well as methodological development. Merging the conclusions of the two studies helped for the development of a method for conducting the Life Cycle Assessment studies on a product-service system based on existing research as well as research gaps. This could include the development of a method for assessing the environmental benefits when establishing a product-service system.

For the environment, the following criteria were established as critical for the determination of environmental fate as well as effects of engineered nanomaterials (Som et al., 2011, [174]): (1) Certain warning of dangerous effects at existing genuine exposure concentrations, (2) a propensity to be dissolved in water, leading to the vanishing of the engineered nanomaterials, and the development of dissolved metal ions, (3) a propensity for sedimentation or agglomeration under normal conditions, (4) destiny in wastewater facility, and (5) steadiness in the course of incineration. The aforementioned conditions comprise of consequences of engineered nanomaterial and the performance of these materials in ecological compartments as well as the technosphere. The dissolution of alumina nanoparticles, zinc oxide nanoparticles, and nano-silver in water enhances the harmful effects. However, the dissolution of nano-silicon dioxide reduces the harmful effects as it caused the vanishing of the nanomaterials and the dissolved silica is not dangerous at environmental concentrations. With the probable exemption of nanosilica, all other nanomaterials can agglomerate intensely in natural waters and are therefore detached from the water system and are less mobile. A majority of engineered nanomaterial seem to be separated in the course of wastewater treatment, but this is based on only limited accessible research, which have employed unrealistically higher concentrations of nanomaterials. A probable exclusion might be carbon nanotubes and silica where a lower separation rate was noted depending on functionalization. For the duration of waste incineration, the carbon-containing nanomaterials might probably be destroyed totally, while metal-oxide or metal particles might persist intact. In general, it can be concluded that mostly silver nanoparticles and zinc oxide nanoparticles could lead to a maximum risk to the environment, however the titanium dioxide nanoparticles also need to be additionally examined. From an environmental perspective and the present utilization of engineered nanomaterials like aluminum dioxide, silicon dioxide, carbon black, and montmorillonite probably cause no or little hazard to the environment (Black, 2013, [178]).

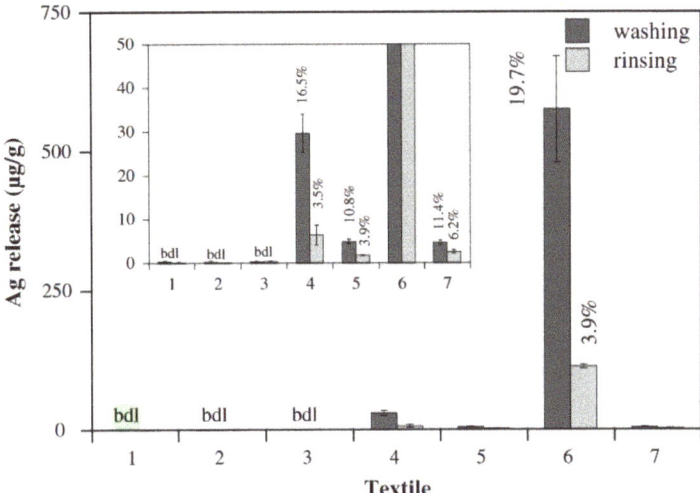

Figure 8. Quantity of silver released from the seven textiles at the time of washing as well as rinsing. The inset presents an expanded outlook of the lower concentration range. The percentage of total silver released is presented on top of the columns. bdl: Below detection limit. Reproduced from (Lorenz et al., 2012, [161]).

3.2. Health Risks of Nanomaterials from Textile Industry

The exposure to engineered nanomaterials from textile materials could happen by means of several pathways: Skin absorption, inhalation, and ingestion (Sahu et al., 2017, [179]), (Alanezi et al., 2018, [180], (Murphy et al., 2020, [181]), (Yu, 2018, [182]), (Abdelrahman et al., 2020, [183]). Maximum exposure to nanomaterials arise for laborers in the textile industries because of continued and prolonged exposure to higher quantities of engineered nanoparticles (Torabifard et al., 2018, [184]). The most frequent routes for the engineered nanoparticle uptake are the skin as well as respiratory track. Subsequent to body entry, nanomaterial accumulates in the spleen, bon, kidney, and liver (Tavares et al., 2017, [185]). The hypothetical model of engineered nanoparticles (ENPs) pathway in the human body and its toxic and harmful effects are shown in Figure 9. Due to the aforementioned reasons, the toxico-kinetics of engineered nanomaterials are presently under numerous studies (de Jong et al., 2017, [186]). Due to the fact that the engineered nanomaterials are smaller than 100 nm, these materials could smoothly penetrate cells (Aryal et al., 2019, [187]). Even though the harmfulness of nanoparticles varies based on their properties (chemical composition, surface energy, charge, shaper, size, and others), they also depend on the living organisms and their diverse DNA covering ratios (Sukhanova et al., 2018, [188]). The major concern for nanoparticle exposure from textiles is by means of skin absorption. The skin is considered to be a superior absorptive material because of the rich supply of blood and tissue macrophages, dendrites, lymph vessels, and various types of sensory nerve endings (Ramachandran, 2014, [189]). The paper by (Filon et al., 2016 [190]) reviewed and analytically assessed evidence on the significance of different skin adsorption paths for engineered nano-objects, nanoparticles, their aggregates as well as agglomerates.

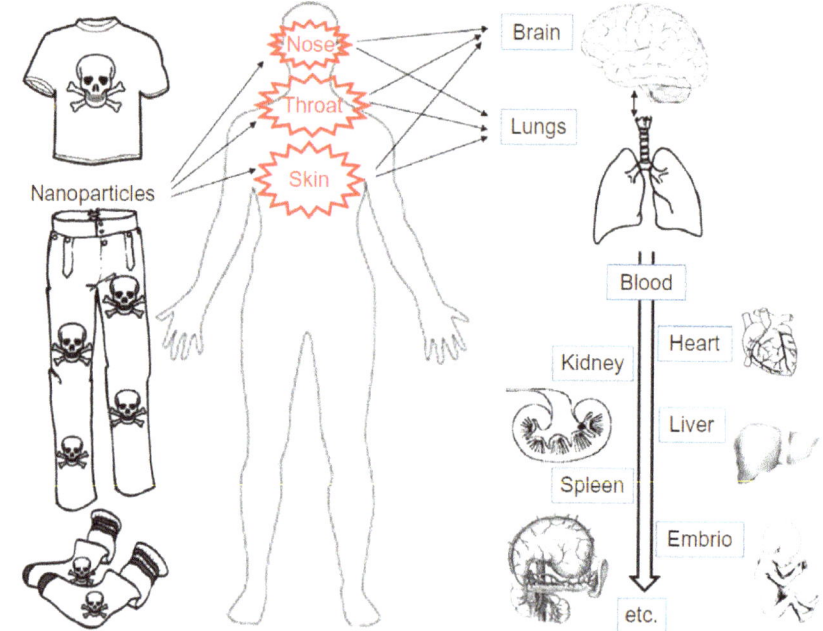

Figure 9. Model of the transport of nanomaterials in human body. Reproduced from (Rezic et al., 2012, [191]).

Several studies have confirmed that nanomaterials can lead to adversarial toxic effects in living beings (Exbrayat et al., 2015, [192]), (Roberto et al., 2019, [193]), and DNA damage (Grumezescu et al., 2017, [194]). Studies on DNA damage have received more research attention because of associations with cancer, neurological diseases, and ageing. In the study by (Grumezescu et al., 2017 [194]), the damage of DNA caused by different nanomaterials will be evaluated with respect to DNA damage products, types as well as detection methods. The results of the study carried out by (Nallanthighal et al., 2017, [195]) suggest that a human being with genetic polymorphisms as well as mutations in 8-Oxoguanine DNA glycosylase 1 might have enhanced susceptibility to silver nanoparticle-mediated DNA damage. The factors that regulate the possible toxicity of engineered nanoparticles are biodegradation, biodistribution, biocompatibility, inflammation, as well as interference with cells and a regular functioning of organs (Adabi et al., 2017, [196]). The aforementioned factors are associated with the reactivity, composition, shape, and size of the engineered nanoparticles.

The textiles are categorized on the basis of usage in products for babies, products with no direct skin contact, products with direct skin contact, and decorative materials. The boundaries for toxic as well as allergenic metals and chemicals differ according to the degree of fabric contact with the skin of a consumer and on heavy metal toxicity. Those limits do not involve the entire amount of compounds existing in the fabric, but the part that could be extracted (Yin et al., 2015, [197]). Analogous method must be developed for examining the harmfulness of nanoparticles on textiles, as their impact on the environment and human health is presently unpredictable. For now, humans are exposed to the emission of nanoparticles from textiles, textile industries, and textile laundries in a cycle we cannot monitor and control (Figure 10).

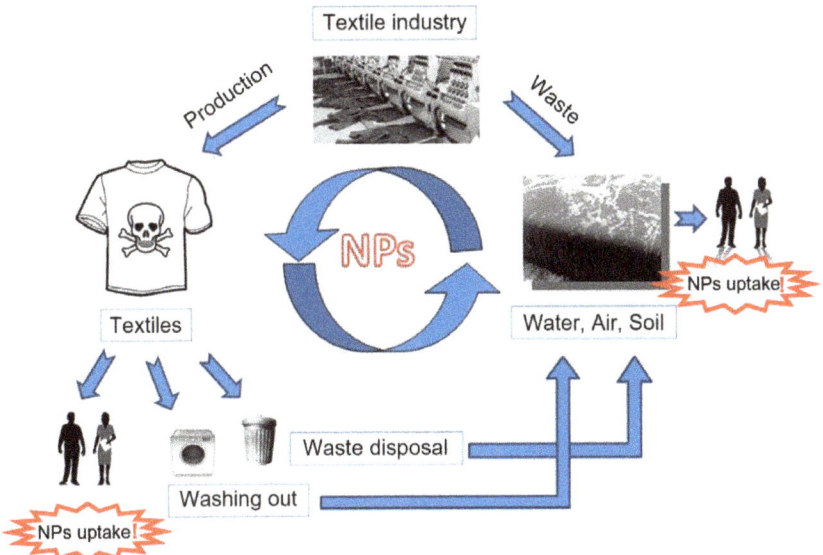

Figure 10. The discharge of nanoparticles (NPs) from textile materials and textile industry into surroundings and uptake by human body. Reproduced from (Yin et al., 2015, [197]).

Therefore, the dangerous effects of nanotextiles against human health and the environment have not been broadly demonstrated yet and accessible data were extremely debatable, being contingent upon the stability, porosity, shape, size, dosage level, and end use of nanomaterials. Consequently, the obtainable reports are not much consistent and could not be compared. Further studies on the conceivable dangers in the utilization of nanofinishing in textile industries must be performed. An absence of guidelines for the control of nano-based treatments is acknowledged therefore, there might be greater concern of the supervisory establishments in the coming years. Ecolabeling is needed when nano-sized materials are included in textile clothing, particularly with direct human contact. Efforts to suggest eco-friendlier fabrication steps utilizing green chemistry, a development of nanomaterials with lesser toxicity, utilization of in-situ preparation techniques resulting in lower effluent, as well as increased durability of the nanomaterials bonded to the textile substrates will also be broadly made in forthcoming studies.

4. Approaches for Assessing the Nanomaterial Toxicity

The nanomaterial toxicity examinations might be carried out using in vivo (live) organisms, like rodents, fishes, microcrustaceans, and several other animals or/and cell cultures (in vitro). Different normalized toxicological examinations are obtainable for measuring the natural response of a living organism to a chemical. On the other hand, there is no standard existing for the assessment of nanomaterial toxicity, which impedes the evaluation of results as well as understanding about its toxicity. A majority of research carried out up to now are revisions of standard procedures utilized for other materials (Ju-Nam et al., 2008, [198]). Even though certain nominal associations of assays are recommended, (Drasler et al., 2017, [199]) defined that there exists no typical assessment protocol because of the extensive range of physico-chemical properties that the nanomaterials could contribute.

Animal experiments are highly prognostic for human effect however there are restrictions, primarily due to the biochemical and physiological dissimilarities among the species. Furthermore, there exist an increasing legal and public demand that morally supports the replacement of animal analysis for substitutes not based on in vivo testing. Novel concepts of testing are based on approaches with the primary culture of human cells as well as permanent cultures of cell lines, due to the fact they provide reliable, cheap, and efficient results (Drasler et al., 2017, [199]). In the following section, we discuss certain major evaluation techniques, developed both in vitro and in vivo, for properly characterizing nanomaterial toxicity.

4.1. In Vivo Techniques

In vitro assessments have augmented significantly, however in vivo confirmation is still needed for understanding as well as interpreting results. Additionally, animal testing was also associated with the NanoTEST venture, whose objective was to recognize the effects on the physiology of verified living organisms. Presently, the Organization for Economic Co-operation and Development (OECD) offers certain examination strategies on which biomarkers must be utilized for every testing organism (Juillerat-Jeanneret et al., 2015, [200]).

Generally, there are several studies on human toxicity by means of rodent models, while only limited research exists in vivo dealing with available nanomaterial ecotoxicity. Additionally, a majority of those observed in the studies reflect the influence of nanomaterials on aqueous organisms, due to the fact that marine and continental waters are the chief reception compartments. As a general rule, rodents (Wistar rat and mice), fish (example, Danio rerio), mollusks (example, *Lymnaea stagnalis*), microcrustaceans (example, *Ceriodaphnia dubia, D. pulex, Daphnia magna*), nematodes (example, *Caenorhabditis elegans*), algae (example, *Raphidocelis subcapitata*), and bacteria (example, *Aliivibrio fischeri*) are mostly utilized testing organisms for the assessment of acute toxicity.

4.2. In Vitro Techniques

As stated by (Drasler et al., 2017, [199]), assays could be carried out using eternal cell lines or primary cultures. As stated by these authors, cell lines are preferred as they provide increased stability as well as homogeneity, which favors the consistency in test results, particularly in preliminary examinations. For more precise examinations, the same investigators indorse the usage of three-dimensional co-cultures for better understanding the action mechanisms of nano-sized materials on tissues. In order to assess nanomaterial toxicity, the utilization of epithelial cell lines (lung, skin, etc.) is typically designated because the aforementioned cells demonstrate characteristics of actual barriers against destructive agents and are consequently the foremost to suffer the effect of these compounds (Rothen-Rutishauser et al., 2012, [201]). Conversely, certain strains might not be receptive to the impacts of nanomaterials and here the primary cultures might be better designated.

In the case of in vitro comet assay using the mammalian cell culture, Collins and his team members (Collins et al., 2017, [202]) put forward certain recommendations: (i) Utilize non-cytotoxic concentrations; (ii) select the type of cell as per the exposure scenario; (iii) identify both long (24 h) and short (2–3 h) tests for obtaining a clear knowledge of the action mode of the nanomaterial; and (iv) identify if the genotoxic destruction demonstrated is a consequence of the direct effect with DNA or because of DNA oxidation. As stated by (Catalán et al., 2017, [203]), the limitations as well as relevance of mutagenicity/genotoxicity assays must be considered while selecting the most suitable monitoring technique. As per the aforementioned study, examinations considered in the assessment must be based on three classes: (1) DNA damage, (2) chromosomal destruction, and (3) gene mutation. According to OECD guidelines (OECD, 2014, [204]), for selecting a test and evaluating the genotoxicity of a nanoform, the solubility, absorption, exposure, metabolites, as well as other derivatives must be taken into account, along with possible side effects.

5. Environmental Risk Assessment—Case Studies

In a study performed by (Voelker et al., 2015, [205]), a standard environmental risk assessment of silver nanomaterials applied in textiles has been carried out. Environmental exposure scenarios were developed for three distinct categories of textiles equipped with silver nanomaterials. Based on these scenarios, the predicted environmental concentrations were deduced for sewage treatment plants and for the environmental compartments such as surface water, sediment, and soil. The information on ecotoxicology were obtained from different analysis on earthworms, chironomids, macrophytes, duckweed, fish, daphnids, algae, cyanobacteria, activated sludge, terrestrial plants, and soil microorganisms. Emission information on silver nanomaterials NM-300K from textiles were obtained from washing experiments. The environmental risk assessment performed was based on the specifications defined in the European Chemicals Agency (ECHA) guidance on information requirements as well as chemical safety assessment. Depending on the selected scenarios as well as preconditions, no environmental hazard of the silver nanomaterials NM-300K discharged from textiles was noticed. Under conservative assumptions, a risk quotient for surface water pointed out that the marine compartment might be influenced by a higher emission of silver nanomaterials to the surroundings because of the higher sensitivity of marine life to silver. Depending on the effective retention of silver nanomaterials in sewage sludge and the continuing application of sewage sludge on farm land it is suggested to introduce a threshold for total silver concentration in sewage sludge. With regards to the potential risk mitigation measures, it is highlighted that one should directly introduce silver nanomaterials into the textile fiber due to the fact that this would lessen the discharge of silver nanomaterials extremely in the course of washing. If this is not conceivable because of technical restrictions or some other reason, then the inclusion of a threshold level controlling the discharge of silver nanomaterials from textiles is recommended. It should be noted that the aforementioned specific study is a case study that is only valid for examined silver nanomaterial NM-300K and its possible application in textiles.

The study conducted by (Yasin et al., 2019, [206]) elaborated on certain points for technical textile waste. Initially, the Life Cycle Assessment method for end-of-life is feasible if the waste treatment depends on the technical textile functionality instead of common textile waste. Secondly, this end-of-life study confirmed that the Life cycle assessment results of any technical textile product at its disposal are also case dependent and must not be considered the same as collective textile waste, in spite of environmental correspondence being considered or not. Figure 11 demonstrates the life cycle of a textile product system as well as its environmental interventions at different phases. The life cycle assessment "gate-to-grave" method was used for studying two technical textiles with different functionalities however with the same weight, one is a silver nanoparticle-treated polyester and the other one is flame retardant-treated wool. They were examined for having an improved understanding of environmental parity, particularly in their usage phase as well as at the "end-of-life" phase. Ten-midpoint categories were employed for analyzing the environmental impacts at the time of the use phase and end-of-life phase of both technical textiles. With regard to the technical textile curtain recycling, both use functionality substances, flame retardant (for its flame retardancy), and silver nanoparticles (for its antimicrobial properties). Their life-cycle impact perspectives can be different with their functionality lost in the application phase, for example, substantial loss of silver nanoparticles as the time of laundering, as compared to the well-bond flame retardants to the fibers. This will increase the environmental cost of one technical textile (silver nanoparticle treated) in use phase imposing severe wastewater treatment. In the same way, the behavior of other functionality substance on technical textile (flame retardant treated) needs different considerations for either end-of-life, incineration, or landfill. Thus, the overall results indicated that in the use phase, the life cycle impact of technical textiles is upfront and changes with the variation in number of washes, the types of applied attributional substances, as well as the rate of release. At the "end-of-life" phase, it has been noted that there is no relationship between the two types of technical textiles with respect to environmental impacts.

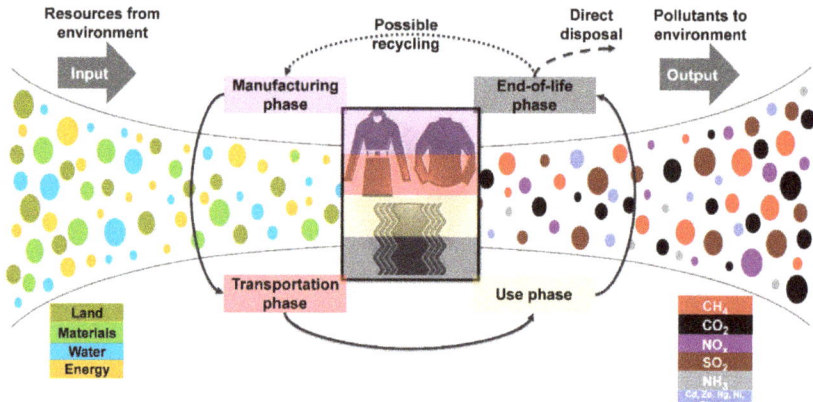

Figure 11. Life cycle of a textile product system as well as its environmental interventions at different phases. Reproduced from (Yasin et al., 2019, [206]).

6. Conclusions

The state-of-the-art nanotechnology has been successfully utilized in several fields for the well-being of mankind. On the other hand, any innovative unverified technology is accompanied with certain drawbacks. There exists concerns about the conceivable hazardous impacts of nanomaterials on the environment as well as human health. There are numerous justifications for believing that the application of nanomaterials is increasing. Up until now, nanotechnology has been considerably used to develop several technologies in addition to industrial sectors, of which includes textiles. Conversely, due to the absence of suitable disposal policies, the level of nanomaterials in the surroundings is persistently developing.

Nanomaterials are promising materials for the functionalization of textiles and fibers. Nanotechnology contributes novel solutions for developing newly advanced textiles with multiple functionalities like antistatic properties, wear/tear resistance, UV protection, flame retardancy, antimicrobial property, water/oil/dirt repellency, self-cleaning ability, and wrinkle free. Despite the fact that there are still certain issues with their utilization in commercial applications, there are many substantial scientific studies that report their appropriateness for advanced functional textile applications. In addition to this progress, it should be noted that the nanotechnology field is still a developing area with several challenges as well as prospects, unsolved difficulties, and commercial opportunities. Our exposure to nanoparticles from textile materials is developing, however, unfortunately this is not supplemented with suitable awareness or/and limits prescribed by safety regulation in regards to toxicological information. It can be anticipated that this situation might vary in our immediate future. Modern science, particularly analytical chemistry with the progresses in instrumental analytics is contributing an extensive range of various techniques that could be utilized for monitoring nanoparticles that exist on textiles as well as in textile wastewaters. Research works must be carried out for preventing possible human health risks for users, inclusive of the adult population, but particularly for babies and children.

Thus, it can be concluded that nanomaterials offer both benefits as well as risks. Currently, nanomaterials are present in certain commercial products, some of which are not labelled while some are labelled, and there will certainly be more to establish. Therefore, it is significant that the textile industries have adequate knowledge about nanomaterials so as to make safe choices. The conclusions of the current paper are anticipated to increase consciousness on the possible influence of nanomaterials containing textile wastes and the significance of better regulations in regards to the ultimate disposal of these wastes.

Author Contributions: Conceptualization, S.J.Z. and H.S.; investigation, H.S.; resources, H.S.; writing—original draft preparation, H.S.; writing—review and editing, H.S.; supervision, S.J.Z.; project administration, S.J.Z.; funding acquisition, S.J.Z. All authors have read and agreed to the published version of the manuscript.

Funding: Qatar University's IRCC research program through grant number IRCC-2019-004.

Acknowledgments: The authors gratefully acknowledge the support from Qatar University's IRCC research program through grant number IRCC-2019-004 for this research.

Conflicts of Interest: The authors declare no conflict of interest.

References

1. Feynman, R.P. There's plenty of room at the bottom. *Eng. Sci.* **1959**, *23*, 22–36.
2. Taniguchi, N. On the basic concept of nanotechnology. In *Proceedings of the International Conference on Production Engineering, Tokyo, 1974*; Japan Society of Precision Engineering: Tokyo, Japan, 1974.
3. Yang, G.; Park, S.J. Deformation of single crystals, polycrystalline materials, and thin films: A Review. *Materials* **2019**, *12*, 2003. [CrossRef] [PubMed]
4. Khan, I.; Saeed, K.; Khan, I. Nanoparticles: Properties, applications and toxicities. *Arab. J. Chem.* **2019**, *12*, 908–931. [CrossRef]
5. Verma, A.; Arif, R.; Jadoun, S. Synthesis, Characterization, and Application of Modified Textile Nanomaterials. *Front. Text. Mater. Polym. Nanomater. Enzym. Adv. Modif. Tech.* **2020**, 167–187.
6. Kumar, R.; Chauhan, M.; Sharma, N.; Chaudhary, G.R. Toxic effects of nanomaterials on environment. In *Environmental Toxicity of Nanomaterials*; CRC Press: Boca Raton, FL, USA, 2018; pp. 1–20.
7. Mishra, R.; Militky, J.; Arumugam, V. Nanotechnology in Textiles: Theory and Application. Characterization of nanomaterials in textiles. In *Nanotechnology in Textiles: Theory and Application*; Woodhead Publishing: Cambridge, UK, 2018.
8. Kumar, V.; Dasgupta, N.; Ranjan, S. (Eds.) *Environmental Toxicity of Nanomaterials*; CRC Press: Boca Raton, FL, USA, 2018.
9. Leong, D.T. *4.10 Nanosafety Issues of Nanomaterials*; Elsevier: Amsterdam, the Netherlands, 2017.
10. Johnston, L.J.; Gonzalez-Rojano, N.; Wilkinson, K.J.; Xing, B. Key challenges for evaluation of the safety of engineered nanomaterials. *NanoImpact* **2020**, *18*, 100219. [CrossRef]
11. Mirshafiee, V.; Osborne, O.J.; Sun, B.; Xia, T. Safety Concerns of Industrial Engineered Nanomaterials. In *Handbook of Nanomaterials for Industrial Applications*; Elsevier: Amsterdam, the Netherlands, 2018; pp. 1063–1072.
12. Karim, M.E. Functional nanomaterials: Selected occupational health and safety concerns. In *Handbook of Functionalized Nanomaterials for Industrial Applications*; Elsevier: Amsterdam, The Netherlands, 2020; pp. 995–1006.
13. Dobrovolskaia, M.A.; McNeil, S.E. (Eds.) *Handbook of Immunological Properties of Engineered Nanomaterials*; World Scientific: Singapore, 2013; Volume 1.
14. Ahmad, F.; Abubshait, S.A.; Abubshait, H.A. Untargeted Metabolomics for Achilles heel of Engineered Nanomaterials' risk assessment. *Chemosphere* **2020**, *262*, 128058. [CrossRef]
15. Kawai, S.; Niwano, M.; Sato, M. A risk assessment framework for self-management of poorly soluble low toxic nanomaterials. *Heliyon* **2019**, *5*, e02165. [CrossRef]
16. Auffan, M.; Masion, A.; Mouneyrac, C.; de Garidel-Thoron, C.; Hendren, C.O.; Thiery, A.; Rose, J. Contribution of mesocosm testing to a single-step and exposure-driven environmental risk assessment of engineered nanomaterials. *NanoImpact* **2019**, *13*, 66–69. [CrossRef]
17. Oomen, A.G.; Steinhäuser, K.G.; Bleeker, E.A.; van Broekhuizen, F.; Sips, A.; Dekkers, S.; Sayre, P.G. Risk assessment frameworks for nanomaterials: Scope, link to regulations, applicability, and outline for future directions in view of needed increase in efficiency. *NanoImpact* **2018**, *9*, 1–13. [CrossRef]
18. Schulte, P.A.; Kuempel, E.D.; Drew, N.M. Characterizing risk assessments for the development of occupational exposure limits for engineered nanomaterials. *Regul. Toxicol. Pharmacol.* **2018**, *95*, 207–219. [CrossRef]
19. Iavicoli, I.; Leso, V.; Ricciardi, W.; Hodson, L.L.; Hoover, M.D. Opportunities and challenges of nanotechnology in the green economy. *Environ. Health* **2014**, *13*, 78. [CrossRef]
20. McKenzie, L.C.; Hutchison, J.E. Green nanoscience. *Chim. Oggi* **2004**, *22*, 30–33.

21. Hutchison, J.E. Greener nanoscience: A proactive approach to advancing applications and reducing implications of nanotechnology. *ACS Nano* **2008**, *2*, 395–402. [CrossRef] [PubMed]
22. Bamoharram, F.F. Role of polyoxometalates as green compounds in recent developments of nanoscience. *Synth. React. Inorg. Met.-Org. Nano-Met. Chem.* **2011**, *41*, 893–922. [CrossRef]
23. Karst, D.; Yang, Y. Potential Advantages and Risks of Nanotechnology for Textiles. *AATCC Rev.* **2006**, *6*, 44–48.
24. Jatoi, A.S.; Khan, F.S.A.; Mazari, S.A.; Mubarak, N.M.; Abro, R.; Ahmed, J.; Baloch, H.; Sabzoi, N. Current applications of smart nanotextiles and future trends. In *Nanosensors and Nanodevices for Smart Multifunctional Textiles*; Elsevier: Amsterdam, the Netherlands, 2021; pp. 343–365.
25. Darwesh, O.M.; Ali, S.S.; Matter, I.A.; Elsamahy, T. Nanotextiles waste management: Controlling of release and remediation of wastes. In *Nanosensors and Nanodevices for Smart Multifunctional Textiles*; Elsevier: Amsterdam, the Netherlands, 2021; pp. 267–286.
26. Schoden, F. Ecological and sustainable smart nanotextiles. In *Nanosensors and Nanodevices for Smart Multifunctional Textiles*; Elsevier: Amsterdam, the Netherlands, 2021; pp. 287–320.
27. Yilmaz, N.D. Introduction to smart nanotextiles. In *Smart Textiles: Wearable Nanotechnology*; Wiley: Hoboken, NJ, USA, 2018; Volume 1.
28. Ehrmann, A.; Nguyen, T.A.; Tri, P.N. *Nanosensors and Nanodevices for Smart Multifunctional Textiles*; Elsevier: Amsterdam, The Netherlands, 2020.
29. Abdullaeva, Z. Nanomaterials for Clothing and Textile Products. In *Nanomaterials in Daily Life*; Springer: Cham, Switzerland, 2017; pp. 111–132.
30. Riaz, S.; Ashraf, M.; Hussain, T.; Hussain, M.T.; Younus, A. Fabrication of robust multifaceted textiles by application of functionalized TiO_2 nanoparticles. *Colloids Surf. A Physicochem. Eng. Asp.* **2019**, *581*, 123799. [CrossRef]
31. Almeida, L.; Ramos, D. Health and safety concerns of textiles with nanomaterials. In *IOP Conference Series: Materials Science and Engineering*; IOP Publishing: Bristol, UK, 2017; Volume 254, p. 102002.
32. Brown, P.; Stevens, K. (Eds.) *Nanofibers and Nanotechnology in Textiles*; Elsevier: Amsterdam, the Netherlands, 2007.
33. Radetić, M. Functionalization of textile materials with silver nanoparticles. *J. Mater. Sci.* **2013**, *48*, 95–107. [CrossRef]
34. Ibrahim, N.A. Nanomaterials for antibacterial textiles. In *Nanotechnology in Diagnosis, Treatment and Prophylaxis of Infectious Diseases*; Academic Press: Cape Town, South Africa, 2015; pp. 191–216.
35. Sundarrajan, S.; Chandrasekaran, A.R.; Ramakrishna, S. An update on nanomaterials-based textiles for protection and decontamination. *J. Am. Ceram. Soc.* **2010**, *93*, 3955–3975. [CrossRef]
36. Afzali, A.; Maghsoodlou, S. Engineering nanotextiles: Design of textile products. *Nanostructured Polym. Blends Compos. Text.* **2016**, 1–40. [CrossRef]
37. Sharon, M. Smart Nanofabrics for Defense. In *Nanotechnology in the Defense Industry: Advances, Innovation, and Practical Applications*; Scrivener Publishing LLC: Beverly, MA, USA, 2019; pp. 235–273.
38. El-Naggar, M.E.; Shaarawy, S.; Hebeish, A.A. Multifunctional properties of cotton fabrics coated with in situ synthesis of zinc oxide nanoparticles capped with date seed extract. *Carbohydr. Polym.* **2018**, *181*, 307–316. [CrossRef] [PubMed]
39. Xue, C.H.; Wu, Y.; Guo, X.J.; Liu, B.Y.; Wang, H.D.; Jia, S.T. Superhydrophobic, flame-retardant and conductive cotton fabrics via layer-by-layer assembly of carbon nanotubes for flexible sensing electronics. *Cellulose* **2020**, *27*, 3455–3468. [CrossRef]
40. Gadkari, R.R.; Ali, S.W.; Joshi, M.; Rajendran, S.; Das, A.; Alagirusamy, R. Leveraging antibacterial efficacy of silver loaded chitosan nanoparticles on layer-by-layer self-assembled coated cotton fabric. *Int. J. Biol. Macromol.* **2020**, *162*, 548–560. [CrossRef] [PubMed]
41. Elsayed, E.M.; Attiab, N.F.; Alshehria, L.A. Innovative Flame Retardant and Antibacterial Fabrics Coating Based on Inorganic Nanotubes. *ChemistrySelect* **2020**, *5*, 2961–2965. [CrossRef]
42. Mejía, M.L.; Zapata, J.; Cuesta, D.P.; Ortiz, I.C.; Botero, L.E.; Galeano, B.J.; Escobar, N.J.; Hoyos, L.M. Properties of antibacterial nano textile for use in hospital environments. *Rev. Ing. Biomédica* **2017**, *11*, 13–19. [CrossRef]
43. Banerjee, B. (Ed.) *Rubber Nanocomposites and Nanotextiles: Perspectives in Automobile Technologies*; Walter de Gruyter GmbH & Co KG: Berlin, Germany, 2019.

44. Jadoun, S.; Verma, A.; Arif, R. Modification of Textiles via Nanomaterials and Their Applications. In *Frontiers of Textile Materials: Polymers, Nanomaterials, Enzymes, and Advanced Modification Techniques*; Scrivener Publishing LLC: Beverly, MA, USA, 2020; pp. 135–152.
45. Gokarneshan, N.; Chandrasekar, P.T.; Suvitha, L. Advances in Nanotextile Finishes—An Approach Towards Sustainability. In *Textiles and Clothing Sustainability*; Springer: Singapore, 2017; pp. 1–56.
46. Perera, S.; Bhushan, B.; Bandara, R.; Rajapakse, G.; Rajapakse, S.; Bandara, C. Morphological, antimicrobial, durability, and physical properties of untreated and treated textiles using silver-nanoparticles. *Colloids Surf. A Physicochem. Eng. Asp.* **2013**, *436*, 975–989. [CrossRef]
47. Ferraris, S.; Perero, S.; Miola, M.; Vernè, E.; Rosiello, A.; Ferrazzo, V.; Fokine, M. Chemical, mechanical and antibacterial properties of silver nanocluster/silica composite coated textiles for safety systems and aerospace applications. *Appl. Surf. Sci.* **2014**, *317*, 131–139. [CrossRef]
48. Bashari, A.; Shakeri, M.; Shirvan, A.R.; Najafabadi, S.A.N. Functional finishing of textiles via nanomaterials. In *Nanomaterials in the Wet Processing of Textiles*; Scrivener Publishing LLC: Beverly, MA, USA, 2018; pp. 1–70.
49. Riaz, S.; Ashraf, M.; Hussain, T.; Hussain, M.T.; Rehman, A.; Javid, A.; Iqbal, K.; Basit, A.; Aziz, H. Functional finishing and coloration of textiles with nanomaterials. *Coloration Technol.* **2018**, *134*, 327–346. [CrossRef]
50. Faccini, M.; Vaquero, C.; Amantia, D. Development of protective clothing against nanoparticle based on electrospun nanofibers. *J. Nanomater.* **2012**, *2012*, 892894. [CrossRef]
51. Lund, A.; van der Velden, N.M.; Persson, N.K.; Hamedi, M.M.; Müller, C. Electrically conducting fibres for e-textiles: An open playground for conjugated polymers and carbon nanomaterials. *Mater. Sci. Eng. R Rep.* **2018**, *126*, 1–29. [CrossRef]
52. Silva, I.O.; Ladchumananandasivam, R.; Nascimento, J.H.O.; Silva, K.K.O.; Oliveira, F.R.; Souto, A.P.; Felgueiras, H.P.; Zille, A. Multifunctional chitosan/gold nanoparticles coatings for biomedical textiles. *Nanomaterials* **2019**, *9*, 1064. [CrossRef] [PubMed]
53. Shabbir, M.; Kaushik, M. Engineered nanomaterials: Scope in today's textile industry. In *Handbook of Nanomaterials for Manufacturing Applications*; Elsevier: Amsterdam, The Netherlands, 2020; pp. 249–263.
54. Ul-Islam, S.; Butola, B.S. (Eds.) *Nanomaterials in the Wet Processing of Textiles*; John Wiley & Sons: Hoboken, NJ, USA, 2018.
55. Singh, M.; Vajpayee, M.; Ledwani, L. Eco-friendly Surface Modification and Nanofinishing of Textile Polymers to Enhance Functionalisation. In *Nanotechnology for Energy and Environmental Engineering*; Springer: Cham, Switzerland, 2020; pp. 529–559.
56. Yetisen, A.K.; Qu, H.; Manbachi, A.; Butt, H.; Dokmeci, M.R.; Hinestroza, J.P.; Yun, S.H. Nanotechnology in textiles. *ACS Nano* **2016**, *10*, 3042–3068. [CrossRef] [PubMed]
57. Joshi, M.; Adak, B. Sector 6. Nanotechnology-based Textiles: A Solution for Emerging Automotive Sector. In *Rubber Nanocomposites and Nanotextiles*; De Gruyter: Berlin, Germany, 2018; pp. 207–266.
58. Haji, A.; Mousavi Shoushtari, A.; Mazaheri, F.; Tabatabaeyan, S.E. RSM optimized self-cleaning nano-finishing on polyester/wool fabric pretreated with oxygen plasma. *J. Text. Inst.* **2016**, *107*, 985–994. [CrossRef]
59. Ghosh, G.; Sidpara, A.; Bandyopadhyay, P.P. High efficiency chemical assisted nanofinishing of HVOF sprayed WC-Co coating. *Surf. Coat. Technol.* **2018**, *334*, 204–214. [CrossRef]
60. Osman, E. Nanofinished Medical Textiles and Their Potential Impact to Health and Environment. In *Nanoparticles and Their Biomedical Applications*; Springer: Singapore, 2020; pp. 127–145.
61. Gokarneshan, N.; Velumani, K. Significant Trends in Nano Finishes for Improvement of Functional Properties of Fabrics. In *Handbook of Renewable Materials for Coloration and Finishing*; Scrivener Publishing LLC: Beverly, MA, USA, 2018; pp. 387–434.
62. Ghosh, S.; Smith, T.; Rana, S.; Goswami, P. *Nanofinishing of Textiles for Sportswear*; World Textile Information Network: Dublin, Ireland, 2020.
63. Radetić, M.; Marković, D. Nano-finishing of cellulose textile materials with copper and copper oxide nanoparticles. *Cellulose* **2019**, *26*, 8971–8991. [CrossRef]
64. Nguyen-Tri, P.; Nguyen, T.A.; Carriere, P.; Ngo Xuan, C. Nanocomposite coatings: Preparation, characterization, properties, and applications. *Int. J. Corros.* **2018**, *2018*. [CrossRef]
65. Joshi, M.; Khanna, R.; Shekhar, R.; Jha, K. Chitosan nanocoating on cotton textile substrate using layer-by-layer self-assembly technique. *J. Appl. Polym. Sci.* **2011**, *119*, 2793–2799. [CrossRef]

66. Smole, M.S.; Stakne, K.; Bele, M.; Jamnik, J.; Hribernik, S.; Ribitsch, V. Nanocoatings for textile. In Proceedings of the 3rd International Textile Clothing & Design Conference, Dubrovnik, Croatia, 8–11 October 2006; pp. 8–11.
67. Peng, L.; Chen, W.; Su, B.; Yu, A.; Jiang, X. CsxWO$_3$ nanosheet-coated cotton fabric with multiple functions: UV/NIR shielding and full-spectrum-responsive self-cleaning. *Appl. Surf. Sci.* **2019**, *475*, 325–333. [CrossRef]
68. Temesgen, A.G.; Turşucular, Ö.F.; Eren, R.; Ulcay, Y. Novel Applications of Nanotechnology in Modification of Textile Fabrics Properties and Apparel. *Int. J. Adv. Multidiscip. Res.* **2018**, *5*, 49–58.
69. Almetwally, A.A.; El-Sakhawy, M.; Elshakankery, M.H.; Kasem, M.H. Technology of nano-fibers: Production techniques and properties—Critical review. *J. Text. Assoc.* **2017**, *78*, 5–14.
70. Nayak, R.; Khandual, A. 18 Nanotextiles and Recent Developments. In *Fibres to Smart Textiles: Advances in Manufacturing, Technologies, and Applications*; CRC Press: Boca Raton, FL, USA, 2019.
71. Naeem, M.A.; Alfred, M.; Saba, H.; Siddiqui, Q.; Naveed, T.; Shahbaz, U.; Wei, Q. A preliminary study on the preparation of seamless tubular bacterial cellulose-electrospun nanofibers-based nanocomposite fabrics. *J. Compos. Mater.* **2019**, *53*, 3715–3724. [CrossRef]
72. Qi, K.; Zhou, Y.; Ou, K.; Dai, Y.; You, X.; Wang, H.; He, J.; Qin, X.; Wang, R. Weavable and stretchable piezoresistive carbon nanotubes-embedded nanofiber sensing yarns for highly sensitive and multimodal wearable textile sensor. *Carbon* **2020**, *170*, 464–476. [CrossRef]
73. Tebyetekerwa, M.; Xu, Z.; Yang, S.; Ramakrishna, S. Electrospun nanofibers-based face masks. *Adv. Fiber Mater.* **2020**, *2*, 161–166. [CrossRef]
74. Montazer, M.; Harifi, T. Conductive nanofinishes for textiles. In *Nanofinishing Textile Materials*; Elsevier: Amsterdam, the Netherlands, 2018; pp. 241–263.
75. Malhotra, B.D.; Ali, M.A. *Nanomaterials for Biosensors: Fundamentals and Applications*; William Andrew: Norwich, NY, USA, 2017.
76. Attia, N.F.; Morsy, M.S. Facile synthesis of novel nanocomposite as antibacterial and flame retardant material for textile fabrics. *Mater. Chem. Phys.* **2016**, *180*, 364–372. [CrossRef]
77. Göcek, İ. Functionalization of textile materials with nanoclay incorporation for improved characteristics. *Politek. Derg.* **2019**, *22*, 509–522. [CrossRef]
78. Korkmaz, N.; Alay Aksoy, S. Enhancing the performance properties of ester-cross-linked cotton fabrics using Al$_2$O$_3$-NPs. *Text. Res. J.* **2016**, *86*, 636–648. [CrossRef]
79. Dogan, O.; Dag, R. Application of nano coating (SiO$_2$) on textile products. *J. Chem. Chem. Eng.* **2017**, *11*, 82–85.
80. Verbič, A.; Gorjanc, M.; Simončič, B. Zinc oxide for functional textile coatings: Recent advances. *Coatings* **2019**, *9*, 550. [CrossRef]
81. Abbas, M.; Iftikhar, H.; Malik, M.H.; Nazir, A. Surface coatings of TiO$_2$ nanoparticles onto the designed fabrics for enhanced self-cleaning properties. *Coatings* **2018**, *8*, 35. [CrossRef]
82. Xu, Q.; Xie, L.; Diao, H.; Li, F.; Zhang, Y.; Fu, F.; Liu, X. Antibacterial cotton fabric with enhanced durability prepared using silver nanoparticles and carboxymethyl chitosan. *Carbohydr. Polym.* **2017**, *177*, 187–193. [CrossRef]
83. Vigneshwaran, N.; Arputharaj, A. Functional Finishing of Cotton Textiles Using Nanomaterials. In *Advances in Functional Finishing of Textiles*; Springer: Singapore, 2020; pp. 43–56.
84. Erdem, N.; Cireli, A.A.; Erdogan, U.H. Flame retardancy behaviors and structural properties of polypropylene/nano-SiO$_2$ composite textile filaments. *J. Appl. Polym. Sci.* **2009**, *111*, 2085–2091. [CrossRef]
85. Attia, N.; Ahmed, H.; Yehia, D.; Hassan, M.; Zaddin, Y. Novel synthesis of nanoparticles-based back coating flame-retardant materials for historic textile fabrics conservation. *J. Ind. Text.* **2017**, *46*, 1379–1392. [CrossRef]
86. Shariatinia, Z.; Javeri, N.; Shekarriz, S. Flame retardant cotton fibers produced using novel synthesized halogen-free phosphoramide nanoparticles. *Carbohydr. Polym.* **2015**, *118*, 183–198. [CrossRef]
87. Kundu, C.K.; Song, L.; Hu, Y. Nanoparticles based coatings for multifunctional Polyamide 66 textiles with improved flame retardancy and hydrophilicity. *J. Taiwan Inst. Chem. Eng.* **2020**, *112*, 15–19. [CrossRef]
88. Rivero, P.J.; Urrutia, A.; Goicoechea, J.; Arregui, F.J. Nanomaterials for functional textiles and fibers. *Nanoscale Res. Lett.* **2015**, *10*, 501. [CrossRef]
89. Ortelli, S.; Malucelli, G.; Blosi, M.; Zanoni, I.; Costa, A.L. NanoTiO$_2$@DNA complex: A novel eco, durable, fire retardant design strategy for cotton textiles. *J. Colloid Interface Sci.* **2019**, *546*, 174–183. [CrossRef]

90. Ortelli, S.; Malucelli, G.; Cuttica, F.; Blosi, M.; Zanoni, I.; Costa, A.L. Coatings made of proteins adsorbed on TiO$_2$ nanoparticles: A new flame retardant approach for cotton fabrics. *Cellulose* **2018**, *25*, 2755–2765. [CrossRef]
91. Carosio, F.; Alongi, J.; Frache, A.; Malucelli, G.; Camino, G. Textile flame retardancy through surface-assembled nanoarchitectures. In *Fire and Polymers VI: New Advances in Flame Retardant Chemistry and Science*; American Chemical Society: Washington, DC, USA, 2012; pp. 327–341.
92. Saleemi, S.; Naveed, T.; Riaz, T.; Memon, H.; Awan, J.A.; Siyal, M.I.; Xu, F.; Bae, J. Surface Functionalization of Cotton and PC Fabrics Using SiO$_2$ and ZnO Nanoparticles for Durable Flame Retardant Properties. *Coatings* **2020**, *10*, 124. [CrossRef]
93. Ali, S.W.; Basak, S.; Shukla, A. Nanoparticles: A potential alternative to classical fire retardants for textile substrates. In *Handbook of Nanomaterials for Manufacturing Applications*; Elsevier: Amsterdam, the Netherlands, 2020; pp. 265–278.
94. Butola, B.S. (Ed.) *Advances in Functional and Protective Textiles*; Woodhead Publishing: Cambridge, UK, 2020.
95. Li, Y.; Wang, B.; Sui, X.; Xie, R.; Xu, H.; Zhang, L.; Zhong, Y.; Mao, Z. Durable flame retardant and antibacterial finishing on cotton fabrics with cyclotriphosphazene/polydopamine/silver nanoparticles hybrid coatings. *Appl. Surf. Sci.* **2018**, *435*, 1337–1343. [CrossRef]
96. Sharma, V.; Basak, S.; Rishabh, K.; Umaria, H.; Ali, S.W. Synthesis of zinc carbonate nanoneedles, a potential flame retardant for cotton textiles. *Cellulose* **2018**, *25*, 6191–6205. [CrossRef]
97. Norouzi, M.; Zare, Y.; Kiany, P. Nanoparticles as effective flame retardants for natural and synthetic textile polymers: Application, mechanism, and optimization. *Polym. Rev.* **2015**, *55*, 531–560. [CrossRef]
98. Yazhini, K.B.; Prabu, H.G. Study on flame-retardant and UV-protection properties of cotton fabric functionalized with ppy–ZnO–CNT nanocomposite. *RSC Adv.* **2015**, *5*, 49062–49069. [CrossRef]
99. Fanglong, Z.; Qun, X.; Qianqian, F.; Rangtong, L.; Kejing, L. Influence of nano-silica on flame resistance behavior of intumescent flame retardant cellulosic textiles: Remarkable synergistic effect? *Surf. Coat. Technol.* **2016**, *294*, 90–94. [CrossRef]
100. Dhineshbabu, N.R.; Bose, S. UV resistant and fire retardant properties in fabrics coated with polymer based nanocomposites derived from sustainable and natural resources for protective clothing application. *Compos. Part B Eng.* **2019**, *172*, 555–563. [CrossRef]
101. Sedighi, A.; Montazer, M.; Mazinani, S. Fabrication of electrically conductive superparamagnetic fabric with microwave attenuation, antibacterial properties and UV protection using PEDOT/magnetite nanoparticles. *Mater. Des.* **2018**, *160*, 34–47. [CrossRef]
102. Kathirvelu, S.; D'souza, L.; Dhurai, B. UV protection finishing of textiles using ZnO nanoparticles. *Indian J. Fibre Text. Res.* **2009**, *34*, 267–273.
103. Becheri, A.; Dürr, M.; Nostro, P.L.; Baglioni, P. Synthesis and characterization of zinc oxide nanoparticles: Application to textiles as UV-absorbers. *J. Nanopart. Res.* **2008**, *10*, 679–689. [CrossRef]
104. Fouda, A.; Saad, E.L.; Salem, S.S.; Shaheen, T.I. In-Vitro cytotoxicity, antibacterial, and UV protection properties of the biosynthesized Zinc oxide nanoparticles for medical textile applications. *Microb. Pathog.* **2018**, *125*, 252–261. [CrossRef]
105. Tsuzuki, T.; Wang, X. Nanoparticle coatings for UV protective textiles. *Res. J. Text. Appar.* **2010**, *14*, 9–20. [CrossRef]
106. Çakır, B.A.; Budama, L.; Topel, Ö.; Hoda, N. Synthesis of ZnO nanoparticles using PS-b-PAA reverse micelle cores for UV protective, self-cleaning and antibacterial textile applications. In *Colloids and Surfaces A: Physicochemical and Engineering Aspects*; Elsevier: Amsterdam, the Netherlands, 2012; Volume 414, pp. 132–139.
107. Farouk, A.; Textor, T.; Schollmeyer, E.; Tarbuk, A.; Grancacic, A.M. Sol-gel-derived inorganic-organic hybrid polymers filled with zno nanoparticles as an ultraviolet protection finish for textiles. *AUTEX Res. J.* **2010**, *10*, 58–63.
108. Radetić, M. Functionalization of textile materials with TiO$_2$ nanoparticles. *J. Photochem. Photobiol. C Photochem. Rev.* **2013**, *16*, 62–76. [CrossRef]
109. Attia, N.F.; Moussa, M.; Sheta, A.M.; Taha, R.; Gamal, H. Synthesis of effective multifunctional textile based on silica nanoparticles. *Prog. Org. Coat.* **2017**, *106*, 41–49. [CrossRef]
110. Attia, N.F.; Moussa, M.; Sheta, A.M.; Taha, R.; Gamal, H. Effect of different nanoparticles based coating on the performance of textile properties. *Prog. Org. Coat.* **2017**, *104*, 72–80. [CrossRef]

111. Lee, S. Developing UV-protective textiles based on electrospun zinc oxide nanocomposite fibers. *Fibers Polym.* **2009**, *10*, 295–301. [CrossRef]
112. Dhineshbabu, N.R.; Bose, S. Smart textiles coated with eco-friendly UV-blocking nanoparticles derived from natural resources. *ACS Omega* **2018**, *3*, 7454–7465. [CrossRef]
113. Noorian, S.A.; Hemmatinejad, N.; Navarro, J.A. Ligand modified cellulose fabrics as support of zinc oxide nanoparticles for UV protection and antimicrobial activities. *Int. J. Biol. Macromol.* **2020**, *154*, 1215–1226. [CrossRef]
114. Ahmed, H.B.; El-Hawary, N.S.; Emam, H.E. Self-assembled AuNPs for ingrain pigmentation of silk fabrics with antibacterial potency. *Int. J. Biol. Macromol.* **2017**, *105*, 720–729. [CrossRef]
115. Bu, Y.; Zhang, S.; Cai, Y.; Yang, Y.; Ma, S.; Huang, J.; Yang, H.; Ye, D.; Zhou, Y.; Xu, W.; et al. Fabrication of durable antibacterial and superhydrophobic textiles via in situ synthesis of silver nanoparticle on tannic acid-coated viscose textiles. *Cellulose* **2019**, *26*, 2109–2122. [CrossRef]
116. Hebeish, A.; El-Naggar, M.E.; Fouda, M.M.; Ramadan, M.A.; Al-Deyab, S.S.; El-Rafie, M.H. Highly effective antibacterial textiles containing green synthesized silver nanoparticles. *Carbohydr. Polym.* **2011**, *86*, 936–940. [CrossRef]
117. Xue, C.H.; Chen, J.; Yin, W.; Jia, S.T.; Ma, J.Z. Superhydrophobic conductive textiles with antibacterial property by coating fibers with silver nanoparticles. *Appl. Surf. Sci.* **2012**, *258*, 2468–2472. [CrossRef]
118. Zhang, F.; Wu, X.; Chen, Y.; Lin, H. Application of silver nanoparticles to cotton fabric as an antibacterial textile finish. *Fibers Polym.* **2009**, *10*, 496–501. [CrossRef]
119. Budama, L.; Çakır, B.A.; Topel, Ö.; Hoda, N. A new strategy for producing antibacterial textile surfaces using silver nanoparticles. *Chem. Eng. J.* **2013**, *228*, 489–495. [CrossRef]
120. Liu, H.; Lv, M.; Deng, B.; Li, J.; Yu, M.; Huang, Q.; Fan, C. Laundering durable antibacterial cotton fabrics grafted with pomegranate-shaped polymer wrapped in silver nanoparticle aggregations. *Sci. Rep.* **2014**, *4*, 5920. [CrossRef]
121. Perelshtein, I.; Applerot, G.; Perkas, N.; Guibert, G.; Mikhailov, S.; Gedanken, A. Sonochemical coating of silver nanoparticles on textile fabrics (nylon, polyester and cotton) and their antibacterial activity. *Nanotechnology* **2008**, *19*, 245705. [CrossRef] [PubMed]
122. Zhang, G.; Liu, Y.; Gao, X.; Chen, Y. Synthesis of silver nanoparticles and antibacterial property of silk fabrics treated by silver nanoparticles. *Nanoscale Res. Lett.* **2014**, *9*, 216. [CrossRef] [PubMed]
123. Attia, N.F.; Elashery, S.E.; Oh, H. Nanomaterials-based antibacterial textiles. In *Nanosensors and Nanodevices for Smart Multifunctional Textiles*; Elsevier: Amsterdam, The Netherlands, 2020; pp. 135–147.
124. Ullah, N.; Yasin, S.; Abro, Z.; Liu, L.; Wei, Q. Mechanically robust and antimicrobial cotton fibers loaded with silver nanoparticles: Synthesized via Chinese holly plant leaves. *Int. J. Text. Sci.* **2014**, *3*, 1–5.
125. Ouadil, B.; Amadine, O.; Essamlali, Y.; Cherkaoui, O.; Zahouily, M. A new route for the preparation of hydrophobic and antibacterial textiles fabrics using Ag-loaded graphene nanocomposite. In *Colloids and Surfaces A: Physicochemical and Engineering Aspects*; Elsevier: Amsterdam, The Netherlands, 2019; Volume 579, p. 123713.
126. Dizaj, S.M.; Lotfipour, F.; Barzegar-Jalali, M.; Zarrintan, M.H.; Adibkia, K. Antimicrobial activity of the metals and metal oxide nanoparticles. *Mater. Sci. Eng. C* **2014**, *44*, 278–284. [CrossRef]
127. Wu, Y.; Yang, Y.; Zhang, Z.; Wang, Z.; Zhao, Y.; Sun, L. Fabrication of cotton fabrics with durable antibacterial activities finishing by Ag nanoparticles. *Text. Res. J.* **2019**, *89*, 867–880. [CrossRef]
128. Toh, H.S.; Faure, R.L.; Amin, L.B.M.; Hay, C.Y.F.; George, S. A light-assisted in situ embedment of silver nanoparticles to prepare functionalized fabrics. *Nanotechnol. Sci. Appl.* **2017**, *10*, 147. [CrossRef]
129. Vasantharaj, S.; Sathiyavimal, S.; Saravanan, M.; Senthilkumar, P.; Gnanasekaran, K.; Shanmugavel, M.; Pugazhendhi, A. Synthesis of ecofriendly copper oxide nanoparticles for fabrication over textile fabrics: Characterization of antibacterial activity and dye degradation potential. *J. Photochem. Photobiol. B Biol.* **2019**, *191*, 143–149. [CrossRef]
130. El-Naggar, M.E.; Shaheen, T.I.; Zaghloul, S.; El-Rafie, M.H.; Hebeish, A. Antibacterial activities and UV protection of the in situ synthesized titanium oxide nanoparticles on cotton fabrics. *Ind. Eng. Chem. Res.* **2016**, *55*, 2661–2668. [CrossRef]
131. 131 Jia, L.; Huang, X.; Tao, Q. Enhanced hydrophilic and antibacterial efficiencies by the synergetic effect TiO_2 nanofiber and graphene oxide in cellulose acetate nanofibers. *Int. J. Biol. Macromol.* **2019**, *132*, 1039–1043. [CrossRef] [PubMed]

132. Asif, A.K.M.A.H.; Hasan, M.Z. Application of nanotechnology in modern textiles: A review. *Int. J. Curr. Eng. Technol.* **2018**, *8*, 227–231.
133. Katiyar, P.; Mishra, S.; Srivastava, A.; Prasad, N.E. Preparation of TiO_2–SiO_2 Hybrid Nanosols Coated Flame-Retardant Polyester Fabric Possessing Dual Contradictory Characteristics of Superhydrophobicity and Self Cleaning Ability. *J. Nanosci. Nanotechnol.* **2020**, *20*, 1780–1789. [CrossRef]
134. Montazer, S.T.M. Denim Fabric with Flame retardant, hydrophilic and self-cleaning properties conferring by in-situ synthesis of silica nanoparticles. *Cellulose* **2020**, *27*, 6643–6661.
135. Das, I.; De, G. Zirconia based superhydrophobic coatings on cotton fabrics exhibiting excellent durability for versatile use. *Sci. Rep.* **2015**, *5*, 18503. [CrossRef]
136. Joshi, M.; Bhattacharyya, A.; Agarwal, N.; Parmar, S. Nanostructured coatings for super hydrophobic textiles. *Bull. Mater. Sci.* **2012**, *35*, 933–938. [CrossRef]
137. Wang, R.; Wang, X.; Xin, J.H. Advanced visible-light-driven self-cleaning cotton by $Au/TiO_2/SiO_2$ photocatalysts. *ACS Appl. Mater. Interfaces* **2010**, *2*, 82–85. [CrossRef]
138. Hebeish, A.A.; Abdelhady, M.M.; Youssef, A.M. TiO_2 nanowire and TiO_2 nanowire doped Ag-PVP nanocomposite for antimicrobial and self-cleaning cotton textile. *Carbohydr. Polym.* **2013**, *91*, 549–559. [CrossRef]
139. Karimi, L.; Yazdanshenas, M.E.; Khajavi, R.; Rashidi, A.; Mirjalili, M. Using graphene/TiO_2 nanocomposite as a new route for preparation of electroconductive, self-cleaning, antibacterial and antifungal cotton fabric without toxicity. *Cellulose* **2014**, *21*, 3813–3827. [CrossRef]
140. Landi, S., Jr.; Carneiro, J.; Soares, O.S.; Pereira, M.F.; Gomes, A.C.; Ribeiro, A.; António, M.F.; Pier, P.; Neves, I.C. Photocatalytic performance of N-doped TiO_2nano-SiO_2-HY nanocomposites immobilized over cotton fabrics. *J. Mater. Res. Technol.* **2019**, *8*, 1933–1943. [CrossRef]
141. Xu, B.; Ding, J.; Feng, L.; Ding, Y.; Ge, F.; Cai, Z. Self-cleaning cotton fabrics via combination of photocatalytic TiO_2 and superhydrophobic SiO_2. *Surf. Coat. Technol.* **2015**, *262*, 70–76. [CrossRef]
142. Joshi, M.; Adak, B. Comprehensive Nanoscience and Nanotechnology (Second Edition). In *Advances in Nanotechnology based Functional, Smart and Intelligent Textiles: A Review*; Elsevier: Amsterdam, the Netherlands, 2019.
143. Priyadarshinirajkumar, A.; Raja, N.V. Innovative herbal nanofinishing on cotton fabric. *Int. J. Fiber Text. Res.* **2015**, *5*, 44.
144. Haque, M. Nano Fabrics in the 21st century: A review. *Asian J. Nanosci. Mater.* **2019**, *2*, 120–256, 131–148.
145. Tripathi, R.; Narayan, A.; Bramhecha, I.; Sheikh, J. Development of multifunctional linen fabric using chitosan film as a template for immobilization of in-situ generated CeO_2 nanoparticles. *Int. J. Biol. Macromol.* **2019**, *121*, 1154–1159. [CrossRef] [PubMed]
146. Hezavehi, E.; Shahidi, S.; Zolgharnein, P. Effect of dyeing on wrinkle properties of cotton cross-linked by Butane Tetracarboxylic Acid (BTCA) in presence of Titanium Dioxide (TiO_2) Nanoparticles. *Autex Res. J.* **2015**, *15*, 104–111. [CrossRef]
147. Uğur, Ş.S.; Bilgiç, M. A novel approach for improving wrinkle resistance and flame retardancy properties of linen fabrics. *Bilge Int. J. Sci. Technol. Res.* **2017**, *1*, 79–86.
148. Memon, H.; Wang, H.; Yasin, S.; Halepoto, A. Influence of incorporating silver nanoparticles in protease treatment on fiber friction, antistatic, and antibacterial properties of wool fibers. *J. Chem.* **2018**, *2018*. [CrossRef]
149. Zhang, F.; Yang, J. Preparation of nano-ZnO and its application to the textile on antistatic finishing. *Int. J. Chem.* **2009**, *1*, 18. [CrossRef]
150. Hassan, M.M. Enhanced colour, hydrophobicity, UV radiation absorption and antistatic properties of wool fabric multi-functionalised with silver nanoparticles. In *Colloids and Surfaces A: Physicochemical and Engineering Aspects*; Elsevier: Amsterdam, the Netherlands, 2019; Volume 581.
151. Hossain, M.; Biswas, A.K.; Parvez, S.; Bain, S. Advanced functionalization of textiles by nanofinishing: A review. In Proceedings of the International Conference on Mechanical, Industrial and Materials Engineering, Rajshahi, Bangladesh, 1–3 November 2013; pp. 503–507.
152. Yadav, A.; Prasad, V.; Kathe, A.A.; Raj, S.; Yadav, D.; Sundaramoorthy, C.; Vigneshwaran, N. Functional finishing in cotton fabrics using zinc oxide nanoparticles. *Bull. Mater. Sci.* **2006**, *29*, 641–645. [CrossRef]
153. Harifi, T.; & Montazer, M. Application of nanotechnology in sports clothing and flooring for enhanced sport activities, performance, efficiency and comfort: A review. *J. Ind. Text.* **2017**, *46*, 1147–1169. [CrossRef]

154. Shalaby, M.N.; Saad, M.M. Advanced Material Engineering and Nanotechnology for Improving Sports Performance and Equipment. *Int. J. Psychosoc. Rehabil.* **2020**, *24*, 1–9.
155. Yu, L. A Study on Decomposition of Nanoparticle Finished Textiles. *Int. J. Nanosci. Nanoeng.* **2018**, *4*, 80.
156. Kausar, A. Textile Nanocomposite of Polymer/Carbon Nanotube. *Am. J. Nanosci. Nanotechnol. Res.* **2018**, *6*, 28–35.
157. Rovira, J.; Domingo, J.L. Human health risks due to exposure to inorganic and organic chemicals from textiles: A review. *Environ. Res.* **2019**, *168*, 62–69. [CrossRef]
158. Montazer, M.; Harifi, T. *Nanofinishing of Textile Materials*; Woodhead Publishing: Cambridge, UK, 2018.
159. Köhler, A.R.; Som, C. Risk preventative innovation strategies for emerging technologies the cases of nano-textiles and smart textiles. *Technovation* **2014**, *34*, 420–430. [CrossRef]
160. Geranio, L.; Heuberger, M.; Nowack, B. The behavior of silver nanotextiles during washing. *Environ. Sci. Technol.* **2014**, *43*, 8113–8118. [CrossRef]
161. Lorenz, C.; Windler, L.; von Goetz, N.; Lehmann, R.P.; Schuppler, M.; Hungerbühler, K.; Nowack, B. Characterization of silver release from commercially available functional (nano) textiles. *Chemosphere* **2012**, *89*, 817–824. [CrossRef]
162. Rather, L.J.; Zhou, Q.; Ganie, S.A.; Li, Q. Environmental Profile of Nano-finished Textile Materials: Implications on Public Health, Risk Assessment, and Public Perception. In *Advances in Functional Finishing of Textiles*; Springer: Singapore, 2020; pp. 57–83.
163. Wagener, S.; Dommershausen, N.; Jungnickel, H.; Laux, P.; Mitrano, D.; Nowack, B.; Schneider, G.; Luch, A. Textile functionalization and its effects on the release of silver nanoparticles into artificial sweat. *Environ. Sci. Technol.* **2016**, *50*, 5927–5934. [CrossRef]
164. Kulthong, K.; Srisung, S.; Boonpavanitchakul, K.; Kangwansupamonkon, W.; Maniratanachote, R. Determination of silver nanoparticle release from antibacterial fabrics into artificial sweat. *Part. Fibre Toxicol.* **2010**, *7*, 8. [CrossRef]
165. Kim, J.B.; Kim, J.Y.; Yoon, T.H. Determination of silver nanoparticle species released from textiles into artificial sweat and laundry wash for a risk assessment. *Hum. Ecol. Risk Assess. Int. J.* **2017**, *23*, 741–750. [CrossRef]
166. von Goetz, N.; Lorenz, C.; Windler, L.; Nowack, B.; Heuberger, M.; Hungerbuhler, K. Migration of Ag-and TiO_2-(Nano) particles from textiles into artificial sweat under physical stress: Experiments and exposure modeling. *Environ. Sci. Technol.* **2013**, *47*, 9979–9987. [CrossRef] [PubMed]
167. Spielman-Sun, E.; Zaikova, T.; Dankovich, T.; Yun, J.; Ryan, M.; Hutchison, J.E.; Lowry, G.V. Effect of silver concentration and chemical transformations on release and antibacterial efficacy in silver-containing textiles. *NanoImpact* **2018**, *11*, 51–57. [CrossRef]
168. Stefaniak, A.B.; Duling, M.G.; Lawrence, R.B.; Thomas, T.A.; LeBouf, R.F.; Wade, E.E.; Abbas Virji, M. Dermal exposure potential from textiles that contain silver nanoparticles. *Int. J. Occup. Environ. Health* **2014**, *20*, 220–234. [CrossRef] [PubMed]
169. Balakumaran, M.D.; Ramachandran, R.; Jagadeeswari, S.; Kalaichelvan, P.T. In vitro biological properties and characterization of nanosilver coated cotton fabrics–An application for antimicrobial textile finishing. *Int. Biodeterior. Biodegrad.* **2016**, *107*, 48–55. [CrossRef]
170. Milošević, M.; Radoičić, M.; Šaponjić, Z.; Nunney, T.; Deeks, C.; Lazić, V.; Mitrić, M.; Radetić, T.; Radetić, M. In situ photoreduction of Ag+-ions by TiO_2 nanoparticles deposited on cotton and cotton/PET fabrics. *Cellulose* **2014**, *21*, 3781–3795. [CrossRef]
171. Cucurachi, S.; Rocha, C.F.B. Life-cycle assessment of engineered nanomaterials. In *Nanotechnology in Eco-efficient Construction*; Woodhead Publishing: Cambridge, UK, 2019; pp. 815–846.
172. Patnaik, A.; Patnaik, S. (Eds.) *Fibres to Smart Textiles: Advances in Manufacturing, Technologies, and Applications*; CRC Press: Boca Raton, FL, USA, 2019.
173. Mitrano, D.M.; Limpiteeprakan, P.; Babel, S.; Nowack, B. Durability of nano-enhanced textiles through the life cycle: Releases from landfilling after washing. *Environ. Sci. Nano* **2016**, *3*, 375–387. [CrossRef]
174. Som, C.; Wick, P.; Krug, H.; Nowack, B. Environmental and health effects of nanomaterials in nanotextiles and facade coatings. *Environ. Int.* **2011**, *37*, 1131–1142. [CrossRef]
175. Harifi, T.; Montazer, M. In situ synthesis of iron oxide nanoparticles on polyester fabric utilizing color, magnetic, antibacterial and sono-Fenton catalytic properties. *J. Mater. Chem. B* **2014**, *2*, 272–282. [CrossRef]
176. Temizel-Sekeryan, S.; Hicks, A.L. Global environmental impacts of silver nanoparticle production methods supported by life cycle assessment. *Resour. Conserv. Recycl.* **2020**, *156*, 104676. [CrossRef]

177. Piontek, F.M.; Müller, M. Literature reviews: Life cycle assessment in the context of product-service systems and the textile industry. *Procedia CIRP* **2018**, *69*, 758–763. [CrossRef]
178. Black Carbon. Screening Assessment for the Challenge. In *Chemical Abstracts Service Registry Number 1333–86–4*; Environment Canada: Ottawa, ON, Canada, 2013.
179. Sahu, S.C.; Hayes, A.W. Toxicity of nanomaterials found in human environment: A literature review. *Toxicol. Res. Appl.* **2017**, *1*, 2397847317726352. [CrossRef]
180. Alanezi, A.M. Impact of Pollution Generated by the Textile Industry on Health and Environment. *J. Univ. Stud. Incl. Res.* **2018**, *2*, 160–176.
181. Murphy, F.; Tchetchik, A.; Furxhi, I. Reduction of Health Care-Associated Infections (HAIs) with Antimicrobial Inorganic Nanoparticles Incorporated in Medical Textiles: An Economic Assessment. *Nanomaterials* **2020**, *10*, 999. [CrossRef] [PubMed]
182. Yu, L. An Overview on the Fate of Fast Fashion and Nanoparticle Treated Textiles. *Res. J. Nanosci. Eng.* **2018**, *2*, 30–33.
183. Abdelrahman, M.S.; Nassar, S.H.; Mashaly, H.; Mahmoud, S.; Maamoun, D.; El-Sakhawy, M.; Khattab, T.A.; Kamel, S. Studies of polylactic acid and metal oxide nanoparticles-based composites for multifunctional textile prints. *Coatings* **2020**, *10*, 58. [CrossRef]
184. Torabifard, M.; Arjmandi, R.; Rashidi, A.; Nouri, J.; Mohammadfam, I. Inherent health and environmental risk assessment of nanostructured metal oxide production processes. *Environ. Monit. Assess.* **2018**, *190*, 73. [CrossRef]
185. Tavares, A.J.; Poon, W.; Zhang, Y.N.; Dai, Q.; Besla, R.; Ding, D.; Ouyang, B.; Li, A.; Chen, J.; Zheng, G.; et al. Effect of removing Kupffer cells on nanoparticle tumor delivery. *Proc. Natl. Acad. Sci. USA* **2017**, *114*, E10871–E10880. [CrossRef]
186. de Jong, W.H.; Oomen, A.G.; Tran, L.; Chaudhry, Q.; Lefebvre, D.E. Engineered Nanoparticles and Food: Exposure, Toxicokinetics, Hazards and Risks. In *Nanotechnologies in Food*; Royal Society of Chemistry: Milton, CA, USA, 2017; pp. 200–227.
187. Aryal, S.; Park, H.; Leary, J.F.; Key, J. Top-down fabrication-based nano/microparticles for molecular imaging and drug delivery. *Int. J. Nanomed.* **2019**, *14*, 6631. [CrossRef]
188. Sukhanova, A.; Bozrova, S.; Sokolov, P.; Berestovoy, M.; Karaulov, A.; Nabiev, I. Dependence of nanoparticle toxicity on their physical and chemical properties. *Nanoscale Res. Lett.* **2018**, *13*, 44. [CrossRef]
189. Ramachandran, M.S. *Heart and Toxins*; Elsevier: Amsterdam, the Netherlands, 2014.
190. Filon, F.L.; Bello, D.; Cherrie, J.W.; Sleeuwenhoek, A.; Spaan, S.; Brouwer, D.H. Occupational dermal exposure to nanoparticles and nano-enabled products: Part I—Factors affecting skin absorption. *Int. J. Hyg. Environ. Health* **2016**, *219*, 536–544. [CrossRef] [PubMed]
191. Rezić, I. Engineered nanoparticles in textiles and textile wastewaters. In *Comprehensive Analytical Chemistry*; Elsevier: Amsterdam, the Netherlands, 2012; Volume 59, pp. 235–264.
192. Exbrayat, J.M.; Moudilou, E.N.; Lapied, E. Harmful effects of nanoparticles on animals. *J. Nanotechnol.* **2015**, *2015*. [CrossRef]
193. Roberto, M.M.; Christofoletti, C.A. How to Assess Nanomaterial Toxicity? An Environmental and Human Health Approach. In *Nanomaterials-Toxicity, Human Health and Environment*; IntechOpen: London, UK, 2019.
194. Grumezescu, A.M. (Ed.) *Nanoscale Fabrication, Optimization, Scale-Up and Biological Aspects of Pharmaceutical Nanotechnology*; William Andrew: Norwich, NY, USA, 2017.
195. Nallanthighal, S.; Chan, C.; Murray, T.M.; Mosier, A.P.; Cady, N.C.; Reliene, R. Differential effects of silver nanoparticles on DNA damage and DNA repair gene expression in Ogg1-deficient and wild type mice. *Nanotoxicology* **2017**, *11*, 996–1011. [CrossRef]
196. Adabi, M.; Naghibzadeh, M.; Adabi, M.; Zarrinfard, M.A.; Esnaashari, S.S.; Seifalian, A.M.; Faridi-Majidi, R.; Aiyelabegan, H.T.; Ghanbari, H. Biocompatibility and nanostructured materials: Applications in nanomedicine. *Artif. Cells Nanomed. Biotechnol.* **2017**, *45*, 833–842. [CrossRef] [PubMed]
197. Yin, Y.; Yu, S.; Yang, X.; Liu, J.; Jiang, G. Source and pathway of silver nanoparticles to the environment. In *Silver Nanoparticles in the Environment*; Springer: Berlin/Heidelberg, Germany, 2015; pp. 43–72.
198. Ju-Nam, Y.; Lead, J.R. Manufactured nanoparticles: An overview of their chemistry, interactions and potential environmental implications. *Sci. Total Environ.* **2008**, *400*, 396–414. [CrossRef]
199. Drasler, B.; Sayre, P.; Steinhäuser, K.G.; Petri-Fink, A.; Rothen-Rutishauser, B. In vitro approaches to assess the hazard of nanomaterials. *NanoImpact* **2017**, *8*, 99–116. [CrossRef]

200. Juillerat-Jeanneret, L.; Dusinska, M.; Fjellsbø, L.M.; Collins, A.R.; Handy, R.D.; Riediker, M.; NanoTEST Consortium. Biological impact assessment of nanomaterial used in nanomedicine. Introduction to the NanoTEST project. *Nanotoxicology* **2015**, *9* (Suppl. 1), 5–12. [CrossRef] [PubMed]
201. Rothen-Rutishauser, B.; Clift, M.J.; Jud, C.; Fink, A.; Wick, P. Human epithelial cells in vitro–Are they an advantageous tool to help understand the nanomaterial-biological barrier interaction? *Euro NanoTox Lett.* **2012**, *4*, 1–20. [CrossRef]
202. Collins, A.; El Yamani, N.; Dusinska, M. Sensitive detection of DNA oxidation damage induced by nanomaterials. *Free Radic. Biol. Med.* **2017**, *107*, 69–76. [CrossRef]
203. Catalán, J.; Stockmann-Juvala, H.; Norppa, H. A theoretical approach for a weighted assessment of the mutagenic potential of nanomaterials. *Nanotoxicology* **2017**, *11*, 964–977. [CrossRef]
204. OECD. *Draft Guidance Document on the Use of OECD Test Guidelines on Genotoxicity Testing for Manufactured Nanomaterials*; Organization for Economic Co-operation and Development: Berlin, Germany, 2014; pp. 1–3. Available online: http://www.oecd.org/env/ehs/testing/Draft_GD_nano_and_genotox_rev2.pdf (accessed on 9 September 2020).
205. Voelker, D.; Schlich, K.; Hohndorf, L.; Koch, W.; Kuehnen, U.; Polleichtner, C.; Kussatz, C.; Hund-Rinke, K. Approach on environmental risk assessment of nanosilver released from textiles. *Environ. Res.* **2015**, *140*, 661–672. [CrossRef] [PubMed]
206. Yasin, S.; Sun, D. Propelling textile waste to ascend the ladder of sustainability: EOL study on probing environmental parity in technical textiles. *J. Clean. Prod.* **2019**, *233*, 1451–1464. [CrossRef]

Publisher's Note: MDPI stays neutral with regard to jurisdictional claims in published maps and institutional affiliations.

 © 2020 by the authors. Licensee MDPI, Basel, Switzerland. This article is an open access article distributed under the terms and conditions of the Creative Commons Attribution (CC BY) license (http://creativecommons.org/licenses/by/4.0/).

Review

Textile-Integrated Thermocouples for Temperature Measurement

Waleri Root, Thomas Bechtold * and Tung Pham

Research Institute for Textile Chemistry/Physics, University of Innsbruck, Hoechsterstrasse 73, 6850 Dornbirn, Austria; waleri.root@uibk.ac.at (W.R.); tung.pham@uibk.ac.at (T.P.)
* Correspondence: thomas.bechtold@uibk.ac.at

Received: 9 December 2019; Accepted: 28 January 2020; Published: 31 January 2020

Abstract: The integration of conductive materials in textiles is key for detecting temperature in the wearer´s environment. When integrating sensors into textiles, properties such as their flexibility, handle, and stretch must stay unaffected by the functionalization. Conductive materials are difficult to integrate into textiles, since wires are stiff, and coatings show low adhesion. This work shows that various substrates such as cotton, cellulose, polymeric, carbon, and optical fiber-based textiles are used as support materials for temperature sensors. Suitable measurement principles for use in textiles are based on resistance changes, optical interferences (fiber Bragg grating), or thermoelectric effects. This review deals with developments in the construction of temperature sensors and the production of thermocouples for use in textiles. The operating principle of thermocouples is based on temperature gradients building up between a heated and a cold junction of two conductors, which is converted to a voltage output signal. This work also summarizes integration methods for thermocouples and other temperature-sensing techniques as well as the manufacture of conductive materials in textiles. In addition, textile thermocouples are emphasized as suitable and indispensable elements in sensor concepts for smart textiles.

Keywords: textiles; temperature sensor; conductivity; coatings; deposition; thermocouple

---

1. Introduction

Extreme exposure of the human body to high temperature can cause severe effects such as heat illness. For appropriate heat monitoring, thermometers, metallic electrodes, or sensor chips can measure human temperature directly in the doctor's office. However, their permanent adherence to the skin, especially during outdoor activities, can lead to the wearer´s discomfort and in particular to skin irritation. This can be avoided by using textile-based temperature sensors, which can detect first signs of a heat illness outside the doctor´s office.

Heat-related illness occurs when the body stores more heat than it can release, which is accompanied by symptoms as heat stroke, heat cramps, and heat exhaustion [1,2]. Avoiding heat illness is particularly important in sports and the mining industry [2,3], which can be achieved by monitoring the temperature of the human body by textile sensors [4–8].

Integrated sensors should exhibit significant flexibility and low weight characteristics [9], which is important for health care applications such as body temperature measurement [10]. In health care, the body temperature is detected on the human skin [9]. This data can be used to investigate wound-healing processes, assessing patient comfort, or monitoring temperature development during sleep [9]. The integration of thermocouples into textile structures is a straightforward solution for temperature monitoring [9].

A thermocouple consists of two different conductive materials, which are connected at one point/form a closed circuit. One connection point is termed the measuring junction and the second can be regarded as a reference junction [11]. The thermocouple develops a voltage between two different materials of wires that can be used to measure temperature. Due to temperature changes, a voltage is generated between the different materials. Therefore, output voltage is related to the change in temperature [12]. This effect was first described by Seebeck in 1826. He discovered that a current flowed in a closed circuit between two dissimilar wires when two junctions are exposed to different temperatures [11,13,14]. The output voltage (ΔU) is calculated by Equation (1):

$$\Delta U = \alpha \times \Delta T \tag{1}$$

where α is the difference in the Seebeck coefficient of the two metal conductors, and ΔT is the temperature difference between the cold and hot junction [12].

In 1834, Peltier observed a current flow when a junction of two different wires was cooled or heated [11,13,14]. Twenty years later, in 1854, Lord Kelvin (W. Thomson) concluded that the current flow results from a temperature gradient in the conductor [11,13,14]. Figure 1 shows the model of a thermocouple, which uses two different conductors, a signal detection, and signal processing unit. The dotted boxes around the measuring and reference junctions show that these regions are isothermal. Isothermal regions do not contribute to the voltage detection (ΔU). The blue and red marked regions are set to different temperatures [13].

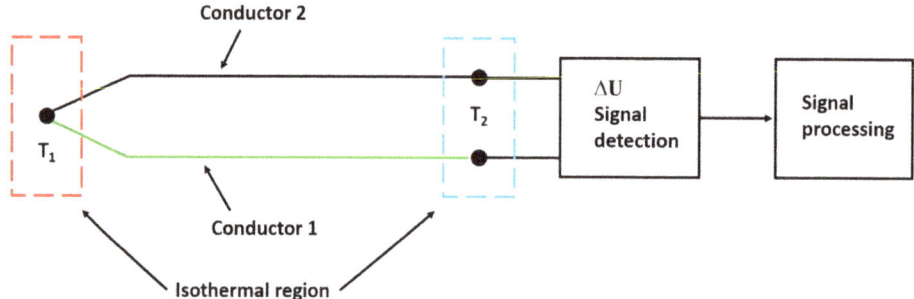

Figure 1. The model of thermocouple detection and signal processing according to [13].

Based on this model, a thermocouple pair can be constructed. A thermocouple pair generates a voltage when two junctions are set at different temperatures (Figure 2). The change of temperature at one junction leads to a voltage change across the thermocouple pair, which is proportional to the change in temperature.

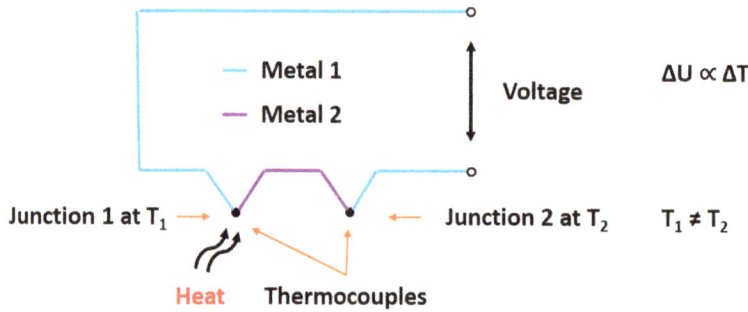

Figure 2. The first junction is heated to T_1 while the second junction stays at temperature T_2. According to [15], this results in an analog voltage signal.

Temperatures have been measured using various thermocouple assemblies such as thermocouple tips, [16], multipoint thermocouples [17], and a combination of dissimilar metal wires (platinum–rhodium alloy) [18]. Thermocouples can also be formed by printing techniques using iron, nickel, and copper inks [19]. The latest reports show that thermocouples can also be formed using two conductors of the same material with different material thicknesses [20–22].

Recent literature reports the invention of single-metal thermocouples consisting of different conductor widths, which are used for temperature detection [20–22].

Over the last three decades, the interest in the integration of temperature measurement systems into textiles has significantly grown. In Figure 3, this evolution is shown by the number of concepts for "Textile thermocouple", "Temperature measurements in textiles", and "Temperature sensors in textiles" published in the years 1990, 1995, 2000, 2005, 2010, and 2019.

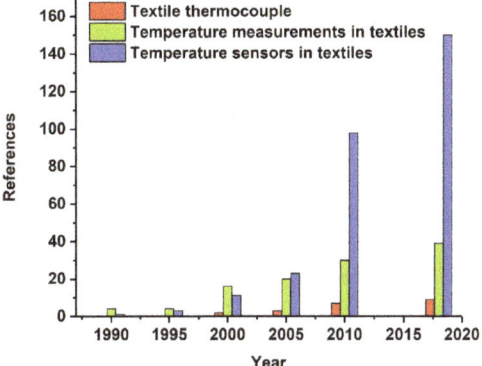

Figure 3. Number of concepts for the integration of temperature measurement systems and thermocouples into textiles published until 2019. The SciFinder database was used for the literature search with the key words textile thermocouples, temperature measurements in textiles, and temperature sensors in textiles.

Regarding the concept of textile thermocouples, the slight increase in publications shows that there is no precise definition of textiles thermocouples in literature. In general, thermocouples can be integrated into textiles by various techniques:

- Screen printing of conductive polymers [23]
- Sputter deposition of metal stripes [24]
- Soldering of metal wires [25]
- Using electro-conductive glues [26]
- Interweaving of metal wires [27]

For the ease of the reader finding the summary of five techniques, we have summarized them in Table 1, which permit having a quick overview of the methods.

Ziegler and Frydrysiak defined that textile thermocouples may be manufactured from thermoelectrodes consisting of functionalized textiles of woven, non-woven and knitted fabric threads, twisted multifilaments, yarns, and fibers. The functionalization of textiles can be conducted with conductive nanoparticles or electro-conductive polymers [26]. A general definition of textile thermocouples used in this review could be stated as:

- Textile thermocouples detect changes in temperature and consist of an indispensable conductive textile matrix with a textile character.

Table 1. Thermocouples used for temperature measurement in textiles.

Materials Used	Technique of Incorporation in Textiles	Reference	Limitations	Advantages
Cu, constantan, PES/CO fabric	Weaving	[28]	Conductive	Thermal insulation
Cu, constantan	Weaving	[27]	Stiffness	Direct application in T-shirts
Polyacrylonitrile thread, steel thread, polyamide thread, polyacrylonitrile yarn, steel fibers, graphite non-woven	Electrical conductive glue	[26]	Stiffness	Direct application in T-shirts
Stainless steel and constantan	Embroidery	[9]	Stiffness	Direct application into mattress
Poly(3,4-ethylendioxythiophene): poly(4 styrenesulfonate), polyaniline	Screen printing	[23]	Conductive	Sensitivity of 10 µV/K
Cu, constantan	Attached on polyester foam	[29]	Stiffness	Adapts to any textile structure
Cu-Ni wires, constantan wires	Soldering	[25]	Stiffness	Direct application in fire fighter gloves

Textile thermocouples should combine the flexibility and light weight of textiles with the conductive property of the conductor material, which can be defined as a truly textile thermocouple.

In this review, various integration methods and conductor materials for the construction of thermocouples in textiles will be described. This review discusses how textiles serve as appropriate carrier materials for the integration of temperature sensors. Different aspects of manufacturing conductive textiles will be shown. An outlook is given, which emphasizes the advantages and limitations of thermocouples in textiles.

2. Concepts of Thermocouple Construction in Textiles

Different thermocouples have been used to measure temperature on woven, non-woven, and knitted textiles.

Figure 4a shows the construction of five thermocouple pairs, which consist of five aluminum conductor strips and a large copper-coated cellulose fabric as a second conductor. Using the copper-coated cellulose textile as a conductor material makes the thermocouple construction more flexible compared to metal wires. The size of the copper-coated cellulose textile can be varied, which allows the positioning of additional thermocouples independently (Figure 4b). This thermocouple construction needs only one conductor as a sensing line. Figure 4c shows a scheme of electron flow in thermoelectric materials. It describes the formation of a temperature difference across a conductor when two junctions (regions) are set to different temperatures. The hot junction (region) generates more free electrons compared to the cold junction. Thus, an electron flow occurs from the hot to the cold junction (region) [12].

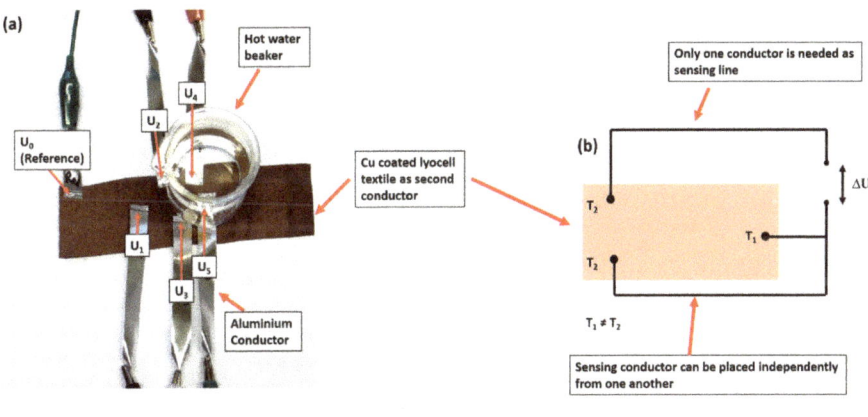

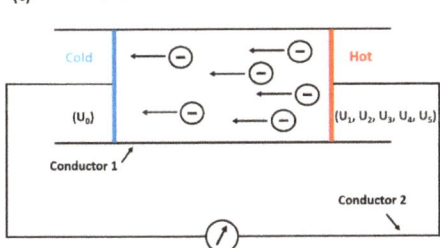

Figure 4. The construction of five thermocouple pairs (**a**), the description of electron flow in thermoelectric materials (**b**) according to [12], and (**c**) an electrical circuit. U_0 is the reference junction and U_1, U_2, U_3, U_4, and U_5 are measuring junctions.

Thermocouples were manufactured from conductive poly(3,4-ethylenedioxythiophene), poly(4styrenesulfonate) (PEDOT-PSS) and polyaniline by screen printing on woven cotton textiles [23]. In addition, thermocouples were used to detect resistivity and temperature as a function of time (up to 35 h). Thermocouple assemblies made from PEDOT-PSS and polyaniline showed a Seebeck coefficient of 18 µV/K comparing to 15 µV/K copper polymer assemblies [23]. In a further composition, thermocouples were manufactured from several textiles such as polyacrylonitrile staple fibers, steel staple fibers, a silver-coated polyamide thread, a knitted steel fabric, a woven polyacrylonitrile fabric, and a graphite non-woven textile [26]. The electrical signal generated from thermocouples was used to measure the temperature in the range of 30 to 120 °C [26]. The thermocouple was constructed from L-shaped copper and constantan (Cu/Ni) stripes on polypropylene textile, which were formed by magnetron sputter deposition. Comparison with a commercial thermocouple indicated no difference in temperature detection [24]. Temperature sensors were constructed from copper–nickel wire thermocouples, which were soldered onto a firefighter´s glove [25].

Thermocouple sensors have been manufactured from wires to monitor the thermal situation in socks and gloves [9]. The body heat regulation was monitored by a sensor-based platinum array outside of the garment [9]. Thermocouple sensors were manufactured from copper and constantan wires and were used to detect temperature at 12 different locations in T-shirts [27]. Consequently, a temperature distribution depending on the garment´s size and a distance from the body could be measured [27].

Copper-coated textiles can be used as flexible and lightweight conductor materials in a thermocouple array (Figure 5). The number of conductive lines can be reduced to measuring junctions (red spots), and a reference junction (green spot) can be formed by the attachment of five aluminum conductors (U_0, U_1, U_2, U_3, U_4, and U_5).

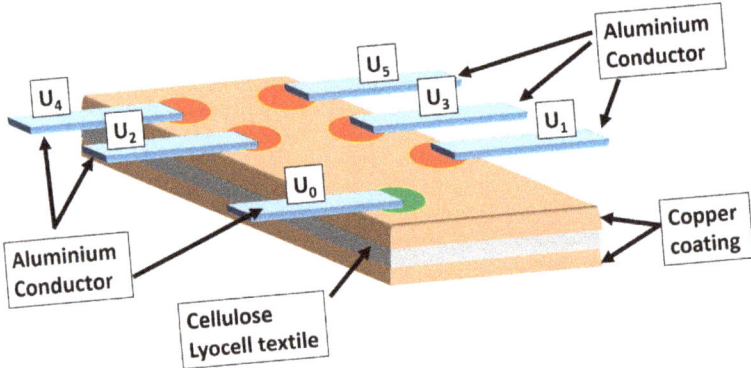

Figure 5. Copper-coated cellulose textiles used as a conductor matrix for temperature measurement.

Thermocouples were used to measure heat flux through polyester and polyester/cotton fabric with different weaves (plain, satin, and twill) [28]. The fabric´s temperature was detected at thermocouple points, which were related to reference points at room temperature [29]. Table 1 discloses various implementations of thermocouples in textiles and their influence on the flexibility of the entire structure, which became stiffer, especially by gluing and soldering.

The thermocouples shown in Table 1 in this summary were made from metal wires, fibers, and yarns, which increase the textile´s weight and reduce its flexibility. The incorporation techniques mentioned often (Table 1) integrate thermocouples in textiles, thus resulting in discomfort and increased stiffness. Further scientific work is needed to deal with these inconveniences and to manufacture a truly textile thermocouple.

3. Other Strategies for Temperature Measurement in Textiles

There are several strategies to measure temperature in textiles such as Positive Temperature Coefficient (PTC), Negative Temperature Coefficient (NTC), Resistance Temperature Detector (RTD), and fiber Bragg grating (FGB). These strategies determine the shape of temperature sensors, which are manufactured by weaving, lithography, adsorption, screen printing, embroidery, knitting, and gluing, using conductive carbon paints and chemical vapor deposition (CVD).

Temperature measurements were conducted with a PTC resistive temperature sensor. It was manufactured as a thin film capacitor with gas/humidity sensitive polymers on a 50 µm Kapton substrate, on which two electrode lines were formed by lithography. The temperature sensor was woven into a textile (width = 45 mm, length = 200 mm) [30]. PTC sensors also were manufactured from an activated carbon fiber cloth by an electrothermal swing adsorption method [31]. A PTC sensitive polyamide foil (KAPTON) was manufactured by screen printing carbon polymer composites, while polyethylene and rubber were used as binder materials. The PTC-sensitive foil-detected temperature increase from 30 to 42 °C as a function of resistance [32]. PTC sensors were used to investigate the heating properties of 40 µm embroidered flexible polyurethane-coated copper filaments. PTC sensors on cotton fabrics recorded the increase in temperature as a function of resistance on embroidered PU-Cu composites. [33].

The temperature in textiles was detected by NTC sensors, which were manufactured from thermosensitive polyvinylidene fluoride (PVDF) fibers of 2 to 6 cm in length and 0.15 mm in diameter. Their active sensor area was formed by thermosensitive, polymer conductive pastes, which were manufactured from multiwall carbon nanotubes (MWCNT) and poly(methylmethacrylate) (PMMA) [34]. In another report, the sensitivity of conductive fabrics to different temperatures was investigated by cotton and silver yarns. The sensitivity was related to the fabric's resistance, which was measured between two brass blocks (500 g each) at a pressing force of 25 N in an oven [35]. Five sensing yarns were incorporated into 2 mm diameter channels on a knitted sock and measured temperature on the skin. One temperature-sensing yarn consisted of a copper wire, six polyester yarns, one NTC thermistor, and a polymer resin, which were processed by a flatbed knitting machine [36].

In a further method, the temperature in textiles was measured by RTD sensors consisting of Kapton and Ti/Au conductors. The sensors were produced in a commercial band weaving process. The sensors were glued with conductive epoxy to metal strips, which were connected to a measurement device [37]. Another RTD detector was manufactured from nylon-6 after electrospinning and functionalized with multiwalled carbon nanotubes (MWCNTs) and polypyrrole (PPy). The samples were treated in the pyrrole vapor for 48 h and connected to two copper electrodes with conductive carbon paint [38]. RTDs were also made from a 100 nm platinum-coated plastic strips of 67.5 mm length and 500 µm width. The strips were woven in Kapton textiles at a distance of 200 µm and were used for temperature measurements [39]. In a further construction, RTD sensors were manufactured from nickel, copper, and tungsten wires on a temperature sensor fabric. These wires were knitted in the middle of a polyester fabric using a flatbed knitting machine. In the design, electrical short circuits were avoided, and a resistance of 3 to 130 ohms was determined during temperature measurements [40]. Flexible temperature and humidity sensors were made from graphene woven fabrics (GWF) on flexible polydimethylsiloxane (PDMS) films, which were deposited by CVD. The change in resistance, which was recorded as a function of temperature from 20 to 60 °C, demonstrated the use as a sensor [41].

A further method of temperature detection in textiles is the use of reflected wavelengths, which was caused by angular deformation. This deformation was used during temperature measurement in fiber optical sensors, which are based on the fiber Bragg grating (FBG) method and are woven in socks. The FBG sensors are made of silica core and plastic substrates, which provide the material a durability and light weight [42]. As an example, the body temperature can be detected by FBG sensors, which are woven into fabrics and embedded into a polyester resin [10]. The FBG method was used to investigate the temperature from 30 to 70 °C of liquid mixtures (water/glycerin) with a negative thermo-optic coefficient of -5×10^{-4} °C^{-1}. The FBG fiber was placed in an aqueous solution with the mixture solution. The Bragg wavelength of the FBG fiber was measured while heating the solution [43]. The structural state of textiles can be monitored when the beam IR laser (1064 nm) impinges on the surface, leading to a thermal gradient of 100 °C. The FBG temperature sensors measured radiation on the polymer surface, which can be used for flame or energy attack detection [44]. In addition to the FBG method, plastic optical fibers (POFs) were used as temperature sensors. During that method, the temperature was measured as a function of the intensity, which was caused by the thermal bending of the fiber. The POF sensors consisted of polymethyl methacrylate and fluorinated polymers, which were used as core and cladding [45].

The POF was used in chirped fiber Bragg grating sensors of 10 mm length, which indicated a sensitivity of −191.4 pm/°C. These sensors measured temperatures along the grating length, which were designed for biomedical treatments and thermotherapies [46]. For biomedical application, POF was used due to the rapid production of POF grating devices, which worked below 248 nm and 266 nm UV wavelengths. This led to the manufacture of chirped POF-FBG sensors with a higher sensitivity and better biocompatibility compared to silica-based sensors [47]. In biomechanical investigations, multiple FBG sensors showed a sensitivity of 10.6 pm/°C, which were connected in serial on textiles. This connection formed a temperature sensor network with multiple points, with which temperature values from 20 to 130 °C were measured [48].

Different techniques and materials are summarized next, which permit temperature measurements in textiles (Table 2).

Table 2. Temperature measurement techniques in textiles.

Measurement Technique	Set Up	Reference	Advantages
PTC	Cr/Au metal electrode structure	[30]	Sensitivity of 1.175 Ω/°C
PTC	Au/Cu contacts as flexible thermistor on the Kapton foil	[32]	Elastic conductive paste enhance flexibility
PTC	Polyurethane-coated copper filaments used as temperature-sensing textile	[33]	Construction of circuits
PTC	Activated carbon fiber cloth used as heating textile clamped between stainless steel electrodes	[31]	Measure temperature up to 200 °C
NTC	Two brass blocks placed on conductive fabric	[35]	Fast measurement
NTC	Multiwalled carbon nanotube coated poly(methylmethacrylate) yarn placed on metal clamps	[34]	Measure temperature up to 850 °C
NTC	Sensor yarns connected to microcontroller	[36]	Fast sensor integration
RTD	Conductive metal thread connected to sensor unit	[37]	Detect temperature, relative humidity
RTD	Copper electrodes	[38]	Nanocomposites enhance flexibility
RTD	Single platinum metal sensor stripes woven into textile	[39]	Fabrication of 120 sensors on one substrate
RTD	Platinum wire embedded into polyester fabric as sensing element	[40]	Knitting method similar to standard industrial process
RTD	Graphene woven fabrics and polydimethylsiloxane used as temperature sensing unit	[41]	Detect temperature, humidity
FBG	Sensors embedded into fabric by cobalt naphthenate and methyl ethyl ketone peroxide resin mixtures	[10]	Temperature sensitivity 150 pm/°C
FBG	Cladding-etched fibers used as temperature-sensing substrates	[43]	Glycerin/water compensate Bragg wavelength shift
FBG	Flexible optic fiber sensor embedded in sock reflecting the infrared light to the infrared detector	[42]	Measure temperature, pressure, joint angles
FBG	Sensors woven into a carbon fiber fabric	[44]	12 sensors measure temperature on surface and through the thickness

4. Aspects of Manufacturing

4.1. Integration of Electrically Conductive Elements

The central element to integrate sensors in textiles is to achieve stable connections. Figure 6 shows an output of a literature search on manufacturing techniques in textiles. The concepts are highlighted with green for "conductive printing on textiles", blue for "conductive deposition on textiles", and red for "conductive coating on textiles". The research dynamics of the three approaches increased over the period of 29 years, which indicates the growing interest in techniques to miniaturize textile sensors.

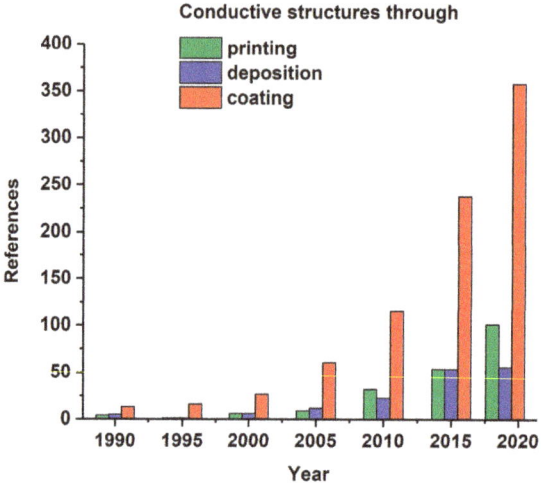

Figure 6. Research activity in coating, deposition, and printing processes to manufacture conductive structure in textiles.

In general, the coating or deposition of conductive materials has more advantages than the soldering, welding, and weaving of metallic wires or sensors in textiles. The classification of products was related to the aspect by which the coating or deposition of materials can be conducted on large textile areas and spatially defined structures such as yarns or fibers. The coating or deposition of conductive materials leads to a conductive thin layer formation on substrates compared to soldering. Thus, coated conductive textiles are flexible and have a higher motion of freedom compared to soldered textiles. The scientist makes the choice of textile materials because it is more likely that different materials and textile structures will be used in each scientific field. The properties of conductive materials are determined by the firm or loose textile structures, their swelling properties, and the amount of reactive groups.

The advantages of coated or deposited materials are the low weight and the thin coating thickness. Thus, conductive coatings better retain the flexibility, bending, and stretch properties of textiles compared to rigid metallic wires or sensors. Consequently, the combination of conductive coatings and deposits on textiles also contributes to the construction of truly miniaturized textile thermocouples. A truly textile thermocouple was described in the Introduction.

The manufacturing of electrically conductive substrates on textiles can be performed by different techniques e.g., soldering, stapling, and bonding components through conductive adhesives. Soldered substrates do not withstand the bending of textiles. Stapled substrates increase the wear and tear of textiles due to the rigid structure and reduce the freedom of movement. The connection between textiles and conductive substrates can be made by a flexible conductive material [49].

Representative examples for such materials are bicomponent fibers that include poly(vinylidene fluoride) as the sheath material, carbon black, and high-density polyethylene as the core material.

Bicomponent yarns were made during the melt spinning process with two screw extruders consisting of the core and sheath material [50]. Conductive core–sheath yarns of copper core filament and cotton sheath were manufactured through the Dref-3 friction spinning method. The core–sheath yarns were made of copper filaments as a core and cotton fibers as sheath. These yarns showed a resistance of 3 to 28 MΩ and a shielding effect of 760 to 860 MHz at a core sheath ratio of Cu 0.26 gram per meter and cotton 0.13 grams per meter [51]. Elastic conducting inks were made of Ag flakes, fluorine rubber, and fluorine surfactant, which showed a conductivity of 182 Scm^{-1} during stretching. These materials were used as wearable electromyogram sensors to detect the signal activity of the muscles of the forearm. Elastic conductor inks were printed on polyimide stencil masks, which formed flexible conductor wires on the upper side of the textile and an elastic conductor vital electrode on the lower side of the textile [52]. Wearable electronic textiles were created by a lockstitching method and were used in apparel textiles. Conductive assemblies were made by stitching conductive threads (such as silver, aluminum, stainless steel, copper, and carbon) on the surface of a cellulose and stitching thermofusible threads (polyamide, polyolefin, and polyvinyl) on the polyester/elastane [53].

The surface of cotton textiles was rendered conductive by impregnating with carboxylated multiwall carbon nanotubes by dispersion. The cotton fabrics were treated with aqueous NaOH/urea mixture at −10 °C for 1 h and showed a low electrical resistivity of 281 Ω cm [54]. The formation of conductive textiles was manufactured through screen printing of the FeCl$_3$ and by applying high voltage from 5 to 30 kV during the coating of pyrrole by vapor deposition. The high voltage along the polypropylene-coated fabric stabilized pyrrole monomers during vapor deposition [55]. Non-conductive epoxy surfaces were laminated with copper sheets by the pressing method. Afterwards, these materials were activated with stannous/palladium chloride particles. The epoxy substrates were made conductive after 20 h of electroless copper plating [56]. The formation of conductive tracks of 1.5 and 4.0 mm was achieved on cotton textiles by the reduction of silver nitrate from sodium borohydride during the spray deposition. Subsequently, the silver seeded tracks were plated selectively with copper during the electroless process from aqueous solution [57]. Copper foils were used to form circuits in cloths, which consisted of silk organza fibers. The electrical circuits structure was manufactured by embroidery and by an industrial sewing machine [58].

Additionally, conductive coatings on cotton fabrics were manufactured by the surface activation in NaOH and poly(diallyldimethylammoniumchloride) solution. The activated cotton fabrics were impregnated with NaBH$_4$ in aqueous solution, and afterwards, a silver nitrate solution was added to the fabric. The cotton fabrics were completely coated with silver nanoparticles after the reduction of silver ions by NaBH$_4$ [59].

Coated textiles powered small consumers without the use of metal wires and impart electromagnetic shielding properties by the examples below. Conductive woven cellulose fabrics power a light-emitting diode (LED) at 20 mA. The copper layer was formed after silver seeding through an electroless deposition in alkaline solution comprising a Cu L-tartrate complex and formaldehyde [60].

Figure 7 describes the electroless deposition method of copper on silver seeded cellulose textiles in alkaline solution. The silver seeded textile is dipped into copper sulfate, formaldehyde, and potassium hydrogen L-tartrate solution (Figure 7a). Formaldehyde is a chemical reducing agent, which reduced copper ions on silver seeds from the copper tartrate complex (TH) to metallic copper (Figure 7a,b). When the deposition proceeds (Figure 7c), copper islands are formed on silver seeds, which then grow to a continuous coating [61].

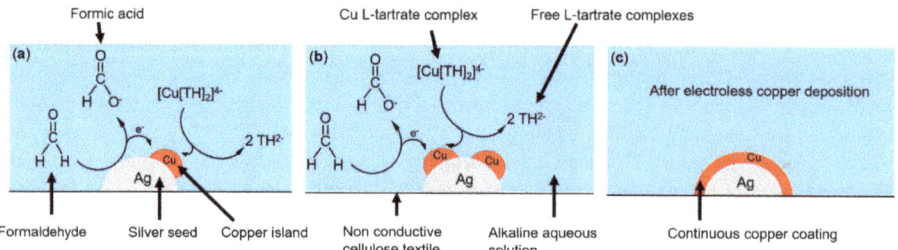

Figure 7. The electroless copper deposition method conducted on cellulose textiles, where the tartrate complex (TH) is a free L-tartrate ligand (**a**). The copper deposition continues on Ag seed (**b**), which leads to the copper layer formation (**c**).

Cotton fabrics imparted conductive properties after the in situ deposition of copper particles and repeated dipping steps in the $CuSO_4$ and $Na_2S_2O_4$. Copper-coated textiles can be used as flexible and light materials. The copper-coated cotton fabrics showed a shielding property of 6 dB, 10 dB, and 13 dB when the fabric was dipped in the copper sulfate solution 50, 100, and 150 times, respectively [62].

A low electrical resistance of textile material can be achieved also by treatment with conductive polymers after impregnating, vapor deposition, and melt mixing methods. Electro-conductive fabrics can be made from wool, cotton, and silver-coated acrylic yarns. Textiles composed of silver-coated wool yarns and silver-coated cotton/acrylic were used as heating elements in textiles [63]. The incorporation of conductive material during fiber formation also leads to polymer fibers with conductive properties.

A non-woven poly(ethylene oxide) (PEO) matrix was mixed with 3 wt % multiwalled carbon nanotubes (MWNT), which formed conductive polymer composites by an electrospinning process. The maximum electrical resistance of PEO/MWNT composites changed when exposed to methanol, dioxan, and toluene vapors [64]. Conductive monofilaments composites were formed from carbon nanotubes (CNT), polypropylene, poly(ε-caprolactone) (PCL), and polypropylen substrates. The materials manufactured from 50%PP/50%PCL/4%CNT composites showed a resistivity of 1.1 Ωm at 154 °C [65].

Coated textiles were used for temperature detection in the range of 15 to 57 °C. Conductive polyamide fabrics of 17% Lycra and 83% Tactel (5 cm × 1 cm) were coated from aqueous solution with poly(3,4-ethylenedioxythiophene)-poly(4-styrenesulfonate) (PEDOT-PSS). PEDOT-PSS-coated fibers were exposed to environmental temperatures of 15 and 45 °C. The electrical resistance of coated fibers decreased with increasing temperature [66]. Conductive polyester yarns were manufactured from copper nanowires and a silicon rubber substrate during a dip-coating. The coated polyester yarns were used as stretchable heating fibers. The composites were woven into a heating fabric and connected to a microcontroller unit to manufacture wearable and smart personal heating systems [67].

4.2. Aspects of Aging

The information about the working property of textile sensors, conductive materials, and protective clothing over their entire lifetime still remains underreported in the literature. For firefighters' protective garments, the aging of materials under environmental conditions reduced due to the low shear resistance even after a short period of time. The mechanical strength of textiles was reduced by up to 80% before a damage was detected visually [68]. The thermal degradation in aramid/basophil firefighter cloth occurred before the optical change was detected. After a convective heat of 80 kW/m^2 and a radiant heat exposure of 40 kW/m^2, the mechanical properties of fabrics in a tensile test decreased by 40% and 60%, respectively (660 N) [69].

The advantages of thin flexible and electric coatings are the good conductivity and the low impact on textile properties such as their handle, flexibility, and density. Possible problems with coatings are their corrosion and insufficient adhesion between the textile and the coating substrate [70]. Polyester fabrics were coated with polypyrrole during incubation in saline substrate for up to two weeks at 37 °C. They exhibited a resistance in the range of 10^3 to 10^4 Ω/square. It was observed that the decrease in electrical conductivity was related to the oxygen uptake during incubation and due to cracking of the coating [71].

Figure 8 shows the results of a literature search on sensor aging containing three different search concepts. The concepts are highlighted with green for "thermal aging of sensors in textiles", blue for "functional aging of sensors in textiles" and red for "aging of temperature sensors in textiles". There are gaps in the literature dynamics of all concepts in the last 29 years, which do not provide a general concept for sensor aging in textiles. During the period of 2000–2019, the total number of references for the concepts of thermal aging of sensors in textiles, functional aging of sensors in textiles, and aging of temperature sensors in textiles were two, five, and four respectively, which indicates low scientific interest in degradation and aging in temperature sensors in textiles.

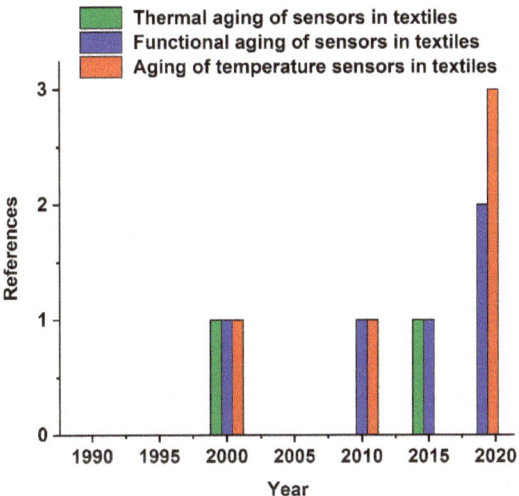

Figure 8. The effects of aging on sensors in textiles.

4.3. Aspects of Life Cycle of Conductive Textiles and their Regulation

Electrically conductive textiles will gain more importance for mass consumer applications. Thus, a new kind of waste will be formed. The market of smart textiles and wearable electronics is estimated to grow from $20 billion in 2015 to $70 billion in 2025 [72], which emphasizes their importance for the mass consumer application. According to the European Commission in 2017, the high potential of wearables on the European market was reported in the orientation paper about smart wearables [73].

The waste difficulties of e-textiles can be overcome by implementing an appropriate eco-design strategy, which include e-textile labeling and the use of compatibility standards [74]. The impact of new waste could cause toxicological stress on human health, the ecosystem, resources, land use, and water use. These negative impacts can be reduced through the life cycle assessment at an early stage of the development, which assesses the potential environmental impact of products and identifies solutions for preventing pollution and decreasing the resource consumption [75].

As an example for a toxicological assessment of a surface of a modified textile, the coating of polyester and cotton fabrics with nano-metal oxides such as CuO and ZnO was studied. Fabrics treated with water and ethanol showed a release of CuO and ZnO nanoparticles up to subtoxic concentrations of 1 µg/mL in A549 cells. At a low concentration up to 10 µg/mL, there was no acute toxicity observed in lung epithelial and macrophage cells compared to an exposure of 100 µg/Ml [76].

Besides the toxicological evaluation, the production of e-textiles in industrial processes has to comply with the legal requirements of European Eco-design, which describe the development of energy-related goods. Future goods design and sustainable material management can be related to the U.S. Environmental Protection Agency, which regulates the life cycle of products during their manufacture [77]. The use of metals for conductive substrates in textiles should be regarded as a metal finishing process, which is conducted by the industry. Consequently, the industry is bound by the laws of regulation for metal finishing such as the Resource Conservation and Recovery Act, the Clean Air Act, and the Clean Water Act (CWA). The CWA includes the Effluent Guidelines and Standards for Metal Finishing and the Effluent Guidelines and Standards for Electroplating. These guidelines and standards are mandatory for facilities dealing with electroplating, coating techniques, electroless plating, printed circuit board production, chemical etching, and milling. The standards determine the concentration of pollutants in wastewater from the above-mentioned processes, which are described in milligrams per m^3 [78].

5. Temperature Sensors and E-Textiles

5.1. Wearable Heaters

Wearable heaters also record temperature profiles as a function of time and can be used in many applications e.g., thermotherapy. In many cases, a combination of heating device and temperature sensor is implement with the aim to control heat generation and to avoid over temperature.

Wearable heaters, which are manufactured from Ag nanofibers (AgNF) on polyethylenterephthalat (PET) and polyimide (PI) by electrospinning, can be affixed to the skin. Heaters were connected at both ends by Cu wires, while the current was applied from the power supply for heat generation. The AgNW (nanowire) heater on the PI substrate shows a considerably stable temperature of 42 °C during a stretching test up to 90%. The use of SiO$_2$ as a passivation layer on AgNW heaters can retard Ag oxidation and allow the detection of temperature up to 250 °C [79].

Wearable and stretchable heaters were made from PEDOT:PSS, polyurethane, and reduced graphene oxide films, which can be applied in thermotherapy. They imparted an electrical conductivity of 18.2 Scm^{-1} and withstood elongation up to 530%. The temperature distribution of composite films was measured in the middle when voltage was applied by two copper wires [80]. Heaters were also manufactured from Ag NWs (nanowires), PEDOT:PSS, and PET materials, which withstood a temperature of 120 °C [81]. Stretchable heaters were also fabricated from graphene fiber (GF). The GFs were embroidered into cotton fabric and withstood finger bending and wrist movement. The temperature was recorded by an infrared camera [82].

Flexible and stretchable heaters were manufactured from carbon nanotubes (CNT), copper foil, and silicon elastomers [83]. Flexible and stretchable heaters were constructed from copper-coated polyacrylonitrile fibers, which can operate at temperature up to 328 °C. These heaters were manufactured from copper-coated fibers by electroplating on glass substrates [84]. Flexible heaters were manufactured from nylon-coated fabric, which was coated with Ag NWs and rubber shape memory polymer during dip-dry and spray coating. Bending, rolling, gripping, and rubbing did not show any damage of the heaters [85].

Stretchable and conductive heaters were manufactured from poly(3,4-ethylenedioxythiophene): poly(4-styrenesulfonate) (PEDOT:PSS) and sodium dodecyl sulfate on cotton and polyurethane fabrics by dip coating. The temperature changes were investigated with a digital thermometer while IR images were recorded with an infrared camera [86]. Stretchable heaters were used in thermotherapy, which were produced from styrene–butadiene–styrene and Ag NW substrates. These substrates formed a mesh by thermal welding and heat treatment [87].

In thermotherapy, stretchable heaters could increase the blood flow near the wrist. The heaters were manufactured from kirigami–aluminum paper, thin elastomers of silicon polymer, and polyethylene terephthalate films, and these could be stretched to 400% at a temperature of 40 °C [88]. Stretchable heaters were also manufactured from copper wire/alumina/polyimide composites. These composites showed a high visible light transmittance up to 91.4% and reached temperatures up to 300 °C. They withstood 100 stretching and relaxation cycles at 30% strain [89]. Stretchable and wearable heaters were manufactured from CuZr and poly(dimethylsiloxane) (PDMS), which could be used at 70% elongation. They were used as portable patch units on human hands and reached temperatures up to 50 °C [90]. Stretchable heaters produced from Ag nanowires and polydimethylsiloxane (PDMS) substrates were used to heat human skin. A constant temperature of 50 °C could be observed up to 40% strain [91].

Temperature measurements were conducted by conductive substrates in textiles, which formed sensors and flexible electronic structures. Flexible electronic circuits were made by coating 35 nm Cr substrates by photolithography and 25 nm Al_2O_3 substrates by atomic layer deposition on Kapton E materials. Electronic circuits were integrated through the commercial weaving process integrated in textiles. They formed woven temperature sensors, which operated in the range of 20 to 100 °C [92]. Flexible and conductive polyester fabrics were manufactured from ploly(3,4-ethylenedioxythiophene):poly(4-styrenesulfonate) (PEDOT:PSS), 15 wt % graphite, and dimethyl sulfoxide mixtures by coating. These fabrics were used as thermoelectric (TE) textiles, which measured temperatures up to 398 K and showed a power of 0.025 $\mu Wm^{-1}K^{-2}$ [93].

Bimodal sensors were used to detect temperature and pressure simultaneously by making use of a piezo-thermoresistive organic conductor and a dialectic substrate. The dielectric substrate was composed of poly(vinylidenefluoride-trifluoro-ethylene) and $BaTiO_3$ nanoparticles. When the human finger pressed on the bimodal sensor, a pressure of up to 0.03 N/mm^2 and a temperature of up to 35 °C were measured [94].

5.2. Sensor Integration in Textiles

Figure 9 shows eight possible application areas, where the integration of sensors in textiles is of interest. The temperature detection already has been investigated in functional garments, sport garments, the automobile industry, medical institutions, security packaging, and the fashion industry. The future seamless compatibility of sensors with textiles will increase their wearing comfort and lead to prototypes, which can be produced on an industrial scale.

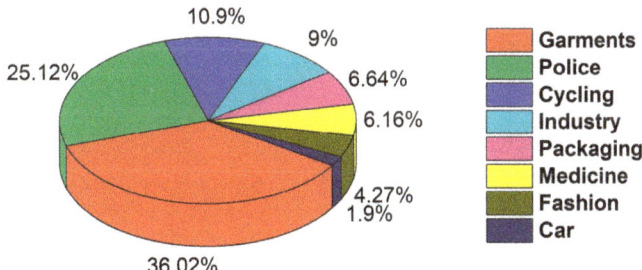

Figure 9. Eight areas for sensor integration in textiles in November 2019.

5.3. Body Sweat/Moisture and Heat Transfer in Textiles

Besides sweat, water content influences the wearer´s comfort in textiles. The presence of water in textiles increased the mass and reduced the heat transfer in sport and protective clothing [95]. Textiles with high water vapor permeability can transfer moisture from the skin through the textile into the environment, which continuously keeps the human body in thermal equilibrium.

Therefore, the transmission of water vapor was recorded as a function of air temperature and relative humidity in polytetrafluoroethyle (PTFE) laminated with nylon fabric, woven cotton fabric, polyester fabric (laminated with polyurethane), and hybrid PTFE membranes. The transmission of water vapor was high at high air temperature and low relative humidity [96].

In addition to the body motion, health condition can be monitored by using biocompatible and stretchable carbon nanotube-based electrodes (CNTs), which are used to detect sweat [97]. Sweat also can be detected by a wearable colorimetric pH sensor, which provides information on the metabolic state and activity of a patient. The collection of sweat in T-shirts was investigated on textile biosensors in health management [98].

Figure 10 shows the increase in the literature on concepts, which are related to thermal effects and energy generation. The concepts are highlighted with green for "thermal insulation in textiles", yellow for "heat transfer in textiles", blue for "textiles exposed to temperature", and red for "energy harvesting in textiles". Energy harvesting in textiles is a new fast growing field. Its role will be significant with the development of miniaturized temperature sensors that seamlessly adapt to textiles.

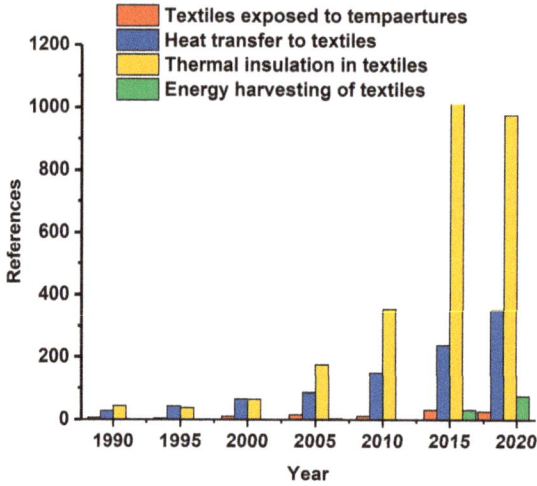

Figure 10. Temperature measurement in textiles and their use for energy generation.

The thermoelectric effect also can be used to generate electrical energy from temperature differences between a human body and the environment.

As an example, the heat of the human body was used to power a flexible thermoelectric glass fabric, which was formed from eight thermocouples consisting of Bi_2Te_3 and Sb_2Te_3 films. It indicated an output voltage of 28 mWg^{-1} ($\Delta T = 50$ K) [99]. The temperature of the human body was detected by polyethylene (PE) and polyethylene oxide (PEO) substrates, which were melt mixed with 40 wt % Ni microparticles. The PEO/PE matrix treated with 40 wt % Ni showed sensitivity as temperature sensors of 0.3 V/°C in the range of 35 to 42 °C compared to 50 wt % [100]. The skin temperature was measured by an embedded wire sensor, which was composed of aluminum carbon epoxy composites. These composites detected a higher skin temperature compared to multiple thermistors [101].

6. Outlook and Future Perspectives

The coating of textiles with metals is a key technology for the miniaturization of low weight textile thermocouples. The metal coating follows the structure of the textile and covers its surface with a thin conductive metallic layer. The advantages of thin conductive coatings are the ability to form different geometries on small surfaces and provide a better flexibility compared to thicker substrates. Combining the conductivity of metal coatings with a fabric's flexibility, light weight, and stretch can provide substantial progress in miniaturized textile thermocouple construction. The combination of textiles with low weight thermocouples will improve the sustainability of the assembly.

Using a thermocouple is a simple way to measure temperature in textiles. Conductive thin-coated textiles can be used for a thermocouple construction, which measures temperature based on an electrical signal. There is a growing demand for miniaturized temperature measuring methods in textiles in the near future.

Besides the functionality of a device, material costs will also determine the selection of conductive parts. The use of silver as a conducive material for the manufacture of wearable heaters can be explained by its high electrical conductivity of 6.3×10^7 Sm^{-1} compared to that of copper, which is 5.9×10^7 Sm^{-1}. Despite the lower cost of copper ($6.7/kg) compared to silver ($510/kg), immediate oxide layer formation on the copper surface makes its application difficult. Conductive PEDOT/PSS substrates (2×10^4 Sm^{-1}) are very expensive ($167,000/kg) and may not be suitable for the large-scale production of flexible substrates [102].

The durability of conductive textile thermocouples during wearing under different weather conditions is still underreported in the current literature. The influence of use and wear conditions on the durability of textile thermocouples is due to effects of moisture and low or elevated temperature. Additionally, the abrasion and mechanical deformation of conductive textile thermocouples increase the rate of degradation, which is often due to limited adhesion between the textile and conductor materials. The future scientific work should focus on the loss conductivity of textile thermocouples during aging and in situ mechanical deformation. Comprehensive scientific work is required to optimize the design, lifetime, and miniaturization of textile thermocouples. This work must include the life cycle assessment of conductive textile thermocouples to prevent hazardous waste, reduce production costs, and provide appropriate strategies for their recycling.

Author Contributions: The authors contributed equally to the manuscript. All authors have read and agreed to the published version of the manuscript.

Funding: This research was funded by the CORNET framework in the project Ambitex –"Textile integrated sensors for monitoring of ambient parameters" (FFG 855282), NanoStretch "Stretchable conductive textiles based on nanostructured templates" (FFG 865927), and TCCV (FFG 860474).

Conflicts of Interest: The authors declare no conflict of interest.

References

1. Bonauto, D.; Anderson, R.; Rauser, E.; Burke, B. Occupational heat illness in Washington State, 1995–2005. *Am. J. Ind. Med.* **2007**, *50*, 940–950. [CrossRef]
2. Donoghue, A.M. Heat illness in the U.S. mining industry. *Am. J. Ind. Med.* **2004**, *45*, 351–356. [CrossRef] [PubMed]
3. Howe, A.S.; Boden, B.P. Heat-Related Illness in Athletes. *Am. J. Sports Med.* **2007**, *35*, 1384–1395. [CrossRef] [PubMed]
4. Xu, X.; Karis, A.J.; Buller, M.J.; Santee, W.R. Relationship between core temperature, skin temperature, and heat flux during exercise in heat. *Graefe's Arch. Clin. Exp. Ophthalmol.* **2013**, *113*, 2381–2389. [CrossRef] [PubMed]
5. Niedermann, R.; Psikuta, A.; Rossi, R.M. Heat flux measurements for use in physiological and clothing research. *Int. J. Biometeorol.* **2014**, *58*, 1069–1075. [CrossRef] [PubMed]
6. Leonov, V. Thermoelectric Energy Harvesting of Human Body Heat for Wearable Sensors. *IEEE Sensors J.* **2013**, *13*, 2284–2291. [CrossRef]
7. Morozumi, Y.; Akaki, K.; Tanabe, N. Heat and moisture transfer in gaps between sweating imitation skin and nonwoven cloth: effect of gap space and alignment of skin and clothing on the moisture transfer. *Heat Mass Transf.* **2012**, *48*, 1235–1245. [CrossRef]
8. Gibson, P.; Charmchi, M. Coupled Heat and Mass Transfer Though Hygroscopic Porous Materials—Application to Clothing Layers. *Sen'I Gakkaishi* **1997**, *53*, 183–194. [CrossRef]
9. Dias, T. *Electronic Textiles: Smart Fabrics and Wearable Technology*, 1st ed.; Dias, T., Ed.; Woodhead: Oxford, UK, 2015; ISBN 9780081002018.
10. Li, H.; Yang, H.; Li, E.; Liu, Z.; Wei, K. Wearable sensors in intelligent clothing for measuring human body temperature based on optical fiber Bragg grating. *Opt. Express* **2012**, *20*, 11740–11752. [CrossRef]

11. Michalski, L.; Eckersdorf, K.; Kucharski, J.; McGhee, J. *Temperature Measurement*, 2nd ed.; John Wiley & Sons LTD: West Sussex, UK, 2001; ISBN 0521433630.
12. Lee, H.S. *Thermal Design: Heat Sinks, Thermoelectrics, Heat Pipes, Compact Heat Exchangers, and Solar Cells*; John Wiley & Sons: Hoboken, NJ, USA, 2010; ISBN 9780470496626.
13. Nicholas, J.V.; White, D.R. *Traceable Temperatures*, 2nd ed.; Wiley: West Sussex, UK, 2001; ISBN 0471492914.
14. Göpel, W.; Hesse, J.; Zemel, J.N. *Thermal Sensors*; VCH: Weinheim, Germany, 1990; Volume 4, ISBN 3527267700.
15. Senturia, D.S.; Wedlock, D.B. *Electronic Circuits and Applications*; Wiley Inte.; John Wiley & Sons, Inc.: New York, NY, USA, 1975; ISBN 0-471-77631-9.
16. McCall, C.I. US 3284247 1966, 1–3. Espacene. Available online: https://worldwide.espacenet.com/patent/search/family/023136070/publication/US3284247A?q=pn%3DUS3284247A (accessed on 2 December 2019).
17. Daily, N.J.; Poteet, F.R.; Rahn, W.M.; Welch, D.L. US 6550963B2 2003. Espacenet. Available online: https://worldwide.espacenet.com/patent/search/family/025288530/publication/US2003016730A1?q=US%206550963B2 (accessed on 12 December 2019).
18. Usher, J.D.; Blaze, J.E.; Phillippi, M.R. US 5071258 1991. Espacenet. Available online: https://worldwide.espacenet.com/patent/search/family/024605372/publication/US5071258A?q=US%205071258%20 (accessed on 2 December 2019).
19. Smith, E.T.; Cooper, L.C. US 20050257822A1 2005. Espacenet. Available online: https://worldwide.espacenet.com/patent/search/family/034970526/publication/US2005257822A1?q=US2005257822 (accessed on 1 December 2019).
20. Liu, H.; Sun, W.; Xu, S. An Extremely Simple Thermocouple Made of a Single Layer of Metal. *Adv. Mater.* **2012**, *24*, 3275–3279. [CrossRef]
21. Sun, W.; Liu, H.; Gong, W.; Peng, L.-M.; Xu, S.-Y. Unexpected size effect in the thermopower of thin-film stripes. *J. Appl. Phys.* **2011**, *110*, 83709. [CrossRef]
22. Basko, D.; Basko, D. A Photothermoelectric Effect in Graphene. *Appl. Phys.* **2011**, *334*, 610–611. [CrossRef] [PubMed]
23. Seeberg, T.M.; Røyset, A.; Jahren, S.; Strisland, F. Printed organic conductive polymers thermocouples in textile and smart clothing applications. In Proceedings of the 2011 Annual International Conference of the IEEE Engineering in Medicine and Biology Society, Boston, MA, USA, 30 August–3 September 2011; pp. 3278–3281.
24. Depla, D.; Segers, S.; Leroy, W.; Van Hove, T.; Van Parys, M. Smart textiles: an explorative study of the use of magnetron sputter deposition. *Text. Res. J.* **2011**, *81*, 1808–1817. [CrossRef]
25. Mrugala, D.; Ziegler, F.; Kostelnik, J.; Lang, W. Temperature Sensor Measurement System for Firefighter Gloves. *Procedia Eng.* **2012**, *47*, 611–614. [CrossRef]
26. Ziegler, S.; Frydrysiak, M. Initial research into the structure and working conditions of textile thermocouples. *Fibres Text. East. Eur.* **2009**, *77*, 84–88.
27. Takatera, M.; Uchiyama, E.; Zhu, C.; Kim, K.; Ishizawa, H. Effect of air gap on apparent temperature of body wearing various sizes of T-shirt. *IOP Conf. Series: Mater. Sci. Eng.* **2017**, *254*, 182012. [CrossRef]
28. Gidik, H.; Bedek, G.; Dupont, D.; Codau, C. Impact of the textile substrate on the heat transfer of a textile heat flux sensor. *Sensors Actuators A: Phys.* **2015**, *230*, 25–32. [CrossRef]
29. Zhu, C.; Takatera, M. A new thermocouple technique for the precise measurement of in-plane capillary water flow within fabrics. *Text. Res. J.* **2014**, *84*, 513–526. [CrossRef]
30. Ataman, C.; Kinkeldei, T.; Quintero, A.V.; Molina-Lopez, F.; Courbat, J.; Cherenack, K.; Briand, D.; Troster, G.; De Rooij, N. Humidity and Temperature Sensors on Plastic Foil for Textile Integration. *Procedia Eng.* **2011**, *25*, 136–139. [CrossRef]
31. Johnsen, D.L.; Rood, M.J. Temperature Control during Regeneration of Activated Carbon Fiber Cloth with Resistance-Feedback. *Environ. Sci. Technol.* **2012**, *46*, 11305–11312. [CrossRef]
32. Bielska, S.; Sibiński, M.; Lukasik, A. Polymer temperature sensor for textronic applications. *Mater. Sci. Eng. B* **2009**, *165*, 50–52. [CrossRef]
33. Roh, J.S.; Kim, S. All-fabric intelligent temperature regulation system for smart clothing applications. *J. Intell. Mater. Syst. Struct.* **2016**, *27*, 1165–1175. [CrossRef]
34. Sibiński, M.; Jakubowska, M.; Sloma, M. Flexible Temperature Sensors on Fibers. *Sensors* **2010**, *10*, 7934–7946. [CrossRef] [PubMed]

35. Pola, T.; Vanhala, J. Resistance Measurements in Conductive Fabrics. *Adv. Mater. Res.* **2011**, *213*, 121–125. [CrossRef]
36. Hughes-Riley, T.; Lugoda, P.; Dias, T.; Trabi, C.L.; Morris, R.H. A Study of Thermistor Performance within a Textile Structure. *Sensors* **2017**, *17*, 1804. [CrossRef] [PubMed]
37. Kinkeldei, T.; Zysset, C.; Cherenack, K.; Troster, G. A textile integrated sensor system for monitoring humidity and temperature. In Proceedings of the 2011 16th International Solid-State Sensors, Actuators and Microsystems Conference, Beijing, China, 5–9 June 2011; Institute of Electrical and Electronics Engineers (IEEE): Piscataway Township, NZ, USA, 2011; pp. 1156–1159.
38. Blasdel, N.J.; Wujcik, E.K.; Carletta, J.E.; Lee, K.S.; Monty, C.N. Fabric nanocomposite resistance temperature detector. *IEEE Sens. J.* **2015**, *15*, 300–306. [CrossRef]
39. Kinkeldei, T.; Zysset, C.; Cherenack, K.; Troester, G. Development and evaluation of temperature sensors for textile integration. In Proceedings of the 2009 IEEE Sensors, Christchurch, New Zealand, 25–28 October 2009; pp. 1580–1583.
40. Husain, M.D.; Kennon, R.; Dias, T. Design and fabrication of Temperature Sensing Fabric. *J. Ind. Text.* **2014**, *44*, 398–417. [CrossRef]
41. Zhao, X.; Long, Y.; Yang, T.; Li, J.; Zhu, H. Simultaneous High Sensitivity Sensing of Temperature and Humidity with Graphene Woven Fabrics. *ACS Appl. Mater. Interfaces* **2017**, *9*, 30171–30176. [CrossRef]
42. Najafi, B.; Mohseni, H.; Grewal, G.S.; Talal, T.K.; Menzies, R.A.; Armstrong, D.G. An Optical-Fiber-Based Smart Textile (Smart Socks) to Manage Biomechanical Risk Factors Associated With Diabetic Foot Amputation. *J. Diabetes Sci. Technol.* **2017**, *11*, 668–677. [CrossRef]
43. Kim, K.T.; Kim, I.S.; Lee, C.-H.; Lee, J. A Temperature-Insensitive Cladding-Etched Fiber Bragg Grating Using a Liquid Mixture with a Negative Thermo-Optic Coefficient. *Sensors* **2012**, *12*, 7886–7892. [CrossRef]
44. Jenkins, R.B.; Joyce, P.; Mechtel, D. Localized Temperature Variations in Laser-Irradiated Composites with Embedded Fiber Bragg Grating Sensors. *Sensors* **2017**, *17*, 251. [CrossRef] [PubMed]
45. Moraleda, A.T.; García, C.V.; Zaballa, J.Z.; Arrue, J. A Temperature Sensor Based on a Polymer Optical Fiber Macro-Bend. *Sensors* **2013**, *13*, 13076–13089. [CrossRef] [PubMed]
46. Korganbayev, S.; Min, R.; Jelbuldina, M.; Hu, X.; Caucheteur, C.; Bang, O.; Ortega, B.; Marques, C.; Tosi, D.; Rui, M. Thermal Profile Detection Through High-Sensitivity Fiber Optic Chirped Bragg Grating on Microstructured PMMA Fiber. *J. Light. Technol.* **2018**, *36*, 4723–4729. [CrossRef]
47. Min, R.; Ortega, B.; Marques, C. Latest Achievements in Polymer Optical Fiber Gratings: Fabrication and Applications. *Photonics* **2019**, *6*, 36. [CrossRef]
48. Xiang, Z.; Wan, L.; Gong, Z.; Zhou, Z.; Ma, Z.; Ouyang, X.; He, Z.; Chan, C.C. Multifunctional Textile Platform for Fiber Optic Wearable Temperature-Monitoring Application. *Micromachines* **2019**, *10*, 866. [CrossRef] [PubMed]
49. Post, E.R.; Orth, M.; Russo, P.R.; Gershenfeld, N. E-broidery: Design and fabrication of textile-based computing. *IBM Syst. J.* **2000**, *39*, 840–860. [CrossRef]
50. Nilsson, E.; Lund, A.; Jonasson, C.; Johansson, C.; Hagström, B. Poling and characterization of piezoelectric polymer fibers for use in textile sensors. *Sensors Actuators A: Phys.* **2013**, *201*, 477–486. [CrossRef]
51. Ramachandran, T.; Vigneswaran, C. Design and Development of Copper Core Conductive Fabrics for Smart Textiles. *J. Ind. Text.* **2009**, *39*, 81–93. [CrossRef]
52. Matsuhisa, N.; Kaltenbrunner, M.; Yokota, T.; Jinno, H.; Kuribara, K.; Sekitani, T.; Someya, T. Printable elastic conductors with a high conductivity for electronic textile applications. *Nat. Commun.* **2015**, *6*, 7461. [CrossRef]
53. Satharasinghe, A.S.; Jayasundara, H.; Vitarana, R.K. US 20160194792A1 2016. Espacenet. Available online: https://worldwide.espacenet.com/patent/search/family/053765507/publication/US2016194792A1?q=US20160194792A1%20 (accessed on 3 December 2019).
54. Li, L.; Fan, T.; Hu, R.; Liu, Y.; Lu, M. Surface micro-dissolution process for embedding carbon nanotubes on cotton fabric as a conductive textile. *Cellulose* **2017**, *24*, 1121–1128. [CrossRef]
55. Tao, X.; Mei-Yi Leung Leung, S.; Chun-Wah Yuen, M.; Kwok, W.-Y.; Ho, H.-L. US 20060148351A1 2006, 1. Espacenet. Available online: https://worldwide.espacenet.com/patent/search/family/036641162/publication/US2006148351A1?q=US20060148351A1%20 (accessed on 3 December 2019).

56. Amelio, W.J.; Lemon, G.; Markovich, V.; Panasik, T.; Sambucetti, C.; Trevitt, D. US 4,448,804 1984. Espacenet. Available online: https://worldwide.espacenet.com/patent/search/family/024156146/publication/US4448804A?q=US4448804%20 (accessed on 1 December 2019).
57. Wills, K.A.; Krzyzak, K.; Bush, J.; Ashayer-Soltani, R.; Graves, J.E.; Hunt, C.; Cobley, A.J. Additive process for patterned metallized conductive tracks on cotton with applications in smart textiles. *J. Text. Inst.* **2017**, *109*, 268–277. [CrossRef]
58. Post, E.R.; Orth, M. Smart fabric, or 'wearable clothing'. In *Digest of Papers*; First International Symposium on Wearable Computers: Cambridge, MA, USA, 1997; pp. 167–168, IEEE Xplore Digital Library; Available online: https://ieeexplore.ieee.org/abstract/document/629937/metrics#metrics (accessed on 3 December 2019).
59. Ashayer-Soltani, R.; Hunt, C.P. WO2014128505A1 SR.pdf 2014. Espacenet. Available online: https://worldwide.espacenet.com/patent/search/family/048092026/publication/WO2014128505A1?q=WO2014128505A1 (accessed on 30 December 2019).
60. Root, W.; Aguiló-Aguayo, N.; Pham, T.; Bechtold, T. Conductive layers through electroless deposition of copper on woven cellulose lyocell fabrics. *Surf. Coatings Technol.* **2018**, *348*, 13–21. [CrossRef]
61. Schlesinger, M.; Paunovic, M. *Modern electroplating*, 5th ed.; John Wiley & Sons, Inc.: Hoboken, NJ, USA, 2010; ISBN 9780470167786.
62. Ali, A.; Baheti, V.; Militky, J.; Khan, Z.; Tunakova, V.; Naeem, S. Copper coated multifunctional cotton fabrics. *J. Ind. Text.* **2018**, *48*, 448–464. [CrossRef]
63. Li, L.; Au, W.M.; Ding, F.; Hua, T.; Wong, K.S. Wearable electronic design: Electrothermal properties of conductive knitted fabrics. *Text. Res. J.* **2014**, *84*, 477–487. [CrossRef]
64. Krucinska, I.; Surma, B.; Chrzanowski, M. Study on Sensing Properties of Electro-spun PEO/MWNT Non-woven Fabric. *Res. J. Text. Appar.* **2010**, *14*, 89–96. [CrossRef]
65. Ferreira, A.; Ferreira, F.; Paiva, M.C. Textile Sensor Applications with Composite Monofilaments of Polymer/Carbon Nanotubes. *Adv. Sci. Technol.* **2012**, *80*, 65–70. [CrossRef]
66. Daoud, W.A.; Xin, J.H.; Szeto, Y.S. Polyethylenedioxythiophene coatings for humidity, temperature and strain sensing polyamide fibers. *Sensors Actuators B: Chem.* **2005**, *109*, 329–333. [CrossRef]
67. Cheng, Y.; Zhang, H.; Wang, R.; Wang, X.; Zhai, H.; Wang, T.; Jin, Q.; Sun, J. Highly Stretchable and Conductive Copper Nanowire Based Fibers with Hierarchical Structure for Wearable Heaters. *ACS Appl. Mater. Interfaces* **2016**, *8*, 32925–32933. [CrossRef]
68. Dolez, P.I.; Vu-Khanh, T. Recent Developments and Needs in Materials Used for Personal Protective Equipment and Their Testing. *Int. J. Occup. Saf. Ergon.* **2009**, *15*, 347–362. [CrossRef]
69. Rossi, R.M.; Bolli, W.; Stämpfli, R. Performance of Firefighters' Protective Clothing After Heat Exposure. *Int. J. Occup. Saf. Ergon.* **2008**, *14*, 55–60. [CrossRef]
70. Stoppa, M.; Chiolerio, A. Wearable Electronics and Smart Textiles: A Critical Review. *Sensors* **2014**, *14*, 11957–11992. [CrossRef] [PubMed]
71. Jiang, X.; Tessier, D.; Zhang, Z. Biostability of electrically conductive polyester fabrics: An in vitro study. *J. Biomed. Mater. Res.* **2002**, *62*, 507–513. [CrossRef] [PubMed]
72. Wearable Technology 2015-2025: Technologies, Markets, Forecast E-Textiles, Wearable Electronics, Medicals Diagnostics/Telemedicine, Smart Glasses, Smart Wristbands and More. By Dr Peter Harrop, Mr James Hayward, Raghu Das and Glyn Holland. IDTechEx. Available online: https://www.idtechex.com/ja/research-report/wearable-technology-2015-2025-technologies-markets-forecasts/427?setlang=ja (accessed on 12 January 2020).
73. Google. Available online: https://ec.europa.eu/digital-single-market/en/news/feedback-stakeholders-smart-wearables-reflection-and-orientation-paper (accessed on 12 January 2020).
74. Köhler, A.R. Challenges for eco-design of emerging technologies: The case of electronic textiles. *Mater. Des.* **2013**, *51*, 51–60. [CrossRef]
75. Rebitzer, G.; Ekvall, T.; Frischknecht, R.; Hunkeler, D.; Norris, G.; Rydberg, T.; Suh, S.; Weidema, B.P.; Pennington, D.W. Life cycle assessment Part 1: Framework, goal and scope definition, inventory analysis, and applications. *Environ. Int.* **2004**, *30*, 701–720. [CrossRef]
76. Mantecca, P.; Kasemets, K.; Deokar, A.; Perelshtein, I.; Gedanken, A.; Bahk, Y.K.; Kianfar, B.; Wang, J. Airborne Nanoparticle Release and Toxicological Risk from Metal-Oxide-Coated Textiles: Toward a Multiscale Safe-by-Design Approach. *Environ. Sci. Technol.* **2017**, *51*, 9305–9317. [CrossRef]

77. Köhler, A.R.; Hilty, L.M.; Bakker, C. Prospective Impacts of Electronic Textiles on Recycling and Disposal. *J. Ind. Ecol.* **2011**, *15*, 496–511. [CrossRef]
78. Kutz, M. *Environmentally Conscious Manufacturing*, 1st ed.; John Wiley & Sons, Inc.: Hoboken, NJ, USA, 2007; ISBN 9780471726371.
79. Jang, J.; Hyun, B.G.; Ji, S.; Cho, E.; An, B.W.; Cheong, W.H.; Park, J.-U. Rapid production of large-area, transparent and stretchable electrodes using metal nanofibers as wirelessly operated wearable heaters. *NPG Asia Mater.* **2017**, *9*, e432. [CrossRef]
80. Zhou, R.; Li, P.; Fan, Z.; Du, D.; Ouyang, J. Stretchable heaters with composites of an intrinsically conductive polymer, reduced graphene oxide and an elastomer for wearable thermotherapy. *J. Mater. Chem. C* **2017**, *5*, 1544–1551. [CrossRef]
81. Ji, S.; He, W.; Wang, K.; Ran, Y.; Ye, C. Thermal Response of Transparent Silver Nanowire/PEDOT:PSS Film Heaters. *Small* **2014**, *10*, 4951–4960. [CrossRef]
82. Wang, R.; Xu, Z.; Zhuang, J.; Liu, Z.; Peng, L.; Li, Z.; Liu, Y.; Gao, W.; Gao, C. Highly Stretchable Graphene Fibers with Ultrafast Electrothermal Response for Low-Voltage Wearable Heaters. *Adv. Electron. Mater.* **2017**, *3*, 1–7. [CrossRef]
83. Li, Y.; Zhang, Z.; Li, X.; Zhang, J.; Lou, H.; Shi, X.; Cheng, X.; Peng, H. A smart, stretchable resistive heater textile. *J. Mater. Chem. C* **2017**, *5*, 41–46. [CrossRef]
84. Jo, H.S.; An, S.; Lee, J.-G.; Park, H.G.; Al-Deyab, S.S.; Yarin, A.L.; Yoon, S.S. Highly flexible, stretchable, patternable, transparent copper fiber heater on a complex 3D surface. *NPG Asia Mater.* **2017**, *9*, e347. [CrossRef]
85. Kim, C.-L.; Lee, J.-J.; Oh, Y.-J.; Kim, D.-E. Smart wearable heaters with high durability, flexibility, water-repellent and shape memory characteristics. *Compos. Sci. Technol.* **2017**, *152*, 173–180. [CrossRef]
86. Yeon, C.; Kim, G.; Lim, J.W.; Yun, S.J. Highly conductive PEDOT:PSS treated by sodium dodecyl sulfate for stretchable fabric heaters. *RSC Adv.* **2017**, *7*, 5888–5897. [CrossRef]
87. Choi, S.; Park, J.; Hyun, W.; Kim, J.; Kim, J.; Lee, Y.B.; Song, C.; Hwang, H.J.; Kim, J.H.; Hyeon, T.; et al. Stretchable Heater Using Ligand-Exchanged Silver Nanowire Nanocomposite for Wearable Articular Thermotherapy. *ACS Nano* **2015**, *9*, 6626–6633. [CrossRef]
88. Jang, N.-S.; Kim, K.-H.; Ha, S.-H.; Jung, S.-H.; Lee, H.M.; Kim, J.-M. Simple Approach to High-Performance Stretchable Heaters Based on Kirigami Patterning of Conductive Paper for Wearable Thermotherapy Applications. *ACS Appl. Mater. Interfaces* **2017**, *9*, 19612–19621. [CrossRef]
89. Li, P.; Ma, J.; Xu, H.; Xue, X.; Liu, Y. Highly stable copper wire/alumina/polyimide composite films for stretchable and transparent heaters. *J. Mater. Chem. C* **2016**, *4*, 3581–3591. [CrossRef]
90. An, B.W.; Gwak, E.J.; Kim, K.; Kim, Y.C.; Jang, J.; Kim, J.Y.; Park, J.U. Stretchable, Transparent Electrodes as Wearable Heaters Using Nanotrough Networks of Metallic Glasses with Superior Mechanical Properties and Thermal Stability. *Nano Lett.* **2016**, *16*, 471–478. [CrossRef]
91. Hong, S.; Lee, H.; Lee, J.; Kwon, J.; Han, S.; Suh, Y.D.; Cho, H.; Shin, J.; Yeo, J.; Ko, S.H. Highly Stretchable and Transparent Metal Nanowire Heater for Wearable Electronics Applications. *Adv. Mater.* **2015**, *27*, 4744–4751. [CrossRef]
92. Cherenack, K.; Zysset, C.; Kinkeldei, T.; Münzenrieder, N.; Tröster, G. Woven Electronic Fibers with Sensing and Display Functions for Smart Textiles. *Adv. Mater.* **2010**, *22*, 5178–5182. [CrossRef]
93. Du, Y.; Xu, J.; Wang, Y.; Lin, T. Thermoelectric properties of graphite-PEDOT:PSS coated flexible polyester fabrics. *J. Mater. Sci. Mater. Electron.* **2017**, *28*, 5796–5801. [CrossRef]
94. Tien, N.T.; Jeon, S.; Il Kim, D.; Trung, T.Q.; Jang, M.; Hwang, B.U.; Byun, K.E.; Bae, J.; Lee, E.; Tok, J.B.H.; et al. A Flexible Bimodal Sensor Array For Simultaneous Sensing of Pressure and Temperature. *Adv. Mater.* **2014**, *26*, 796–804. [CrossRef] [PubMed]
95. Neves, S.; Campos, J.; Mayor, T.S. On the determination of parameters required for numerical studies of heat and mass transfer through textiles – Methodologies and experimental procedures. *Int. J. Heat Mass Transf.* **2015**, *81*, 272–282. [CrossRef]
96. Huang, J.; Chen, Y. Effects of Air Temperature, Relative Humidity, and Wind Speed on Water Vapor Transmission Rate of Fabrics. *Text. Res. J.* **2010**, *80*, 422–428. [CrossRef]
97. Yao, S.; Zhu, Y. Nanomaterial-Enabled Stretchable Conductors: Strategies, Materials and Devices. *Adv. Mater.* **2015**, *27*, 1480–1511. [CrossRef] [PubMed]

98. Luprano, J. Bio-Sensing Textile for Medical Monitoring Applications. *Adv. Sci. Technol.* **2008**, *57*, 257–265. [CrossRef]
99. Kim, S.J.; We, J.H.; Cho, B.J. A wearable thermoelectric generator fabricated on a glass fabric. *Energy Environ. Sci.* **2014**, *7*, 1959. [CrossRef]
100. Jeon, J.; Lee, H.B.R.; Bao, Z. Flexible Wireless Temperature Sensors Based on Ni Microparticle-Filled Binary Polymer Composites. *Adv. Mater.* **2013**, *25*, 850–855. [CrossRef]
101. Ueno, S.; Sawada, S. Correction of the evaporative resistance of clothing by the temperature of skin fabric on a sweating and walking thermal manikin. *Text. Res. J.* **2012**, *82*, 1143–1156. [CrossRef]
102. Wang, D.; Zhang, Y.; Lu, X.; Ma, Z.; Xie, C.; Zheng, Z. Chemical formation of soft metal electrodes for flexible and wearable electronics. *Chem. Soc. Rev.* **2018**, *47*, 4611–4641. [CrossRef]

 © 2020 by the authors. Licensee MDPI, Basel, Switzerland. This article is an open access article distributed under the terms and conditions of the Creative Commons Attribution (CC BY) license (http://creativecommons.org/licenses/by/4.0/).

Article

Effect of Jute Fibres on the Process of MICP and Properties of Biocemented Sand

Christine Ann Spencer [1],*, Leon van Paassen [2] and Henrik Sass [3]

1. School of Engineering, Cardiff University, Cardiff CF24 3AA, UK
2. Center for Bio-Mediated and Bio-Inspired Geotechnics (CBBG), Arizona State University, Tempe, AZ 85287-3005, USA; leon.vanpaassen@asu.edu
3. School of Earth and Ocean Sciences, Cardiff University, Cardiff CF10 3AT, UK; sassh@cardiff.ac.uk
* Correspondence: spencerca1@cardiff.ac.uk; Tel.: +44-749-460-1376

Received: 1 November 2020; Accepted: 24 November 2020; Published: 28 November 2020

Abstract: There has been increasing interest, in the past decade, in bio-mediated approaches to soil improvement for geotechnical applications. Microbially induced calcium carbonate precipitation (MICP) has been investigated as a potentially sustainable method for the strengthening and stabilisation of soil structures. This paper presents the results of a study on the effect of jute fibres on both the MICP process and properties of biocemented sand. Ureolytic *Sporosarcina pasteurii* has been used to produce biocemented soil columns via MICP in the laboratory. Results showed that columns containing 0.75% (by weight of sand) untreated jute fibres had unconfined compressive strengths approximately six times greater on average compared to biocemented sand columns without jute fibres. Furthermore, efficiency of chemical conversion was found to be higher in columns containing jute fibres, as measured using ion chromatography. Columns containing jute had calcimeter measured $CaCO_3$ contents at least three times those containing sand only. The results showed that incorporation of jute fibres into the biocemented sand material had a beneficial effect, resulting in stimulation of bacterial activity, thus sustaining the MICP process during the twelve-day treatment process. This study also explores the potential of jute fibres in self-healing MICP systems.

Keywords: biocementation; MICP; jute fibres; unconfined compressive strength; urea hydrolysis; sustainable geotechnics; self-healing

1. Introduction

Growing interest in sustainable methods of soil improvement has led to increased interest and research in the application of microbially induced calcium carbonate precipitation (MICP). MICP is a biogeochemical process which can be used to improve the mechanical properties of loose, saturated sand, by increasing its strength and stiffness and reducing its tendency to dilate [1]. MICP may be used as an alternative to the traditional Portland cement-based method of soil cementation [2]. MICP may occur through a variety of metabolic pathways, including photosynthesis, ureolysis, ammonification, denitrification and methane oxidation [3], in addition to sulphate reduction and iron reduction [4]. Ureolysis was selected for this study. This process is dependent upon suitable bacteria and cementation media. Ureolysis increases the alkalinity of fluid in soil pore spaces as a result of the degradation of urea to carbonate and ammonium (Equation (1)), and induces calcium carbonate precipitation (Equation (2)) [5].

$$CO(NH_2)_2 + 2H_2O \rightarrow 2NH_4^+ + CO_3^{2-} \quad (1)$$

$$Ca^{2+} + CO_3^{2-} \rightarrow CaCO_{3(S)} \quad (2)$$

Precipitation of calcium carbonate ($CaCO_3$) leads to pore-filling, inter-particle binding and particle roughening; resulting in improved soil strength and stiffness and also reduced permeability [6].

Urease positive *Sporosarcina pasteurii* (formerly *Bacillus pasteurii*) has commonly been used for studies on cementation of granular soil via MICP [3,5,7–9]. *S. pasteurii* is tolerant to a wide pH range. Stocks-fischer et al. [8] reported urease activity between pH 6 and 10, with the optimum environment being slightly alkaline. *Sporosarcina ureae* has also been shown to promote MICP via ureolysis, as demonstrated in preliminary work by Spencer and Sass [10], and by Botusharova [11]; however, the process is much slower when compared to using *S. pasteurii*. Both *S. pasteurii* and *S. ureae* will express urease, regardless of ammonia compound concentrations [12] and are spore-forming.

Recent studies have explored the enhancement of biocemented soils with additional materials such as fibres, these have focussed on mostly synthetic fibres. Fibres utilised to improve strength characteristics have included polyvinyl alcohol (PVA) [13], polypropylene (Fibermesh 150) [14], basalt fibres [15] and carbon fibres [16]. Gao et al. [17] used polypropylene fibres in an MICP based surficial treatment of sand for seepage control. Of these examples only basalt fibres are a natural material, these are however a finite resource.

This study investigates the use of jute fibres, as an innovative approach to enhance the properties of MICP treated sand using sustainable natural fibres. The use of fibres to reinforce soil is an established technique, first proposed by Vidal [18] in 1969. Jute fibres are widely used in building materials, textiles and packaging. Jute is the most common natural fibre cultivated in the world, it is biodegradable and has good tensile strength [19]. Natural fibres, such as jute, are affordable and recyclable. For structural applications, fibres may be premixed with soil during construction or incorporated *in situ* using deep mixing techniques. Fibres can be mixed with soil to construct embankment dams and other water-retaining structures to improve resistance to piping erosion, with fibre content determined by suitable piping tests [20].

This current study, which combines jute fibres and MICP, has been completed as part of an investigation into the feasibility of the use of jute and other absorbent materials to enable autonomous self-healing of biocemented sand via MICP [10]. Self-healing, in the context of construction materials, can be defined as, 'the partial or total recovery of at least one property of a material' [21]. As a further enhancement of a biocemented sand material self-healing is of interest to improve the sustainability of biocemented soil structures and reduce future maintenance and repair costs. Under loading (shear, tension, compression) the calcium carbonate binding between the silica sand particles may fail. The likely failure mechanism being a fracture within the precipitated calcium carbonate or between the precipitate and the silica sand particles [22].

Previous studies by Montoya and Dejong [1] and Botusharova [11] demonstrated that, by injecting the nutrients and precursor chemicals (cementation medium) required for the MICP process into degraded biocemented sand, strength regain could be achieved. To enable biocemented sand to self-heal, provided viable spores of *S. pasteurii* are present, would require a store of this cementation medium within the biocemented sand matrix. The development of a biocemented sand material that can autonomously self-heal is novel and has been inspired by previous studies on self-healing of cementitious materials using immobilisation, as summarised by Spencer and Sass [10]. Having identified jute as a potentially suitable carrier material for immobilisation during preliminary studies, further investigation was required on the effectiveness of jute to store cementation medium and release this to achieve self-healing.

Laboratory experiments have been undertaken to (i) determine the effect of jute fibres on the process of MICP, (ii) quantify effect of jute fibres on strength properties when incorporated into a biocemented sand material, (iii) further investigate potential for use of jute fibres in self-healing geotechnical systems.

2. Materials and Methods

2.1. Soil and Testing of Properties

A fine F60 foundry sand (U.S. Silica, Ottawa, IL, USA) was selected for this study since the optimum grain size for MICP is 0.05 mm to 0.4 mm [23]. Particle size distribution tests were conducted in accordance with ASTM D6913/D6913M–17 [24], to verify the provided product data, in addition to Proctor compaction tests, in accordance with ASTM D1557-12e1 [25], to establish target density for the sand columns. Properties as otherwise reported by U.S. Silica Company (Ottawa, IL, USA) are as per Table 1.

Table 1. Sand properties.

Soil Origin	G_s	ρ (g/cm^3)	Mineralogy	Shape
Ottawa	2.65	1.522	Quartz	Round

Prior to use, the sand was autoclaved at 120 °C for 20 min and oven dried at 105 °C for sterilisation purposes. The F60 sand was found to have negligible calcium carbonate content when tested using a calcimeter, and therefore no further treatment of the sand was required.

2.2. Fibre Preparation

Natural jute fibres (Sunrise Agriculture, Ajmer, Rajasthan, India) were used for this study. The fibres were initially of variable length, as shown in Figure 1 prior to processing. Clumps of fibres were gently brushed and then hand cut to approximately 6 mm. The length of fibres ($n = 20$) averaged 5.88 mm ± 1.60. A scanning electron microscope (SEM; SNE-4500M Plus Tabletop, SEC, Suwon-si, Korea) was used in this study to measure the diameter of the fibres following the production of the biocemented sand columns.

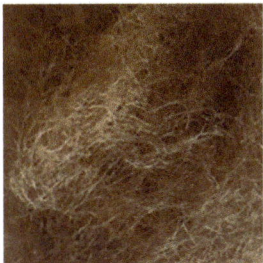

Figure 1. Jute fibres (not to scale).

Prior to use, the jute fibres were washed thoroughly using a sieve and deionised water, followed by autoclaving at 121 °C for 20 min. The fibres were then oven dried at 50 °C. Autoclaving was not considered to have an adverse effect on the fibres given its short duration. Lignocellulosic fibres have been observed to thermally degrade through dehydration, depolymerisation and oxidation when heated [19], dependent upon temperature and duration of heat exposure. Van de Velde and Baetens [26] reported that after exposing flax fibres to 120 °C for up to 2 h no significant decrease in tensile strength was observed.

Pre-Treatment of Jute Fibres

A concentrated cementation medium, CM3, was prepared as per Table 2, to treat fibres to be contained within three of the nine columns prepared. This medium contained the basic chemicals required for the MICP process, urea and calcium chloride in the form of calcium chloride dihydrate, along with Oxoid CM0001 to provide a nutrient source for the bacteria. Oxoid CM0001 (Oxoid Ltd, Basingstoke, UK) is a dehydrated culture medium. The typical 13 g/L solution of Oxoid CM0001 used for the production of liquid broth cultures contains 1 g/L 'Lab-Lemco' beef extract, 2 g/L yeast extract, 5.0 g/L peptone and 5.0 g/L sodium chloride.

After following the fibre preparation procedure outlined above, 1 g quantities of jute fibres were placed onto individual 15 cm × 15 cm squares of stainless-steel mesh with draining trays beneath and sprayed with equal amounts of CM3 (approximately 15 mL) until fully covered with this liquid. The fibres were then placed in a sealed plastic container for 24 h to allow for absorption of the CM3, before oven drying at 50 °C for 48 h. Fibres were removed from the mesh and immediately transferred to sealed sterile containers after drying. Due to the hydrophilic nature of the jute, these fibres will readily absorb moisture once exposed to air. The containers were weighed before and after filling (once fibres had cooled to room temperature) to quantify the amount of solid immobilised CM3 on each set of fibres. A set of six treated fibres were prepared due to expected variation in immobilised quantities of CM3, of which three were selected with closely matching quantities of immobilised CM3 for use in this study.

2.3. Bacteria Culture

Non-pathogenic (ACDP Group 1) *Sporosarcina pasteurii*, commonly found in soil, was obtained from the American Type Culture Collection, Manassas, VA, USA, (ATCC 11859) as a freeze-dried culture and used to produce a stab culture for storage at 4 °C. Bacteria were transferred from the stab culture using a sterile inoculation loop onto plates of Luria–Bertani (LB) agar amended with 20 g/L syringe filtered urea. Growth medium for plates contained 5 g/L yeast extract, 10 g/L tryptone, 10 g/L sodium chloride, 15 g/L agar and 20 g/L urea in deionised water. The inoculated plates, sealed with gas-permeable film, were incubated at 23 °C room temperature for 48 h. Single colonies from the plates were used to inoculate liquid growth medium. Triplicates of 50 mL liquid broth cultures were produced in 250 mL Erlenmeyer flasks, for use as an inoculant for further liquid broth cultures to be used in experiments. The liquid growth medium consisted of 13 g/L autoclave sterilised Oxoid CM0001 and 20 g/L syringe filtered urea in deionised water. Flasks were shaken at 23 °C, 150 rpm until the late-exponential phase of growth was reached after approximately 12 h and then stored at 4 °C. Liquid broth cultures were produced in 50 mL quantities using 250 mL Erlenmeyer flasks for the batch tests, followed by multiples of 150 mL in 500 mL flasks for the column studies for which greater volumes were required.

Bacterial cultures for use in experiments were inoculated using 100 µL liquid broth culture per 50 mL growth medium and aerobically grown at 23 °C, 150 rpm until an optical density at a wavelength of 600 nm (OD_{600}) of 0.9–1.2 was obtained, which equates to approximately 7.5×10^7–1.1×10^8 cells/mL, according to the relationship reported by Ramachandran et al. [27]. For the column studies, freshly grown liquid broth cultures were transferred to 50 mL sterile polypropylene tubes, each containing 35 mL culture, and centrifuged at 5000 rpm for 20 min. The supernatant was then removed, and a sample taken from this to measure the optical density, to take into account any loss of bacteria in the supernatant. The bacteria were then resuspended in a small quantity of phosphate buffered saline (PBS), dispersed using a pipette and transferred to 15 mL centrifuge tubes from which bacteria would be injected in the columns. These bacterial suspensions were made up to 10 mL with additional PBS. Use of PBS ensured the bacteria would not undergo osmotic shock which would otherwise occur in water. Aseptic technique was followed throughout, and involves using lab practices which prevent contamination, to help ensure that the only bacteria present within the culturing flasks and columns was *Sporosarcina pasteurii*.

2.4. Preparation of Cemention Medium

Two variations of the cementation medium (CM) were produced for column treatments, as per Table 2. The basic constituents of the CM as required for the process of MICP are urea and a calcium source. Calcium chloride dihydrate was selected for the calcium source. A slightly higher molarity of urea was used in comparison to calcium chloride dihydrate, since this helps ensure all calcium can be utilised. In addition, a source of nutrients, Oxoid CM0001, was added to the cementation medium to promote ongoing bacterial growth and therefore urease activity during treatment. CM1 consisted of 0.67 M urea and 0.50 M calcium chloride dihydrate, in addition to 3 g/L Oxoid CM0001 and was used as a fixation medium to fix the bacteria to the sand within the columns in addition to initiating MICP. CM2, based on the cementation medium used by Stocks-Fischer et al. [8] and Al Qabany and Soga [28], was as per CM1 with 6 g/L Oxoid CM0001, with 0.187 M ammonium chloride and 0.025 M sodium bicarbonate added. Sodium bicarbonate is added to stabilise the pH of the cementation medium before injections [28] and addition of ammonium chloride was found to help stimulate the MICP process beyond the initial CM injection.

Table 2. Cementation media composition and sterilisation methods.

Precursor Chemicals and Nutrients	CM1 (g/L)	CM2 (g/L)	CM3 (g/L)	Sterilisation Method
Calcium chloride dihydrate ($CaCl_2 \cdot 2H_2O$)	73.51	73.51	147.02	Autoclaved
Urea ($NH_2(CO)NH_2$)	40	40	80	Syringe filtered
Ammonium chloride (NH_4Cl)	0	20	-	Autoclaved
Sodium bicarbonate ($NaHCO_3$)	0	2.12	-	Syringe filtered
Oxoid CM0001 nutrient broth	3	6	12	Autoclaved

The cementation media were prepared using tap water. Results from a batch test conducted as part of this study provided evidence of the beneficial effect of using tap water compared to deionised water. CM1 was produced by first autoclaving a solution containing calcium chloride dihydrate and Oxoid CM0001, into which a solution containing urea was syringe filtered. To prepare 2.0 L of CM2, firstly the ammonium chloride and Oxoid CM0001 were dissolved in 1.6 L tap water. This solution was adjusted to pH 6.0 using 2.0 M HCl prior to then adding the powdered calcium chloride dihydrate. The pH adjustment prevented the calcium precipitating out into the solution. This solution was autoclaved then made up to 2.0 L by adding a solution containing the urea and sodium bicarbonate using a 0.2 µm syringe filter.

2.5. Urease Activity and Batch Test

Urease activity (mM urea hydrolysed/min) is calculated as per the relationship derived by Whiffin [7] below, based on a conductivity assay.

$$Urease\ Activity\ =\ Electrical\ Conductivity \left(\frac{mS}{cm}/min\right) \times 11.11\ (R_2 = 0.9988) \quad (3)$$

Electrical conductivity was measured over five minutes to obtain the average activity per minute, as per Harkes et al. [29]. This process was repeated 3 times for each sample tested and an average taken from the three results.

Specific urease activity (mM urea hydrolysed/min/OD_{600}) is further defined by Whiffin [7] as the amount of urease activity per biomass, as per Equation (4).

$$Specific\ Urease\ Activity = \frac{Urease\ Activity}{Biomass\ (OD_{600})} \quad (4)$$

A batch test was conducted to determine effects on urease activity of (i) bacterial growth in tap water and deionised water, (ii) inoculation of medium with plate cultures or liquid broth culture, (iii) initial pH of growth medium. 50 mL liquid broth cultures were prepared using tap water or deionised water as described above and grown at 23 °C for 14–19 h, to achieve a stationary stage of growth. The nutrient medium was inoculated with either 100 µL of a liquid broth culture grown to 1.0 OD_{600} or with one colony from a plate culture. To test the effect of pH of the nutrient medium on urease activity the pH of the solution of Oxoid CM0001 nutrient broth in water was adjusted prior to autoclaving and adding urea, after which a 1 mL sample was taken to test the pH prior to inoculation.

2.6. Preparation of Columns

Columns were assembled as shown in Figure 2. The column apparatus fabricated for this research was based on that used in previous studies by Botusharova [11], with some enhancement including longer split moulds, to ensure that the columns produced would exceed minimum depth to diameter ratios required for the unconfined compressive strength (UCS) testing and to ensure a secure fit of the bungs. Masses of sand and fibres used within columns were as per Table 3. The jute fibre content of columns was 0.75% by weight of sand. Apparatus was washed with 1% Virkon S Disinfectant solution and rinsed with autoclaved deionised water or autoclaved before use to sterilise.

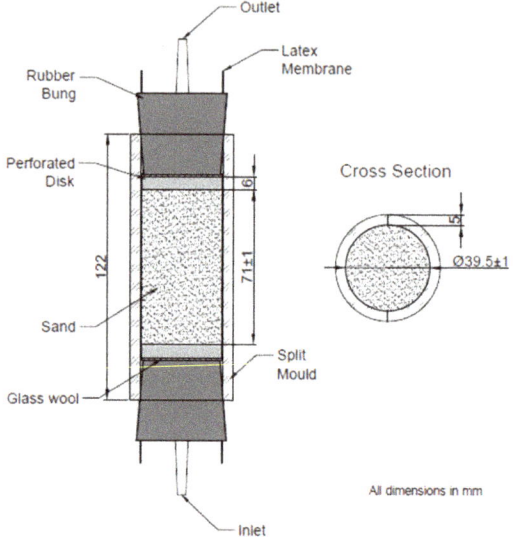

Figure 2. Column assembly.

Table 3. Column contents.

Column ID	Sand, g	Jute, g	Immobilised CM, g
J1	133	1	0
J2	133	1	0
J3	133	1	0
J4	133	1	1.687
J5	133	1	1.681
J6	133	1	1.686
C1	143	0	0
C2	143	0	0
C3	143	0	0

Columns were prepared in triplicates, with J1 to J3 containing sand and fibres, J4 to J6 containing sand and treated fibres and C1 to C3 containing sand only as controls for comparison. Producing the columns in triplicates helps to improve accuracy of results, with results reported as an average from each set of triplicates unless otherwise stated. Each sand column was encased by a 150 mm long, 38 mm diameter, 0.3 mm thick latex membrane, enclosed by clear perspex split moulds of 5 mm thickness and approximately 39 mm inner diameter. To prepare each column, split moulds were secured together using cable ties and the latex sleeve placed inside the mould, with the ends wrapped over the sides of the mould. A perforated 3D printed plastic disk (6 mm depth, 38 mm diameter), wrapped with glass wool was placed into the inlet end of the mould, followed by a bung without the outlet connector in place. The perforated disk was wrapped in glass wool to prevent washout of fine grains of sand during treatments and to provide a gap between the bung and disk. A 3.99 mm × 8 mm silicone tapered plug was inserted into the hole in the inlet bung to retain the CM, into which the sand was deposited using the wet pluviation method combined with vibration. The latex sleeve was wrapped around the inlet bung and this assembly then held upright by a small beaker. A funnel held by a clamp was positioned above the top of the column assembly, leaving 1.5 cm between the funnel tip and column edge. A total of 35 mL of sterile CM1 was poured inside the latex sleeve.

The column assembly was vibrated three times during the filling process, using a vortex mixer, to aid compaction. The second perforated disk, also wrapped with glass wool, was then placed into the top of the column and the second bung inserted. The bung was pressed firmly into the column while holding the lower bung in place, the membrane was then wrapped around the upper bung and secured using rubber O rings to prevent leaking. This column was then transferred to the column frame, as shown in Figure 3. The column frame consisted of 30 × 60 cm pegboard, with metal feet attached, onto which up to six columns could be secured on either side. This process was repeated for each column.

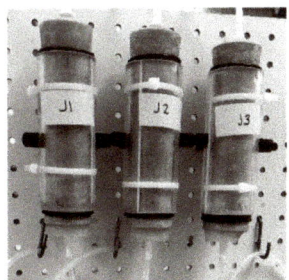

Figure 3. Columns attached to frame.

Prior to addition to the columns, fibres were mixed with sand samples for columns J1 to J6. After weighing the sand and adding fibres a small quantity of CM1 was added, equating to 5% of the total mass of the sand and fibre mixture, to aid mixing. Hydration prior to mixing also helps prevent fibre sand segregation [30]. This was then mixed by hand for approximately five minutes, until an even distribution of fibres was observed. The quantity of liquid added to the sand aided mixing while ensuring that it was still possible to pluviate the sand into the columns. Columns were kept at an ambient room temperature of 23 °C throughout.

2.7. Bacteria Fixing and Biocementation

To achieve biocementation, a two-stage process was applied involving (i) the injection of a bacterial suspension, followed by (ii) injections of cementation medium (treatments), as per Table 4. The first treatment, using CM1, is injected immediately after the bacterial suspension and has the effect of fixing the bacteria within the columns in addition to initiating MICP. Harkes et al. [29] found that when a cementation solution consisting of 1 M equimolar urea and calcium chloride was injected into columns immediately after a bacterial suspension this resulted in 100% retention of bacteria, as determined by optical density measurement of effluent samples. Divalent cationic ions such as Ca^{2+} may enhance the attachment of bacteria to surfaces by reducing electrostatic repulsion [31]. The bacterial suspension was injected upwards into the base of all columns simultaneously using a peristaltic pump and constant 1.5 mL/min pumping rate, followed immediately by one and a half pore volumes of CM1. The outlet tubing from columns was then drained and reconnected. Tubing was closed off with clamps and disconnected from the pump tubing following each treatment. On each of the following four days, one and a half pore volumes of CM2 were pumped through the columns, as per the schedule detailed in Table 4.

Table 4. Columns treatment schedule.

Day	Time Since Prior Injection (h)	Column Injection (1.5 × Pore Volume)	Treatment
0	0	CM1	1
1	16	CM2	2
2	22	CM2	3
3	24	CM2	4
4	24	CM2	5
12	192	Tap Water	None

The timing between treatments had been determined by preliminary studies undertaken as part of this research, and based upon the time taken to deplete the calcium source in columns containing jute, in which this occurred fastest. While injecting CM2 the effluent was collected in a series of 5 mL quantities in 15 mL polypropylene tubes.

2.8. Measurement of Electrical Conductivity, pH and Evaluation of Bacterial Fixing

The measurement of pH and electrical conductivity can provide an indication of the extent of substrate conversion, and thus bacterial or urease activity within columns. The effluent displaced from columns during the injection of new CM was collected in a series of centrifuge tubes in 5 mL quantities. For each 5 mL column effluent collected, the conductivity and pH of the effluent was measured using a Consort multi parameter analyser C3010, pH probe and conductivity probe with temperature compensation. Following the first CM treatment, 1 mL of effluent from the 5–10 mL sample of effluent from each tube was taken to measure optical density, as an indication of biomass concentration, to determine effectiveness of bacteria fixing. Optical density was measured using a spectrophotometer (Hach DR 6000, Loveland, CO, USA) at a wavelength of 600 nm. The first 5 mL was not used since this may include some CM that had been retained in the column outlet after treatment.

2.9. Geochemical Analysis

The calcium ion concentration of column effluent was measured using an Ion chromatograph (IC). Dionex ICS 5000+ Cation analysis was conducted using 20 mM methanesulfonic acid eluent starting concentration, on a Dionex CS12A column, using 112 mA suppressor output. The first 5 mL effluent from each column was discarded at this stage, since this may contain some unreacted substrates that had been held within the column outlet, outlet tubing had otherwise been drained after each treatment. Beyond this point, up to 20 mL the effluent was mixed for each column to obtain a representative average concentration of calcium ions for each column. The effluent beyond this point was not tested, since this may contain the new CM being injected at later stages in the treatment process as pore volume decreases. Effluent samples were tested daily within two hours of collection to ensure accuracy of results.

2.10. Quantification of Calcium Carbonate Precipitate

The calcium carbonate content of the sand columns was quantified using a calcimeter, in accordance with Eijkelkamp [32] and the method of Scheibler, to determine carbonate content based upon a volumetric method. Carbonates in the sample are converted to CO_2 by adding HCl to the sample. The pressure of the CO_2 released causes water in a burette to rise and this difference in level from the start of the test enables calcium carbonate content by mass, $w(CaCO_3)$, to be calculated according to the below equation obtained following calibration, whereby V_1 is the volume of carbon dioxide produced by the reaction of the test portion.

$$w(\text{CaCO}_3) = 0.0045 \times V_1 \tag{5}$$

The result of the above divided by the mass of the sample tested gives the percentage of the sample which contains calcium carbonate. To obtain an estimate of the average calcium carbonate content of each column, as a percentage of the total dry mass, samples of between 4 g and 5 g were taken from the top, centre and base of each column for testing after oven drying at 105 °C. Tests were conducted at constant room temperature of 23 °C.

2.11. Mineral Analysis

A combination of scanning electron microscopy (SEM) and X-ray diffraction (XRD) was used to confirm the presence of calcium carbonate and analyse the characteristics of the mineral precipitate in samples taken from biocemented columns. SEM was used to observe the morphology of the mineral crystals and fibres, and to measure the diameter of the jute fibres in the samples. Samples for SEM were prepared using a sputter coater (MCM-100 Ion Sputter Coater, SEC, Korea) and gold coating. X-ray diffraction (XRD; Malvern Panalytical, Aeris powder diffractometer, Malvern, UK) was used to characterise the mineral crystals observed and confirm the presence of $CaCO_3$.

2.12. Unconfined Compressive Strength Testing

The unconfined compression test was carried out in accordance with 7.2 of BS 1377-7:1990 [33], to determine the unconfined compressive strength (UCS) of biocemented columns using the load frame method. A loading rate of 1.27 mm/min was applied. This test was performed twice, firstly after the biocementation treatment process had been completed (UCS1), and a second time to test for self-healing effects (UCS2). Following the injection of treatment five the cementation medium was left in columns for eight days before UCS1, with the inlet and outlet tubing clamped. The retention time had been longer at this stage due to the significant reduction in calcium ion depletion rate in control columns and also those containing treated fibres. Tested columns were in a saturated condition, with evaporation prevented during testing by the test specimens being encased by the latex membranes and end caps. Membranes were kept in place for this test, perforated discs were removed, and Perspex end caps were used to provide a level testing surface between the column and UCS testing apparatus. Following the initial UCS test, the columns were reconstituted and returned the column assembly, with one and a half pore volumes of autoclaved tap water injected into columns. Samples were left hydrated a further eight days before the UCS test was conducted again. At this stage, the theory that some CM may have been retained, and later leached out of fibres to enable healing was also being tested, following earlier work by Spencer and Sass [10].

3. Results and Discussion

3.1. Urease Activity Batch Test

Urease activity test results, as per Table 5, show that bacterial growth was faster in a medium prepared using tap water as opposed to deionised water, based on measurements of optical density. Consequently, the cementation medium for the column treatments was prepared using tap water. Liquid broth culture (LBC) inoculant resulted in a higher urease activity than inoculation of the growth medium with a single colony plate culture (PC). It is, however, noted that this may be due to inoculum size, and that this can be better controlled by using liquid broth culture as an inoculant. The liquid broth cultures for the columns were prepared using deionised water (DI). Three measurements of urease activity were taken and averaged for each of the samples prepared.

Table 5. Results from urease activity tests.

ID	Water	Inoculant	pH	Culture Time	OD	Electrical Conductivity (mS/cm/min)	Urea Hydrolysed (mM/min)	Specific Urease Activity (mM/min/OD)
1	Tap	PC	8.37	19	0.940	0.56	6.22	6.62
2	DI	PC	7.97	19	0.937	0.41	4.59	4.90
3	Tap	LBC	8.37	19	1.308	0.75	8.30	6.34
4	Tap	LBC	8.37	16	1.222	0.69	7.63	6.24
5	Tap	LBC	8.37	14	1.068	0.57	6.30	5.89
6	DI	LBC	6.46	19	1.005	0.63	6.96	6.93
7	DI	LBC	7.15	19	0.822	0.57	6.37	7.75
8	DI	LBC	7.97	19	1.038	0.55	6.15	5.92
9	DI	LBC	8.53	19	1.134	0.50	5.56	4.90
10	DI	LBC	8.86	19	1.196	0.44	4.89	4.09
11	DI	LBC	9.17	19	1.156	0.41	4.59	3.97

The specific urease activity of the liquid broth culture grown using deionised water peaks at a pH of approximately 7, as shown in Figure 4. This is close to the pH of sample 7 above, this being the sample with unadjusted pH, and, hence, the pH of the growth medium was not adjusted for the following column studies.

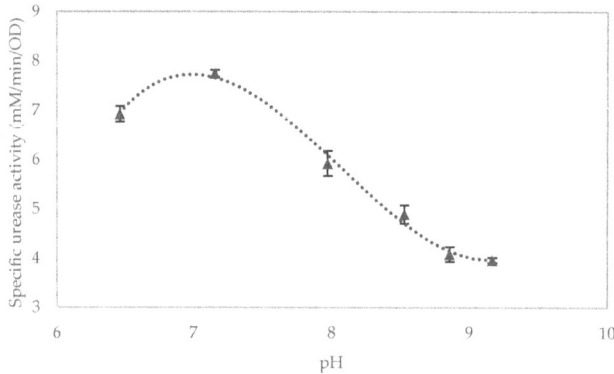

Figure 4. Effect of pH on specific urease activity in liquid broth cultures, with error bars showing standard errors of the means.

3.2. Column Studies

3.2.1. Soil Properties

Figure 5 gives the optimum dry density (target density) of sand, and the sand and jute mixture, for compaction into the columns, this being 1.726 g/cm^3 and 1.700 g/cm^3 respectively. An additional measurement was taken for the test with the sand and jute mixture, as it had been evident during testing that the optimum water content was higher when compared with the sand only test. This had been expected due to absorption of water by the jute fibres. Based on column measurements taken using vernier calipers prior to the first UCS test, columns containing sand only and sand and untreated jute were compacted to 95.7% and 91.4% respectively of their target densities. These results, as an average of the triplicate columns, show that the inclusion of fibres hindered compaction of column contents to some extent.

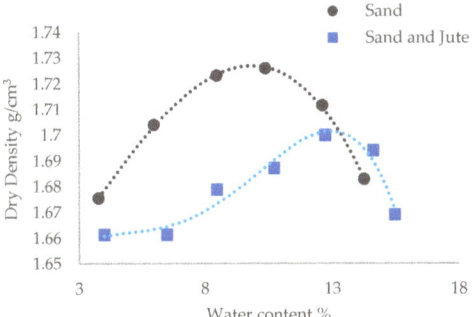

Figure 5. Proctor compaction curves for sand only and fibre and sand mixtures.

Results obtained for particle size distribution, as shown in Figure 6, are in close alignment with those reported by U.S. Silica. The general slope and shape of this distribution curve are described by means of the coefficient of uniformity (C_u) and coefficient of curvature (C_z), with a C_z value of between 1 and 3, indicative of a well-graded soil [34]. The parameters determined from the gradation curve for the F60 sand are given in Table 6. The C_z value of 1.064 indicates that this soil is fairly well-graded. The low C_u value reflects the narrow range of particle sizes.

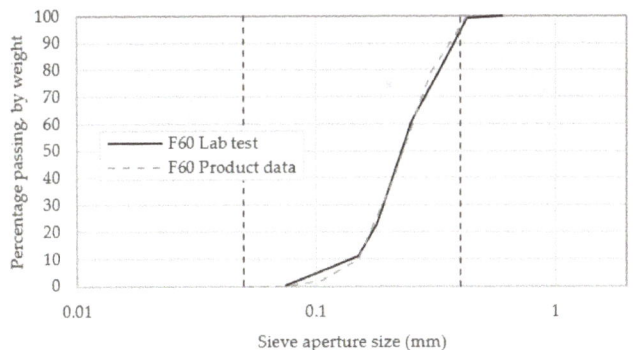

Figure 6. Particle size distribution of F60 sand.

Table 6. Sand parameters.

D_{10}	D_{50}	D_{60}	D_{30}	C_u	C_z
0.140	0.226	0.250	0.193	1.786	1.064

3.2.2. Bacteria Fixation and Initial Activity

The optical density (OD_{600}) of the *S. pasteurii* liquid broth culture grown for the column studies was measured as 0.992 using a spectrophotometer. Following centrifugation, the supernatant optical density was 0.154. Taking into account this loss of bacteria in the supernatant the resulting optical density was 0.840. When using the larger 500 mL Erlenmeyer flasks to produce 150 mL volumes of bacterial cultures, grown for 24 h at 150 rpm and 23 °C, the measured urease activity had been lower than when using the 250 mL flasks to produce the 50 mL cultures. Urease activity of the culture used to inoculate the columns was measured as 4.37 mM/min. The reduced surface area to volume ratio demonstrated the effect of limited oxygen transfer on the urease activity of the culture. The concentration gradient between the oxygen at the surface and within the medium promotes oxygen transfer into the medium [35].

The bacteria appeared to have been successfully fixed by CM1 as shown by the low optical densities of effluent in Table 7, as measured following the injection of CM2. Some bacterial losses were occasionally observed in effluent just after one pore volume of CM had been pumped into the column, when some cloudiness was observed in the effluent. It is noted that should any mineral precipitate be contained within the effluent that this would affect the optical density measurements. However, given the very low optical densities measured, as given in Table 7, this has been deemed to have a negligible effect if any at this stage.

Table 7. Optical density of column effluent following bacteria fixing.

Column	J1	J2	J3	J4	J5	J6	C1	C2	C3
Effluent OD$_{600}$ (5–10 mL)	0.018	0.01	0.015	0.011	0.008	0.011	0.055	0.02	0.044

3.2.3. Distribution of Bacterial Activity

Figure 7a–j shows the breakthrough curves for measured electrical conductivity and pH of effluent displaced during each treatment flush, which are used to analyse the distribution of the reaction products within the columns. These results indicate that the distribution is fairly even for control columns, with more variation in the columns containing jute, as was expected given the jute may absorb some of the bacteria. The results also indicate a slightly lower bacterial activity towards the top of columns (outlet) containing jute following treatment one. This trend was observed to reverse after three CM treatments. The dashed vertical lines in Figure 7a–j represent the interpreted location of column boundaries at the outlet (approx. 5 mL) and inlet (approx. 30 mL) locations.

Conductivity measurements of the effluent from columns show that there is some initial inhibition of MICP activity in columns containing jute fibres (both treated and untreated) during treatment one, as shown in Figure 7a, when compared to the control columns containing sand only, as was similarly observed in aqueous studies reported by Spencer and Sass [10]. However, results from testing of the effluent flushed following subsequent treatments show that for columns containing untreated jute fibres the EC and pH values corresponded to full conversion. The results from effluent tested after treatment three and four show a decline in measured pH and EC from columns containing treated jute compared to the untreated jute. Due to the excess of urea, full conversion of urea would deplete almost all calcium and the remaining solution would be expected to contain about 1.25 mol/L ammonium, 1 mol/L chloride and 0.25 mol/L carbonate/bicarbonate, which according to Van Paassen [36], has an EC of about 125 mS/cm and an expected pH of 8.5 to 9. Incomplete conversion would render lower EC and pH values. The pH and conductivity results in Figure 7 show a similar trend for columns containing treated and untreated fibres, however, the longer error bars for those with treated fibres indicative a greater variability of bacterial activity within these columns.

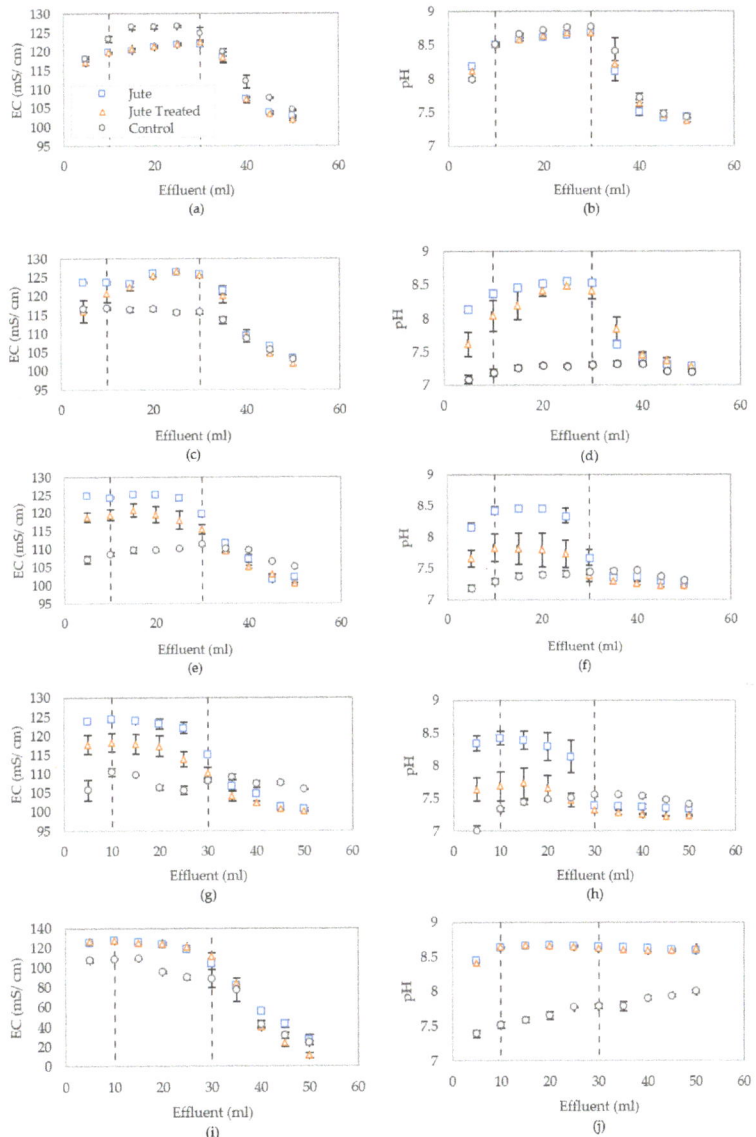

Figure 7. Conductivity and pH of columns effluent measured following treatments 1 (**a,b**), 2 (**c,d**), 3 (**e,f**), 4 (**g,h**) and 5 (**i,j**), with error bars showing standard errors of the means for the triplicates.

3.2.4. Efficiency of Calcium Ion Conversion

The concentration of calcium ions in the effluent has been used as a measure of the efficiency of substrate conversion following CM treatments one to four. The initial calcium ion concentration in the injected CM was 500 mmol, which reduces to an average of 2 mmol across all columns after treatment one. The calcium ion depletion in Figure 8 refers to this reduction in concentration, and has been represented as a cumulative value over time. The concentration of calcium ions in the column effluent shows that the efficiency of conversion of calcium ions to produce calcium carbonate precipitate declines over time for the control columns containing sand only and, to a lesser extent, the columns containing the pretreated jute fibres, between treatments one and four. Where jute has been mixed with the sand the relationship between calcium ion conversion to produce calcium carbonate, i.e., depletion of the calcium ions, in respect of time is almost linear. This clearly demonstrates a beneficial effect of the jute fibres on the MICP process. This effect is likely due to adsorption/absorption of bacteria by the jute fibres, which appears of have had a positive effect on bacterial cell growth/ viability. Cells adsorbed on surfaces replicate and grow into microcolonies [31].

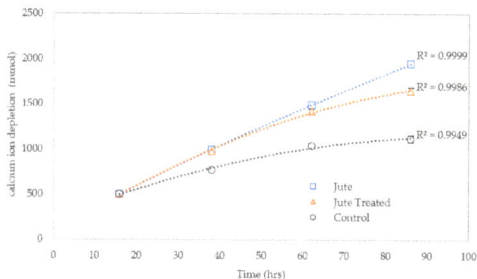

Figure 8. Cumulative reduction in concentration of calcium ions in columns between cementation medium (CM) treatments one and four, with error bars showing standard errors of the means.

Figure 9 shows the chemical conversion efficiency following all five CM treatments, based on measurement of calcium ions in the effluent. Following treatment one the conversion efficiency is near 100% for all columns, despite the slightly reduced urease activity of the bacteria injected into columns compared to the batch study. There is a rise in efficiency following treatment five since columns had been left eight days before the first UCS test and subsequent flushing with tap water and collection of effluent. This increase at this stage is significant for the columns containing treated jute and is indicative of a slower but also sustained MICP process when compared to results for columns containing untreated jute. These results, along with those from Figure 7, are indicative of a lower urease activity in columns J4 to J6, suggesting that there are less viable bacteria in these columns at this stage compared to J1 to J3. Based on results in Table 7, bacteria had been fixed adequately following CM treatment one but the fibre pre-treatment may have rendered these fibres less able to absorb/adsorb bacteria and more bacteria may have instead adhered to the sand particles. The adhesion of bacteria to sand particles in columns J4 to J6 is likely somewhere between that of the sand only and sand and untreated jute columns. Adhesion of bacteria to a surface is affected by the physical properties of the surface and surface chemistry, with topography being the most influential factor on bacterial adhesion [31]. The treatment of the fibres may have resulted in a smoothened outer surface, and will likely have also affected their ability to absorb bacteria.

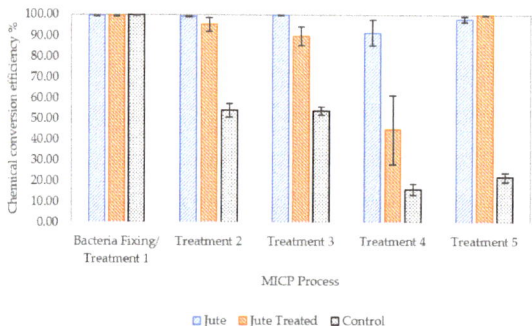

Figure 9. Chemical conversion efficiency, with error bars showing standard error of means.

3.2.5. Unconfined Compressive Strength, and CaCO$_3$ Precipitated

Unconfined compression test results are shown in Figure 10. The average unconfined compressive strength of the three control columns containing sand only was 66 kPa. The average unconfined compressive strength of the columns containing untreated jute and treated jute fibres was 370 kPa, and 320 kPa, respectively. It is observed that on average the unconfined compressive strengths of columns containing untreated fibres are approximately 5.6 times higher than the columns containing sand only. Figure 10a shows a relatively close relationship between the peak unconfined compressive strength results for columns J1 to J3, with the peak strengths all occurring close to 5% strain. This is indicative of good repeatability for these columns. More variation between results is observed in Figure 10b, for columns J4 to J6 containing treated fibres, which appear to fail in a more brittle manner at varying strains between approximately 2.5% and 5%, and have lower residual strengths. For this set of columns, the highest, and also the lowest, UCS is obtained out of the six columns containing fibres. The lower strength for column J5 is attributed to this splitting down the centre during the UCS test, as can be seen in Figure 11e. The highest strength obtained was 520 kPa for column J4 containing treated jute fibres. The variability between results for columns J4 to J6 may have been due to variable absorption or adsorption of bacteria by these fibres. The pre-treatment did not appear to hinder the mixing of fibres with sand.

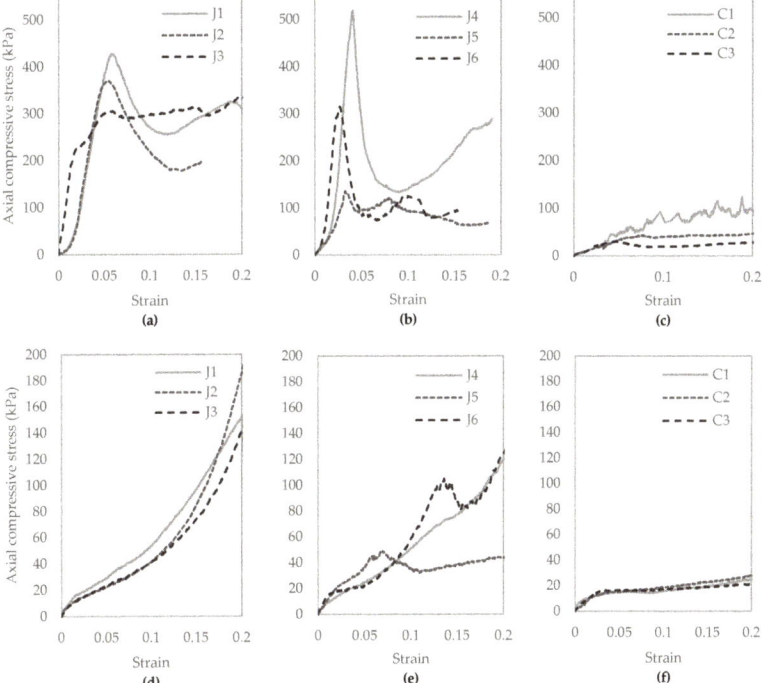

Figure 10. Unconfined compression test results following CM treatment five (**a–c**), and after reconstitution, flushing and saturation with water and eight days curing to test for self-healing (**d–f**).

These results demonstrated the significant contribution to strength of the jute fibres, when compared to controls, as a result of the mechanical properties of the fibres and greater precipitation of calcium carbonate within these columns. The confining effect of the latex membranes will have had a small contribution to strengths obtained, which is assumed to be consistent across all columns tested. Of interest in this study is the comparison between the results for columns tested.

Figure 10d–f is indicative of longer-term strengths of the soil following failure. Between the first set of unconfined compression test results (Figure 10a–c) and second (Figure 10d–f), the column contents contained within the latex membranes were reconstituted, this process itself will have had some effect on material properties and may contribute to some strength reduction. There is a noticeable difference in the trend of the UCS results for the reconstituted samples when comparing Figure 10d–f, with peaks only visible for columns containing treated jute fibres.

Figure 11 shows images of the column samples following the onset of failure during the UCS test. The diagonal shear failure can be clearly seen in most samples. When comparing images, there is a greater inconsistency in observed failure mechanisms for the samples with treated fibres, J4 to J6. J5 was observed to break apart down the centre of the column during testing. The controls typically sheared across the middle third of the test specimen. The columns containing jute had greater resistance to shear failure and the shear failure line appears higher up in the specimen, indicating strengths may have been greater towards the column inlet (base of column).

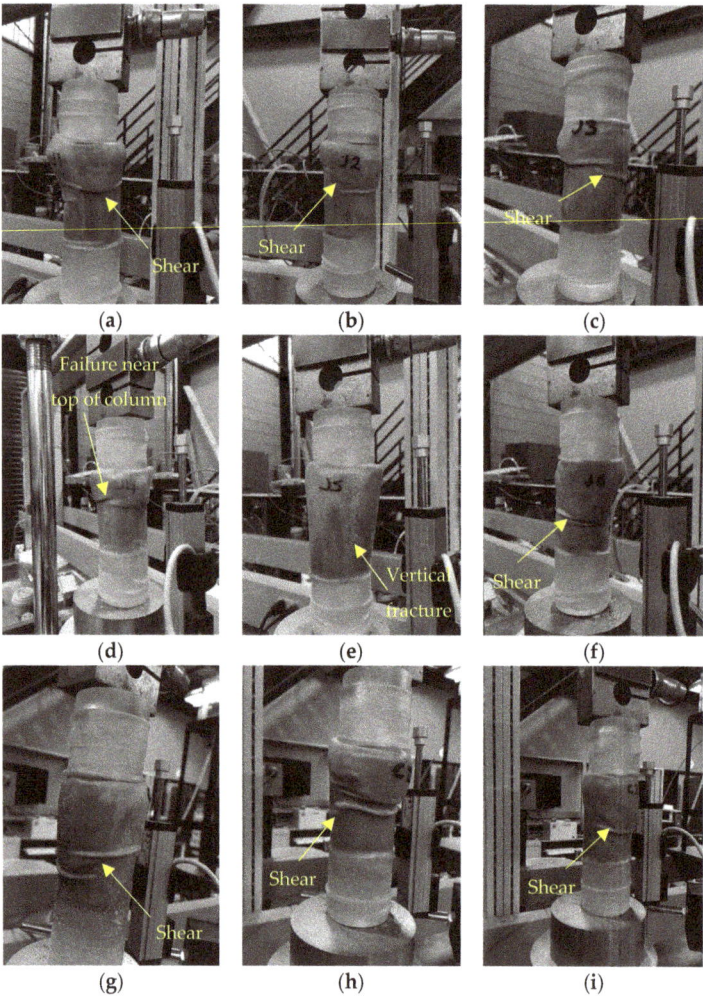

Figure 11. Images of columns J1 to J3 (**a–c**), J4 to J6 (**d–f**) and C1 to C3 (**g–i**), following the onset of failure during unconfined compressive strength testing.

Figure 12 shows the unconfined compression test results following the initial biocementation (peak 1 and residual 1) and after the reconstitution and self-healing test stage (peak 2). The self-healing stage consisted of saturation with sterile tap water and curing over eight days. When the Peak 2 strength is compared with the residual strength from UCS1, the results for two columns (J2 and most notably J6) indicate some strength regain. In accordance with BS 1377-7:1990 the 'Peak 2' unconfined compressive strength has been determined from results at 20% axial strain for columns J1 to J3, J4 and J6 and C1 to C3.

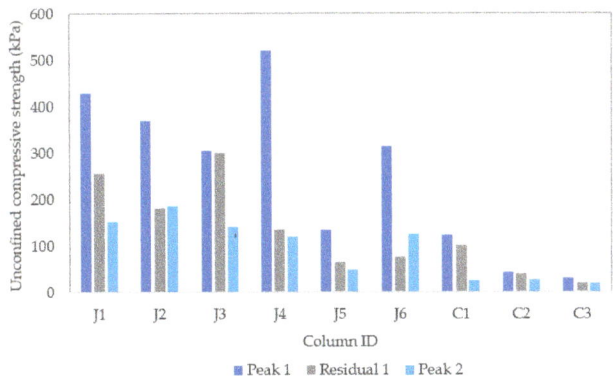

Figure 12. Unconfined compressive strength (UCS) test results.

Following the second UCS test, the columns were removed from the latex membranes and oven dried, to determine moisture contents, followed by measurement of calcium carbonate content, as given in Table 8. Columns containing jute will have had a higher void ratio prior to biocementation since columns containing fibres did not compact quite as well those containing sand only.

Table 8. Moisture and $CaCO_3$ contents of biocemented columns (averages from triplicates).

Columns	Moisture Content %	$CaCO_3$ Content%
J1–J3	18.35 ± 0.34	3.98 ± 0.59
J4–J6	16.82 ± 1.33	4.63 ± 0.31
C1–C3	17.03 ± 0.39	1.08 ± 0.47

The results in Table 8 show that the inclusion of jute and treated jute in the columns had resulted in an increase in calcium carbonate content by 3.69 and 4.33 times on average, respectively, when compared to the columns with no fibres. This increase is significant when compared to studies using synthetic fibres and also shows jute outperforms natural basalt fibres. Choi et al. [13] reported that MICP treated sand specimens containing 0.8% (by weight of sand) PVA fibres had just 1.06 times more calcium carbonate on average than those without fibres. Choi et al. [13] report an average 28.18% unconfined compressive strength increase resulting from PVA fibre additions, although it is noted that there is considerably more variability in the results they obtained. Li et al. [14] found that the UCS of MICP-treated sand with 0.3% (by weight of sand) polypropylene (Fibermesh 150) fibres was 2.4 times higher on average, which reduced to 1.5 times when the fibre percentage increased to 0.4%. Improved results have been achieved using natural basalt fibres. Xiao et al. [15] reported that inclusion of 0.4% basalt fibres in biocemented sand results in a 4.9 times higher unconfined compressive strength on average, and 1.62 times greater calcium carbonate content when compared to specimens with no fibres. Similarly, Xiao et al. [15] reported a UCS reduction to 1.7 times that of sand only specimens when the fibre percentage was doubled to 0.8%.

The greatest amount of calcium carbonate was precipitated within columns J4 to J6, despite the measured reduction in chemical conversion efficiency in these columns following MICP treatments 2 to 4. This is likely due to the leaching of the immobilised cementation medium.

The calcium carbonate contents of the nine individual columns, as determined using a calcimeter, are shown in Figure 13. This analysis relates the calcium carbonate contents to the tested unconfined compressive strengths of the columns. There had been a greater consistency between results for the controls and columns containing untreated jute.

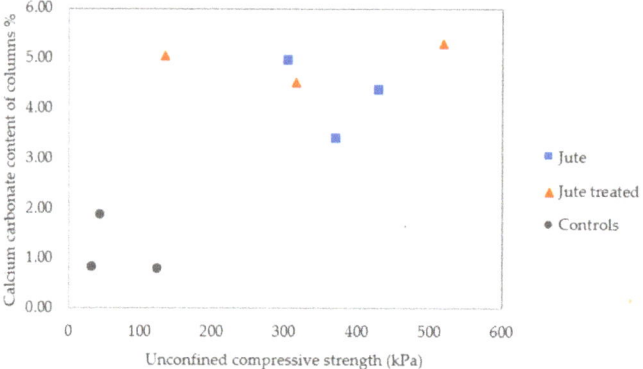

Figure 13. Calcium carbonate content of columns.

3.2.6. Morphology of CaCO$_3$ Precipitate and Jute Fibres

Samples used for this stage of analysis were J1, J4 and C1. Figure 14 shows the distribution of jute fibre diameters in samples taken from J1 ($n = 21$) and J4 ($n = 20$), measured using SEM. These results indicate the treated fibres had swollen and were more variable in diameter.

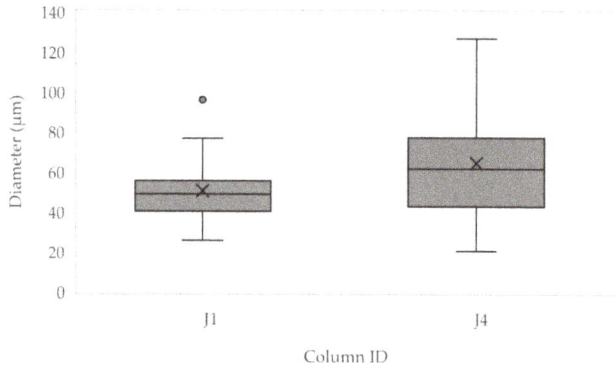

Figure 14. Diameters of jute fibres measured using scanning electron microscopy (SEM).

SEM images of the sample from J1 show that, where there is little to no fibre deterioration observed, as seen in Figure 15a, there is much less CaCO$_3$ precipitate observed on the fibre surface when compared to the visibly deteriorated fibre in Figure 15b. The fibre shown in Figure 15b has significant deterioration both on its surface and at depth, since it is breaking apart and has a much more roughened surface covered in CaCO$_3$ crystals. Rough fibres will be better at filtering and absorbing bacteria. Fewer crystals were generally observed on the treated jute fibres as can be seen in Figure 15c,d. The fractured fibres shown in Figure 15d suggest perhaps greater brittleness of fibres as a result of the pre-treatment process. The samples containing fibres (Figure 15a–d), show fibres and sand particles with a combination of rhombohedral and rounded crystals of calcium carbonate on the surface. The images of samples containing biocemented sand only, Figure 15e,f, show clusters of what appear to be more rhombohedral shaped calcium carbonate on the sand surface and bridging sand particles. It can be observed that the fibre and sand grain in Figure 15b are bonded together by calcium carbonate crystals and that there were generally a greater number of crystals in jute containing samples. This supports the findings of chemical analyses that show the increased efficiency of substrate conversion to form calcium carbonate in the columns containing jute.

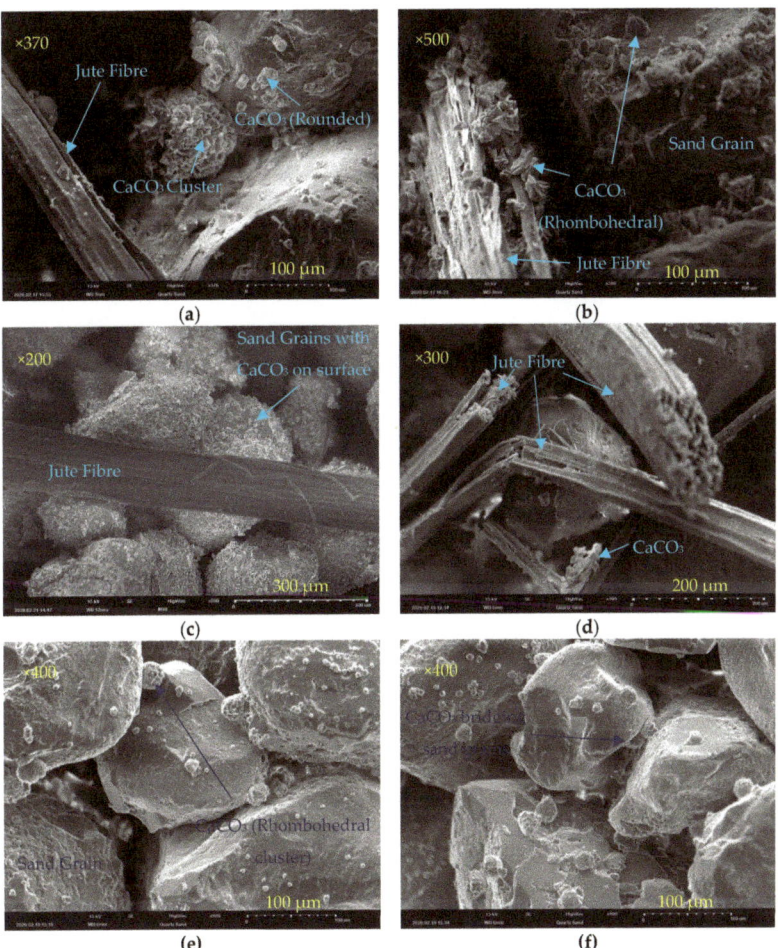

Figure 15. SEM images of samples from biocemented sand columns containing jute (**a**,**b**), treated jute (**c**,**d**), and sand only controls (**e**,**f**).

This spherical shape of some of the crystals observed has been associated with crystals of vaterite [28,36,37], with calcite reported to precipitate in a more rhombic form [37]. There are three polymorphs of anhydrous calcium carbonate: vaterite, aragonite and calcite. Calcite is a more thermodynamically stable form of calcium carbonate than vaterite [38]. Experimental evidence has demonstrated that vaterite can transform to aragonite in 60 min at 60 °C and to calcite in 24 h at room temperature [39]. XRD was performed to verify the crystal morphology of the observed precipitate.

The XRD data was analysed using HighScore Plus, with results shown in Figures 16–18. XRD data have been compared with reference patterns to determine crystalline phases present, with phases identified based upon the closest match between intensity and position of reference patterns and the diffraction peaks. These analyses verify the presence of calcite and vaterite polymorphs of calcium carbonate in all samples tested. Significant peaks for each crystalline phase are shown circled in Figures 16–18. These results indicate that vaterite may be the more dominant of the calcium carbonate polymorphs present within all samples, in particular those with untreated fibres (J1–J3), based upon height of peaks and intensity of the reference pattern. Nawarathna et al. [40] reported that addition of chitosan as an organic additive to enhance MICP promoted the production of vaterite. This suggests that jute as an organic material may be influencing the crystal morphology in a similar manner, leading to the observed dominance of vaterite, due to the physicochemical properties of these fibres.

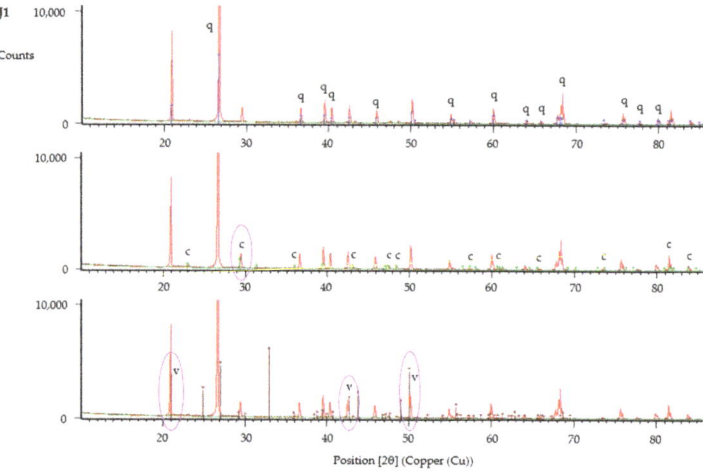

Figure 16. X-ray powder diffraction (XRD) analysis of sample from column J1, showing identified peaks of quartz (q), calcite (c) and vaterite (v).

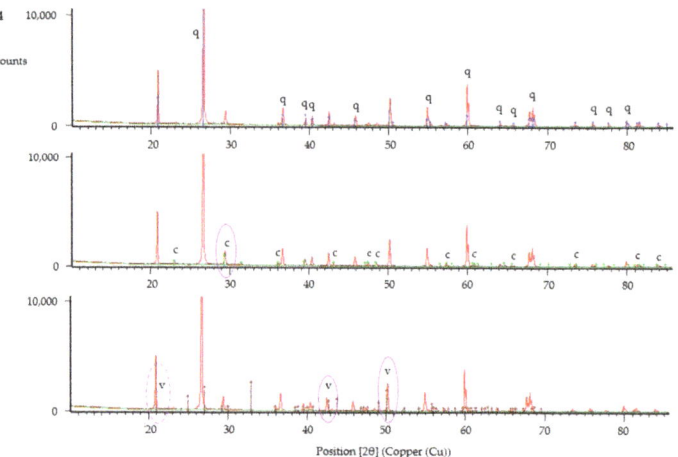

Figure 17. XRD analysis of sample from column J4, showing identified peaks of quartz (q), calcite (c) and vaterite (v). The peak at 21 [°2θ] is identified as vaterite based on results for J1 and C1, however results for J4 alone suggest this could be also be quartz.

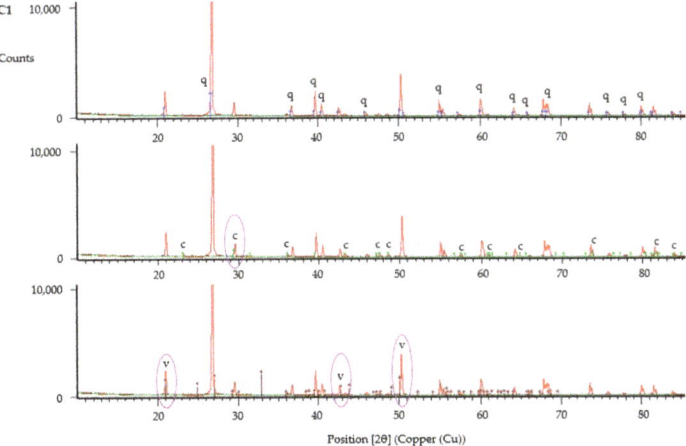

Figure 18. XRD analysis of sample from column C1, showing identified peaks of quartz (q), calcite (c) and vaterite (v).

4. Conclusions

Column studies were undertaken to investigate the effect of jute fibres on both the process of MICP and on properties of biocemented sand. Biocemented sand columns were produced in triplicates, containing sand and untreated jute fibres, sand and treated jute fibres and sand only as controls. The treated fibres immobilised a concentrated cementation medium, with the aim of enabling self-healing via MICP.

The results showed that the incorporation of jute fibres within a biocemented sand material significantly increased the unconfined compressive strength of this material when compared to biocemented sand without the jute fibres. This strength increase results from the contribution to strength properties of not just the fibres themselves, but also the increased amount of calcium carbonate precipitated in the columns containing jute fibres. On the basis of the results obtained the contribution to unconfined compressive strength increase by the fibres alone cannot be ascertained. In addition to increasing strength, the inclusion of fibres had a beneficial effect on the MICP process, improving efficiency of substrate conversion, likely as a result of sustaining the bacterial growth and, hence, urease activity. It is likely that bacteria had been absorbed by the jute and also adsorbed onto the surface of the fibres within the columns, and that this contributed to the positive findings. More investigation would be required to fully understand this effect. Tuson et al. [41] reported that bacterial systems used for sensing and responding to surfaces are still not well understood. Surface roughness of the fibres also appears to have added to this effect given the observed higher density of calcium carbonate crystals observed on the surface of roughened fibres. Renner et al. reported the significant influence of surface topography on bacterial adhesion [31].

A consequence of the fibre inclusions within the biocemented sand sustaining longer term activity of *Sporosarcina pasteurii* bacteria is the enabling of the continuation of the MICP process without the need for multiple injections of bacteria. This could reduce the cost of production of a biocemented sand material and would be beneficial where several treatments of cementation medium are required to achieve a low permeability and/or high strength.

The evaluation of self-healing effects and quantification of this has proved challenging, with only one column containing treated fibres showing any significant potential self-healing capacity. To achieve self-healing via MICP the cementation medium would need to be stored within the biocemented soil matrix for later release. Immobilisation may only be effective if a material embedded within the biocemented soil can retain sufficient cementation medium during the initial MICP treatment process.

Therefore, a material which enables a sufficiently slow release of immobilised chemicals would be required. More testing is required on this aspect and alternatives such as encapsulation explored. A self-healing MICP system may be of particular interest and suitability for seepage control, such as within a dam core or within grouting. For this application, a sufficient amount of cementation medium treatments would need to be applied to achieve a very low permeability, which the addition of jute fibres may help to facilitate. It is then assumed the self-healing MICP process would be activated upon water ingress into this material should micro-cracking occur, to help prevent piping for example.

The results from the set of columns containing the treated fibres may give some insight into effects of pre-treating fibres prior to use in MICP applications. These fibres were subject to chemical treatment with the concentrated cementation medium, and also some additional heat treatment, while these fibres were dried at 50 °C. This is an area which could be explored further and had not been a focus of this study. This study could be further extended using recycled jute fibres. It is expected that surface roughness of recycled fibres may further promote fixing of bacteria to fibres, however, any processing treatment fibres have undergone, contamination of fibres and potential deterioration should also be taken into consideration.

Author Contributions: Conceptualization, C.A.S.; data curation, C.A.S.; formal analysis, C.A.S.; investigation, C.A.S.; methodology, C.A.S.; resources, L.v.P.; supervision, L.v.P. and H.S.; writing—original draft, C.A.S.; writing—review & editing, L.v.P. and H.S. All authors have read and agreed to the published version of the manuscript.

Funding: This research was funded Cardiff University School of Engineering and the Center for Bio-mediated and Bio-inspired Geotechnics (CBBG).

Acknowledgments: The authors acknowledge support by Arizona State University and the Center for Bio-mediated and Bio-inspired Geotechnics, at which this study has been completed as part of the first authors Ph.D. The first Author's Ph.D. stipend is funded by Cardiff University School of Engineering. We further acknowledge the use of facilities within the Eyring Materials Center at Arizona State University.

Conflicts of Interest: The authors declare no conflict of interest.

Abbreviations

The following symbols are used in this paper.

P	Bulk density
G_s	Specific Gravity
D_{10}	Diameter corresponding to 10% finer in the particle-size distribution
D_{50}	Diameter corresponding to 50% finer in the particle-size distribution
D_{60}	Diameter corresponding to 60% finer in the particle-size distribution
D_{30}	Diameter corresponding to 30% finer in the particle-size distribution
C_u	Uniformity Coefficient of uniformity
C_z	Coefficient of curvature

References

1. Montoya, B.M.; Dejong, J.T. Healing of biologically induced cemented sands. *Geotech. Lett.* **2013**, *3*, 147–151. [CrossRef]
2. Park, S.-S.; Choi, S.-G.; Nam, I.-H. Effect of Plant-Induced Calcite Precipitation on the Strength of Sand. *J. Mater. Civ. Eng.* **2014**, *26*, 06014017. [CrossRef]
3. Dhami, N.K.; Alsubhi, W.R.; Watkin, E.; Mukherjee, A. Bacterial Community Dynamics and Biocement Formation during Stimulation and Augmentation: Implications for Soil Consolidaton. *Front. Microbiol.* **2017**, *8*, 1267. [CrossRef] [PubMed]
4. Ersan, Y.C.; de Belie, N.; Boon, N. Microbially induced CaCO3 precipitation through denitrification: An optimization study in minimal nutrient environment. *Biochem. Eng. J.* **2015**, *101*, 108–118. [CrossRef]
5. van Paassen, L.A.; Ghose, R.; van der Linden, T.J.M.; van der Star, W.R.L.; van Loosdrecht, M.C.M. Quantifying Biomediated Ground Improvement by Ureolysis: Large-Scale Biogrout Experiment. *J. Geotech. Geoenvironmental Eng.* **2010**, *136*, 1721–1728. [CrossRef]

6. Khodadadi, T.H.; Kavazanjian, E.; Paassen, L.V.; Dejong, J. Bio-Grout Materials: A Review. *Geotech. Spec. Publ.* **2017**, 1–12. [CrossRef]
7. Whiffin, V.S. Microbial CaCO3 Precipitation for the Production of Biocement. Ph.D. Thesis, Murdoch University, Perth, Australia, 2004.
8. Stocks-Fischer, S.; Galinat, J.K.; Bang, S.S. Microbiological precipitation of CaCO3. *Soil Biol. Biochem.* **1999**, *31*, 1563–1571. [CrossRef]
9. Gomez, M.G.; Martinez, B.C.; DeJong, J.T.; Hunt, C.E.; deVlaming, L.A.; Major, D.W.; Dworatzek, S.M. Field-scale bio-cementation tests to improve sands. *Proc. Inst. Civ. Eng.-Ground Improv.* **2015**, *168*, 206–216. [CrossRef]
10. Spencer, C.A.; Sass, H. Use of carrier materials to immobilise and supply cementation medium for microbially mediated self-healing of biocement. *IOP Conf. Ser. Mater. Sci. Eng.* **2019**, *660*, 012067. [CrossRef]
11. Botusharova, S. Self-Healing Geotechnical Structures Via Microbial Action. Ph.D. Thesis, Cardiff University, Cardiff, UK, 2017.
12. Burbank, M.B.; Weaver, T.J.; Green, T.L.; Williams, B.C.; Crawford, R.L. Precipitation of Calcite by Indigenous Microorganisms to Strengthen Liquefiable Soils. *Geomicrobiol. J.* **2011**, *28*, 301–312. [CrossRef]
13. Choi, S.-G.; Wang, K.; Chu, J. Properties of biocemented, fiber reinforced sand. *Constr. Build. Mater.* **2016**, *120*, 623–629. [CrossRef]
14. Li, M.; Li, L.; Ogbonnaya, U.; Wen, K.; Tian, A.; Amini, F. Influence of Fiber Addition on Mechanical Properties of MICP-Treated Sand. *J. Mater. Civ. Eng.* **2016**, *28*, 04015166. [CrossRef]
15. Xiao, Y.; He, X.; Evans, T.M.; Stuedlein, A.W.; Liu, H. Unconfined Compressive and Splitting Tensile Strength of Basalt Fiber-Reinforced Biocemented Sand. *J. Geotech. Geoenvironmental Eng.* **2019**, *145*, 04019048. [CrossRef]
16. Zhao, Y.; Fan, C.; Ge, F.; Cheng, X.; Liu, P. Enhancing Strength of MICP-Treated Sand with Scrap of Activated Carbon-Fiber Felt. *J. Mater. Civ. Eng.* **2020**, *32*, 04020061. [CrossRef]
17. Gao, Y.; Tang, X.; Chu, J.; He, J. Microbially Induced Calcite Precipitation for Seepage Control in Sandy Soil. *Geomicrobiol. J.* **2019**, *36*, 366–375. [CrossRef]
18. Vidal, H. The Principle of Reinforced Earth. *Highw. Res. Rec.* **1969**. Available online: http://onlinepubs.trb.org/Onlinepubs/hrr/1969/282/282-001.pdf (accessed on 2 September 2020).
19. Vigneswaran, C.; Jayapriya, J. Effect on physical characteristics of jute fibres with cellulase and specific mixed enzyme systems. *J. Text. Inst.* **2010**, *101*, 506–513. [CrossRef]
20. Shukla, S.K. *Fundamentals of Fibre-Reinforced Soil Engineering*; Springer: Singapore, 2017; ISBN 978-981-10-3063-5.
21. Wiktor, V.; Jonkers, H.M. Quantification of crack-healing in novel bacteria-based self-healing concrete. *Cem. Concr. Compos.* **2011**, *33*, 763–770. [CrossRef]
22. DeJong, J.T.; Mortensen, B.M.; Martinez, B.C.; Nelson, D.C. Bio-mediated soil improvement. *Ecol. Eng.* **2010**, *36*, 197–210. [CrossRef]
23. Rebata-Landa, V. Microbial Activity in Sediments: Effects on Soil Behaviour. Ph.D. Thesis, Georgia Institute of Technology, Atlanta, GA, USA, 2007.
24. ASTM D6913/D6913M-17. *Standard Test Methods for Particle-Size Distribution (Gradation) of Soils Using Sieve Analysis*; ASTM International: West Conshohocken, PA, USA, 2017.
25. ASTM D1557-12e1. *Standard Test Methods for Laboratory Compaction Characteristics of Soil Using Modified Effort (56,000 ft-lbf/ft3 (2,700 kN-m/m3))*; ASTM International: West Conshohocken, PA, USA, 2012.
26. Van de Velde, K.; Baetens, E. Thermal and Mechanical Properties of Flax Fibres as Potential Composite Reinforcement. *Macromol. Mater. Eng.* **2001**, *286*, 342–349. [CrossRef]
27. Ramachandran, S.K.; Ramakrishnan, V.; Bang, S.S. Remediation of Concrete Using Microorganisms. *Mater. J.* **2001**, *98*, 3–9. [CrossRef]
28. Al Qabany, A.; Soga, K. Effect of chemical treatment used in MICP on engineering properties of cemented soils. *Geotechnique* **2013**, *63*, 331–339. [CrossRef]
29. Harkes, M.P.; van Paassen, L.A.; Booster, J.L.; Whiffin, V.S.; van Loosdrecht, M.C.M. Fixation and distribution of bacterial activity in sand to induce carbonate precipitation for ground reinforcement. *Ecol. Eng.* **2010**, *36*, 112–117. [CrossRef]
30. Ibraim, E.; Diambra, A.; Russell, A.R.; Muir Wood, D. Assessment of laboratory sample preparation for fibre reinforced sands. *Geotext. Geomembr.* **2012**, *34*, 69–79. [CrossRef]

31. Renner, L.D.; Weibel, D.B. Physicochemical regulation of biofilm formation. *MRS Bull.* **2011**, *36*, 347–355. [CrossRef]
32. Eijkelkamp Soil & Water. *Calcimeter: Manual*; Eijkelkamp Soil & Water, Netherlands; 2018. Available online: https://www.eijkelkamp.com/download.php?file=M0853e_Calcimeter_b21b.pdf (accessed on 20 November 2019).
33. British Standards Institution. *BS 1377-7:1990 - Methods of test for soils for civil engineering purposes. Part 7: Shear strength tests (total stress)*; British Standards Institution: London, UK, 1999.
34. Craig, R.F. *Craig's Soil Mechanics: Seventh Edition*; E & FN Spon: Abingdon, UK, 2004.
35. Doran, P.M. Chapter 10 - Mass Transfer. In *Bioprocess Engineering Principles (Second Edition)*; Doran, P.M., Ed.; Academic Press: London, UK, 2013; ISBN 978-0-12-220851-5.
36. van Paassen, L. Biogrout, ground improvement by microbial induced carbonate precipitation. Ph.D. Thesis, Delft University of Technology, Delft, The Netherlands, 2009.
37. Kawano, J.; Shimobayashi, N.; Kitamura, M.; Shinoda, K.; Aikawa, N. Formation process of calcium carbonate from highly supersaturated solution. *J. Cryst. Growth* **2002**, *237–239*, 419–423. [CrossRef]
38. Ni, M.; Ratner, B.D. Differentiation of Calcium Carbonate Polymorphs by Surface Analysis Techniques–An XPS and TOF-SIMS study. *Surf. Interface Anal. SIA* **2008**, *40*, 1356–1361. [CrossRef]
39. Grasby, S.E. Naturally precipitating vaterite (μ-CaCO3) spheres: Unusual carbonates formed in an extreme environment. *Geochim. Cosmochim. Acta* **2003**, *67*, 1659–1666. [CrossRef]
40. Nawarathna, T.H.K.; Nakashima, K.; Kawasaki, S. Chitosan enhances calcium carbonate precipitation and solidification mediated by bacteria. *Int. J. Biol. Macromol.* **2019**, *133*, 867–874. [CrossRef]
41. Tuson, H.H.; Weibel, D.B. Bacteria-surface interactions. *Soft Matter* **2013**, *9*, 4368–4380. [CrossRef] [PubMed]

Publisher's Note: MDPI stays neutral with regard to jurisdictional claims in published maps and institutional affiliations.

© 2020 by the authors. Licensee MDPI, Basel, Switzerland. This article is an open access article distributed under the terms and conditions of the Creative Commons Attribution (CC BY) license (http://creativecommons.org/licenses/by/4.0/).

Article

Mechanical Behaviors of Flax Fiber-Reinforced Composites at Different Strain Rates and Rate-Dependent Constitutive Model

Dayong Hu [1,2], Linwei Dang [1,2], Chong Zhang [1,2] and Zhiqiang Zhang [3,4,*]

1. Department of Aircraft Airworthiness Engineering, School of Transportation Science and Engineering, Beihang University, Beijing 100191, China; hudayong@buaa.edu.cn (D.H.); danglinwei@buaa.edu.cn (L.D.); chongzhang@buaa.edu.cn (C.Z.)
2. Aircraft/Engine Integrated System Safety Beijing Key Laboratory, Beijing 100191, China
3. Beijing Key Laboratory of Rehabilitation Technical Aids for Old-Age Disability, Key Laboratory of Rehabilitation Technical Aids Analysis and Identification of the Ministry of Civil Affairs, National Research Center for Rehabilitation Technical Aids, Beijing 100176, China
4. Qinhuangdao Institute of National Research Center for Rehabilitation Technical Aids, Qinhuangdao 066000, China
* Correspondence: zhangzhiqiang@nrcrta.cn

Received: 25 January 2019; Accepted: 11 March 2019; Published: 13 March 2019

Abstract: Flax fiber-reinforced composites (FFRCs) exhibit excellent environmentally friendly qualities, such as light weight, low cost, recyclability, and excellent mechanical properties. Understanding the dynamic mechanical behavior of FFRCs could broaden their potential applications in lightweight, crashworthy, and impact-critical structures. This study presents a study on the fabrication of FFRCs by vacuum-assisted resin infusion. The dynamic stress–strain responses of the fabricated specimens at strain rates ranging from 0.006 s^{-1} to 2200 s^{-1} were evaluated using quasi-static tests and the Split–Hopkinson pressure bar (SHPB). The results indicated that the FFRC exhibited superior strain rate sensitivity. Final deformation photographs and scanning electron micrographs clearly revealed the damage evolution of the FFRC specimens, as well as various failure mechanisms, including fiber–matrix debonding, fiber pull-out, and fiber fracture at different strain rates. On the basis of the experimental results, a simplified Johnson–Cook model was established to describe the strain-rate dependent constitutive model of FFRC. The validation of the suggested constitutive model was embedded in the finite element simulations and could well repeat the strain wave observed from the experiment results. Finally, the quasi-static compression and drop-hammer impact of pyramidal lattice structures with FFRC cores were investigated both numerically and experimentally, proving the effectiveness of the simplified Johnson–Cook model. This study could potentially contribute to a deeper understanding of the dynamic mechanical behavior of FFRCs and provide fundamental experimental data for future engineering applications.

Keywords: flax fiber-reinforced composite; strain rate effect; Johnson–Cook model; lattice structure; failure mechanism

1. Introduction

Over the last decades, lightweight, low cost, and recyclability have been the highlighted material properties for automotive and aerospace industries [1]. Synthetic high-performance fiber-reinforced composites were able to overcome the lightweight problem; however, most of them were nondegradable and harmful to the environment. To reduce the environmental impact of composite materials, natural fiber-reinforced composites (NFRCs) with favorable mechanical and economic properties, as well as higher environmental friendliness, have been investigated as potential substitutes for synthetic fibers as reinforcements [1–9]. Extensive research on the mechanical properties of NFRCs have been conducted and compared with those of synthetic fibers. Dittenber and Gangarao [1] reported that the specific modulus of flax fiber was approximately 45 GPa/(g/cm^3), higher than that of glass fiber measuring approximately 29 GPa/(g/cm^3). Joshi et al. [10] showed that flax fibers exhibited specific strength equal to or even higher than that of glass fibers and could be a suitable replacement for glass fibers. Amenini et al. [6] investigated the dynamic characterization of a flax fiber-reinforced polyamide 11 bio-composite. Koh and Madsen [11] studied the failure criteria to accurately predict the strength of NFRCs, and they recommended Hashin and Puck failure theories, as they showed the smallest error compared to experimental data. Ramesh [12] carried out a detailed review on the preparation and properties of flax fiber and its composite materials.

In recent years, NFRCs have been investigated for application in aeronautic and automotive engineering [4,13,14] because of their distinct mechanical properties and low cost. Composites in these fields have been highly susceptible to impact damage induced by striking with foreign objects [13,15,16]. Thus, characterizing the mechanical performance of NFRCs under impact loading conditions is important. Ravandi et al. [17] reported that composites with woven flax fiber exhibited better impact properties compared to those with non-woven fiber. Meredith et al. [18] performed drop hammer experiments at an impact speed of approximately 8.0 m/s to evaluate the energy absorption characteristics of cone tubes made of woven flax and regenerated cellulose textiles; the specific energy absorption of cone tubes varied from 21.2 to 34.2 kJ/kg. Another important finding was that cone tubes made of Biotex flax combined with stiffer resin could obtain a higher specific energy absorption than that of cone tubes made of tougher resin. López–Alba et al. [4] investigated the energy absorption behaviors of NFRC tubes at different impact velocities. They found that the specific energy absorption of NFRC tubes heavily depended on the material parameters, including wall thickness, type of weave material, reinforced textiles, and matrix used. Shishevan et al. [19] assessed the low-velocity impact responses of basalt fiber-reinforced composites at different magnitudes of impact energy—30, 60, 80, 100, 120 and 160 J. In addition, the related key impact parameters, such as maximum contact force, absorbed energy, deflection, and duration were compared with those of carbon fiber-reinforced composites. On the basis of the experimental results, the impact performance of basalt fiber-reinforced composites was greater than that of carbon fiber-reinforced composites because of the higher toughness of basalt fibers. To evaluate the effects of fiber content and fiber orientation on the impact behavior of flax fiber-reinforced polypropylene composites, Rahman et al. [20] measured the impact properties of FFRCs by using the Charpy impact test and drop-weight impact test. Their results indicated that composites with varying fiber orientations exhibited different energy absorption for the in-plane and out-of-plane impact loads. Dhakal et al. [21] evaluated the effects of impactor shape and impact velocity on the dynamic mechanical properties of hemp–unsaturated PE composite under low-velocity impact conditions. The results showed that the specimens impacted by a hemispherical impactor exhibited higher force and absorbed more energy compared with specimens impacted by a conic impactor. In addition, with an increase in impact velocity, the damage to the back face of the specimen was more prominent for the laminates impacted by the hemispherical impactor. Rajaei et al. [22] investigated the effect of preheating on the impact performance of flax fiber-reinforced composite laminates. The impact test results showed that heat exposure at 300 °C reduced the energy absorption of the flax fiber composites. Shen et al. [23] also studied the effect of manufacturing process temperature, exposure temperature and water absorption on the low-velocity impact damage threshold and damage mechanisms of NFRCS. It was found that excessive temperature and water uptake could cause a serious reduction on the impact damage threshold and damage resistance. To further enhance impact properties of flax fibers, Al-Hajaj et al. [24] carried out pendulum impact tests with a range of impact energies (5–40 J) to investigate the effect of hybridization with woven carbon fibers plus flax fibers on impact properties. Results showed that these hybrid composites had superior impact properties compared to pure flax fiber-reinforced epoxy composites, suggesting that hybridization using synthetic and natural fiber could be done successfully. Moreover, under high loading rate conditions, the effect of strain rate on the mechanical properties of NFRCs has received considerable interest and has been extensively studied in metal, foam, and carbon- and glass fiber-reinforced composites [25,26]. Using the Split–Hopkinson pressure bar (SHPB), Omar et al. [27] examined the dynamic properties of pultruded jute and kenaf fiber-reinforced unsaturated polyester composites under different strain rates, nearly reaching 1400 s^{-1}. The compression modulus, compressive strength, and flow stress of both NFRCs were highly sensitive to strain rates. Similar experiments were also performed by Kim et al. [28] to study the dynamic mechanical responses of hemp, hemp/glass hybrid, cellulose, and wheat straw-reinforced polymeric composites at strain rates between 600 and 2400 s^{-1}. These NFRCs exhibited evident strain rate sensitivity. However, studies concerning the effects of strain rate on the mechanical behaviors of NFRCs compared with those of metals and composites, remained relatively inadequate, requiring further studies.

In addition, numerical simulation has become a widely used tool for predicting structural responses, thereby reducing experiment expense and time cost. For NFRCs, several numerical modeling studies can be found in the literature [29–32]. On the basis of the hypotheses of linear–elastic behavior up to failure and strain rate independent behavior, Rubio-López et al. [29] developed a finite element method (FEM) model to predict the low-velocity impact behavior of all-cellulose composite plates. Poilâne et al. [30] proposed a viscoelastoplastic model with eight independent parameters to model the behavior of the unidirectional flax fabric polymer composite. However, these two models failed to consider strain dependency. Rubio-López [31] presented a rheological model to describe the viscoplastic behavior of NFRCs at different strain rates and used this model to successfully analyze the low-velocity impact responses of NFRCs. Numerical prediction was consistent with experimental data conducted with two impactor shapes at different impact energies. However, in their study, the strain rates were considerably low, ranging from 2.08×10^{-4} s^{-1} to 8.33×10^{-3} s^{-1}, which could still be considered under quasi-static loading condition. Consequently, higher strain rates were expected to more accurately describe the dynamic behaviors of NFRCs.

To bridge these gaps, in the present study, flax fiber-reinforced composites (FFRCs) were manufactured by vacuum-assisted resin infusion (VARI). Quasi-static and SHPB experiments were then conducted to obtain mechanical properties in order to investigate the effects of strain rate. Based on experimental results, a simplified Johnson–Cook model, validated by SHPB experiments, was developed to describe both quasi-static and dynamic characteristics of FFRCs. Lastly, the effectiveness of the simplified Johnson–Cook model was further verified by quasi-static and drop-hammer crushing experiments of lattice structures.

2. Materials and Methods

2.1. Fabrication

Flax fiber plain weave fabrics (50% weft yarns per 50% warp yarns: [0°/90°] purchased from Linyi City, Shandong Province, China) and epoxy resin (LY1564/Aradur22962, Huntsman, from Shanghai, China) were selected to fabricate laminated composites by VARI. The diameters of the tows of the fabrics varied from about 0.6 mm to 1.5 mm. The nominal area density of the fabric and the fabric weight fraction were measured to be approximately 230 g/m^2 and 48%, respectively. The quasi-static tension tests of the pure matrix were performed and the stress-strain curves were provided in Figure S1. The viscosity of the resin was 450 mPa·s and the glass transition temperature was 140°C. The operating time of the resin was about 120 min, within which the fabrication should be finished outside the heating oven. The curing cycle was 80 °C for 2 h and 120 °C for 3 h.

VARI is a type of low-cost molding technology for manufacturing large-scale composite structures. The process in this study includes the following four steps:

- First, the mold surface is cleaned with acetone to achieve a perfect surface quality.
- Secondly, the layers of woven fiber dried at 70 °C for 3 h in an oven are laid on a mold sheet and other supplementary materials as shown in Figure 1a.

- Third, the resin is injected into the mold with atmospheric pressure after vacuuming and checking the sealing.
- Last, the resin flow is closed, and the resin is cured in an oven. The composite sheet after demolding is shown in Figure 1b, and the detailed image of the sheet is shown in Figure 1c.

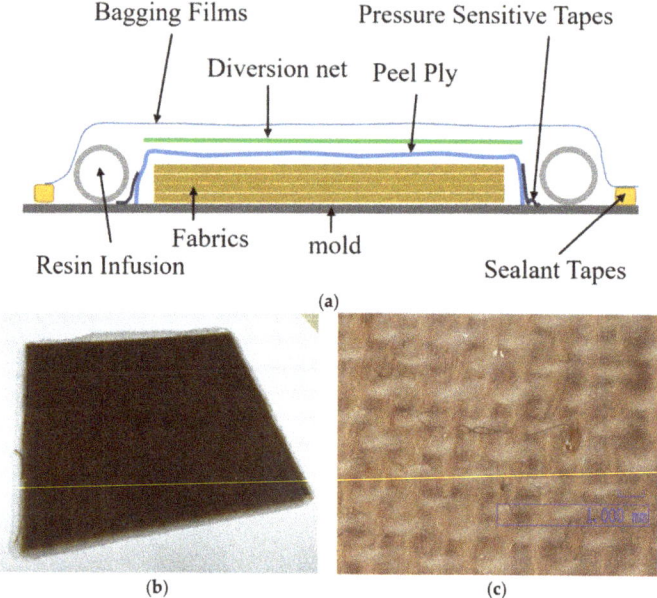

Figure 1. (a) Schematic of woven fiber and other supplementary materials; (b) composite sheet; (c) detailed image of the sheet.

In the present study, the fabricated composite sheet consisted of 20 layers, with a total thickness of 11.0 mm. The measured density of the composite sheet was 1.18 g/cm^3. Cylindrical specimens with measuring 9 mm in diameter and 5 mm in length were adopted for both quasi-static and dynamic compression experiments. For consistency with the fiber orientation of the trusses of the lattice structure in Section 4.3, the same direction with a fiber orientation of 45° was used for the specimens of quasi-static and SHPB compression experiments (Figure 2).

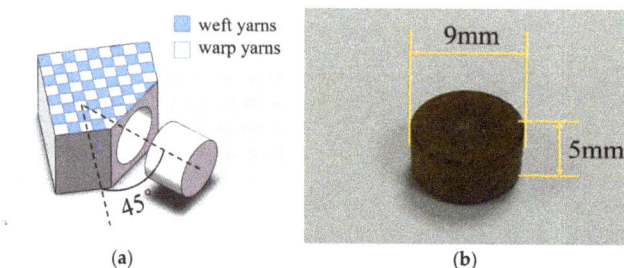

Figure 2. Cutting direction (a) and size (b) of the material test specimen.

2.2. Experimental Set-Up

2.2.1. Quasi-Static Compression Experiment

Quasi-static compression experiments were conducted on a hydraulic servo testing machine INSTRON 8801 (manufactured in Boston, MA, USA) at constant loading rates of 2, 20, and 200 mm/min, corresponding to the nominal strain rates of 0.006, 0.06, and 0.6 s^{-1}, respectively. The temperature was 25 °C, the relative humidity was 20%, and the maximum load capacity of the machine was 100 kN. The specimens were placed centrally between the polished platens. Prior to continuous loading, about 10 N was preloaded to eliminate the clearance. The axial deformation was captured by a non-contacting video extensometer (Instron AVE 2.0, manufactured in Boston, MA, USA) with a precision of 0.5 per thousand during the quasi-static tests.

2.2.2. SHPB Experiment

The SHPB apparatus is widely used to investigate the dynamic behavior of materials. The device typically consists of a striker bar, an incident bar, a transmission bar, and a gas gun, among other things, as shown in Figure 3. The strain gauge, digital storage oscilloscope, and ultrahigh dynamic extensometer are used to calibrate and measure the time history curves of incident, reflected, and transmitted waves. The cylindrical specimen is placed between the incident bar and the transmission bar. When the striker is propelled from the gas gun to impact the incident bar, a compressive elastic wave is generated and propagated through the incident bar. Once the wave reaches the specimen, part of the wave reflects on the interface of the specimen, while the remaining portion passes through the transmission bar. The traveling waves in the incident bar and the transmitted bar can be quantitatively captured by strain gauges mounted on these two bars. Thus, the strain–time histories of the incident, reflected, and transmitted waves can be recorded using the oscilloscope.

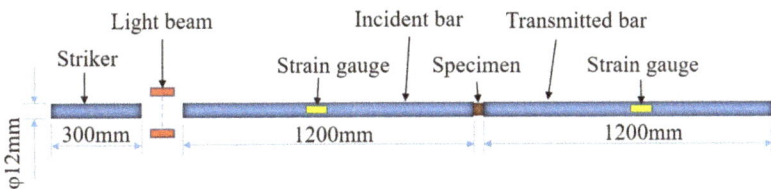

Figure 3. Schematic of the SHPB device.

The engineering stress, engineering strain, and strain rate can be calculated using the following equations [23]:

$$\sigma_e(t) = \frac{EA_0}{A_s}\varepsilon_t(t), \quad (1a)$$

$$\varepsilon_e(t) = -\frac{2C_0}{L_s}\int_0^t \varepsilon_t(t)dt, \quad (1b)$$

$$\dot{\varepsilon}(t) = -\frac{2C_0}{L_s}\varepsilon_r(t), \quad (1c)$$

where E, C_0, and A_0 denote Young's modulus, stress wave speed, and the cross-section area of the incident bar, respectively. A_s and L_s represent the cross-section area and length of the specimen, respectively. $\varepsilon_t(t)$ and $\varepsilon_r(t)$ refer to the amplitude of the transmitted wave and the reflected wave as functions of time t, respectively. $\sigma_e(t)$ and $\varepsilon_e(t)$ denote engineering stress and strain, respectively.

Furthermore, the true stress–strain relationship can be obtained using the following equations:

$$\sigma(t) = \sigma_e(t)(1 - \varepsilon_e(t)), \quad (1d)$$

$$\varepsilon(t) = \ln(1 - \varepsilon_e(t)), \quad (1e)$$

where $\sigma(t)$ and $\varepsilon(t)$ are the true stress and the true strain, respectively.

In the present study, the detailed parameters of the SHPB apparatus were as follows: the striker measured 300 mm in length and 12 mm in diameter; both the incident bar and the transmitted bar were 1200 mm in length and 12 mm in diameter; and the three bars were made of steel. The experiments were conducted at different strain rates: $1300~\text{s}^{-1}$ and $2200~\text{s}^{-1}$. Two repeated experiments were conducted at each strain rate to ensure the repeatability of the experimental results.

3. Experimental Results

In the SHPB tests, the striker bar was launched by varying the gas pressure to achieve different average nominal strain rates. Figure 4 shows the stress–strain curves of the FFRCs under quasi-static and SHPB compression experiments. The experimental results exhibited good repeatability (see Figure 4a–e). In addition, Figure 4f shows that FFRC is strongly affected by strain rates. The yield stress, as well as the flow stress, of the FFRCs markedly increased with an increase in strain rate. Therefore, the FFRCs showed evident strain rate sensitivity. This finding was highly similar to that reported by Omar et al. [27].

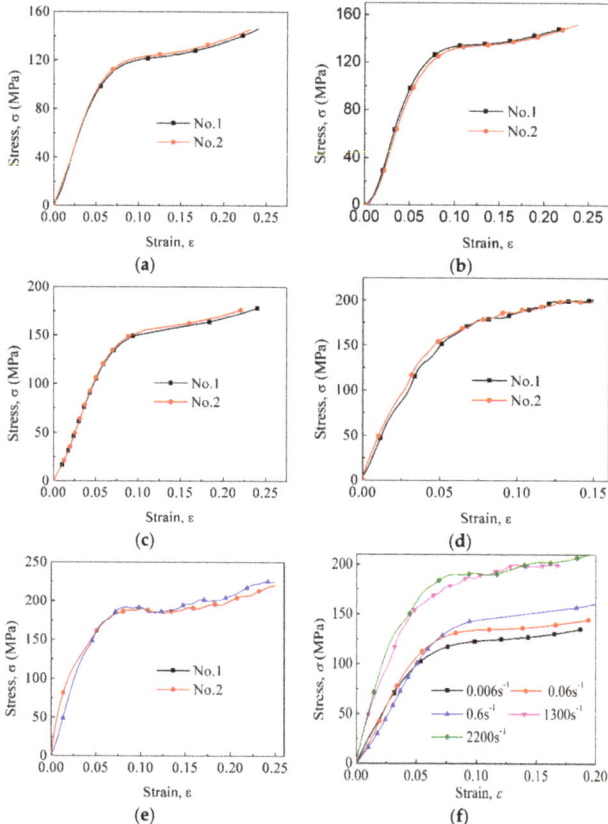

Figure 4. Stress–strain curves at different strain rates: (**a**) $0.006~\text{s}^{-1}$, (**b**) $0.06~\text{s}^{-1}$, (**c**) $0.6~\text{s}^{-1}$, (**d**) $1300~\text{s}^{-1}$, (**e**) $2200~\text{s}^{-1}$. (**f**) Comparison of typical curves.

The representative stress–strain curve in Figure 5 shows the trend of this stress–strain curve can be divided into three distinct stages—elastic region (oa), yield stage (bc), and plastic stage (cd)—similar to that of metal [25]; however, the specimen is made of a fiber-reinforced polymer material. The yield strength, σ_1, of the representative curves were extracted from the figure and listed in Table 1, consistent with the previous study (100–200 MPa) [33]. Evidently, σ_1 increased with an increase in strain rate. For example, when the nominal strain rate reached 1300 s^{-1}, σ_1 markedly increased to 152 MPa, which was about 1.5 times higher than that of 0.006 s^{-1}. The increase in yield strength from 1300 s^{-1} to 2200 s^{-1} was 11.7 MPa.

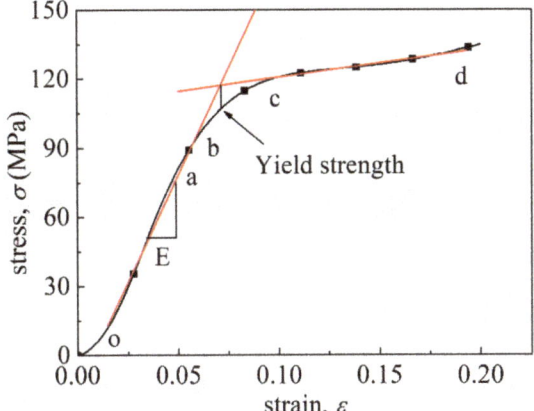

Figure 5. Different stages of a typical stress–strain curve: elastic region (oa), yield stage (bc), and plastic stage (cd).

Table 1. Yield strength of representative curves.

$\dot{\varepsilon}$	0.006 s^{-1}	0.06 s^{-1}	0.6 s^{-1}	1300 s^{-1}	2200 s^{-1}
σ_1	102.0 MPa	112.3 MPa	149.9 MPa	152.0 MPa	163.7 MPa

In addition, Young's modulus exhibited an appreciably increasing trend with an increase in strain rate, which was consistent with reference [27]. They attributed the increase in stiffening to the increase in strain rate, thereby decreasing the molecular mobility of polymer chains. However, dynamic Young's modulus could not be accurately measured by SHPB [34]. In the current study, the value of Young's modulus was not given to avoid inconsistency.

Figure 6a–c shows the final deformation morphologies of the crushed specimens. At a lower strain rate, the specimen only exhibits a reduction in thickness, where no obvious damage in appearance is observed. In the case of 1300 s^{-1}, the margin of the specimen was damaged to a certain extent and compression deformation was evident. In the case of 2200 s^{-1}, the specimen broke into two fragments, along with small cracks, and the fracture angle was approximately 45° (Figure 6c). The magnitude of the shear fracture angle mainly depended on the interfacial bond strength: a strong interface resulted in a larger shear fracture angle, whereas a weak interface generated a small fracture angle [27]. The failure could be inferred to have been initiated by matrix plasticity, followed by cracks passing through the layers of the laminate and forming a shear fracture with an angle of 45°.

The microscopic failure mechanism was analyzed by scanning electron microscopy (SEM, Zeiss Auriga, manufactured in Oberkochen, Germany) to highlight the dominant failure modes at selected locations on the specimens. Prior to SEM observation, the specimens were coated with an ion sputter coater to obtain enhanced conductance. Figure 6d–f presents micrographs of the fractured surface of the crushed specimen at the strain rate of 2200 s^{-1}. Fiber pull-out from the matrix is clearly shown in Figure 6d. Almost no matrix residue could be found on the surface of the fibers (Figure 6d,e). This observation could be attributed to the poor adhesion between the hydrophilic flax fiber and the hydrophobic epoxy matrix [18,35]. As seen in Figure 6e, a crack occurs along the fiber's longitudinal direction, which could be attributed to the shear failure of the fiber when the matrix fractured. The crack also indicated a reduction in the shear strength of the flax fiber. In Figure 6f, flax fragments were stuck to the matrix after fiber pull-out, and superficial flax shavings exhibit partly separated from the fibers—that is, not completely from the reinforcements. This occurrence was highly consistent with the observation of Liang et al. [35] that this could be considered as an additional type of damage mechanism for NFRCs.

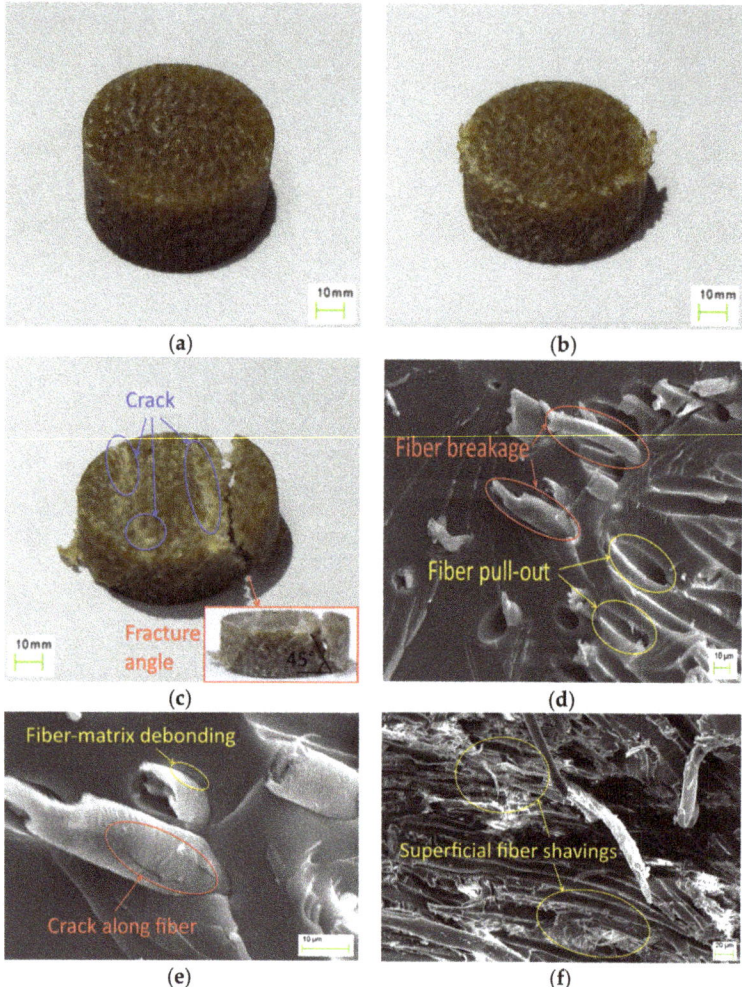

Figure 6. Specimens after experiments (**a**) 0.6 s^{-1}, (**b**) 1300 s^{-1}, (**c**) 2200 s^{-1}; Micrographs of fractured edges: (**d**) fiber breakage and fiber pull-out, (**e**) fiber–matrix debonding and crack along the fiber, (**f**) superficial fiber shavings.

4. Discussion

4.1. Simplified Johnson–Cook Model

The simplified Johnson–Cook model is widely accepted to describe the coupling factors among stress, strain, and strain rate [25,36]. The profile of the stress–strain curves of FFRCs was similar to those of traditional metals with a well-defined Johnson–Cook model. Owing to the difficulty in extracting temperature data and the slight effect of temperature on constitutive behaviors under low impact energy, for the sake of simplicity, only isotropic hardening and strain-rate hardening effects were considered in this study. Therefore, the dynamic behavior of FFRCs can be expressed as

$$\sigma = (A + B\varepsilon^n)\left(1 + C \ln \dot{\varepsilon}^*\right), \tag{2}$$

where σ is the stress; A is the yield stress; B and n represent the effect of strain hardening, respectively; C is the material constant determined by the specific material, representing the strain rate dependence of the material; ε is the equivalent plastic strain and obtained by subtracting the elastic strain from the total strain; $\dot{\varepsilon}$ is the strain rate; and $\dot{\varepsilon}^*$ is the dimensionless plastic strain rate expressed as $\dot{\varepsilon}/\dot{\varepsilon}_0$, where $\dot{\varepsilon}_0 = 0.006$ s^{-1} on the basis of quasi-static experiments.

Naturally, in the quasi-static experiment for $\dot{\varepsilon}^* = 1$, the constitutive model of Equation (2) can be further simplified to

$$\sigma = A + B\varepsilon^n, \tag{3}$$

Taking the logarithm of both sides of Equation (3) may result in the following:

$$\ln(\sigma - A) = \ln B + n \ln \varepsilon, \tag{4}$$

Subsequently, Equation (4) is applied to fit the quasi-static experimental data in logarithmic coordinates by the least square method, such that ln B represents the intercept of the straight line, and n represents the slope. Thus, B and n can be determined using simple mathematical conversion. At room temperature, C can be obtained through the fitting in accordance with Equation (2)

$$\frac{\sigma_2(\dot{\varepsilon})}{\sigma_1} - 1 = C \ln \frac{\dot{\varepsilon}}{\dot{\varepsilon}_0}, \tag{5}$$

where $\sigma_1 = A + B\varepsilon^n$, σ_1 represents the yield stress when the strain rate is 0.006 s^{-1}, and $\sigma_2(\dot{\varepsilon})$ is the yield stress at the strain rate of $\dot{\varepsilon}$.

The simplified Johnson–Cook model (Equation (2)) was used to describe the dynamic rate-dependent constitutive behavior of the FFRCs. The fitting parameters in the constitutive models in accordance with the experimental data are listed in Table 2 and the constitutive relationship was obtained as follow: $\sigma = \left(102.0 + 70.8\varepsilon^{0.416}\right)\left(1 + 0.047 \ln \frac{\dot{\varepsilon}}{0.006}\right)$. The fitting curves of the model and the experimental data are illustrated and compared in Figure 7.

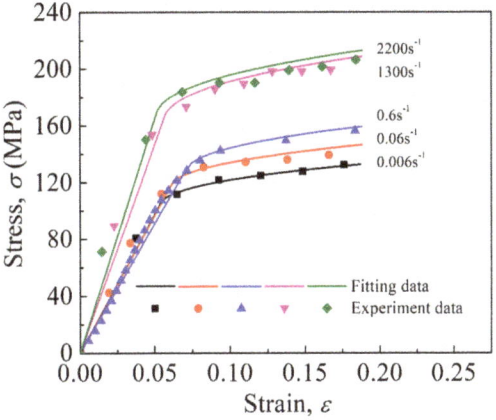

Figure 7. Comparison of the fitting curves of the experimental data.

4.2. Dynamic Wave Verification

To validate the effectiveness of the developed simplified Johnson–Cook model of FFRCs, numerical simulation was conducted using the commercial FEM software ABAQUS 6.13 (Dassault Systemes S.A, Vélizy-Villacoublay, France) to simulate the SHPB experiment.

A three-dimensional FEM model was set up to simulate the SHPB experiments. The FEM model consisted of four components: a striker bar, an incident bar, a transmission bar, and an FFRC specimen, each of which was of the same size as the SHPB apparatus shown in Figure 3. To improve the accuracy of the FEM analysis, an 8-node linear brick with reduced integration and hourglass control (C3D8R) was adopted. The minimum size of all elements was 1 mm. The same material was used for the three components made of steel, with the following measurements: modulus, 190 Gpa; density, 8 g/cm^3, and Poisson ratio, 0.3. The developed simplified Johnson–Cook model was used to simulate the mechanical behavior of the FFRC specimen in Table 2. The time history of the strain wave was obtained from the same location of the strain gauges on the incident and transmission bars with SHPB experiments.

Table 2. Fitting parameters in the simplified Johnson–Cook model.

Simplified Johnson-Cook model	A (MPa)	B (MPa)	n	C	$\dot{\varepsilon}_0$
Value	102.0	70.8	0.416	0.047	0.006

The speed of the striker was varied to obtain the various loading conditions of the tested specimens. The strain wave obtained from the experiments, with its counterpart from the FEM calculations, is shown in Figure 8. Consistency was found, indicating the validity and accuracy of the simplified Johnson–Cook material model. However, a slight general difference still observed, which could be attributed to the following: (a) The influence of thermal softening under impact could not be simulated in ABAQUS Explicit [25]; (2) The geometry of the specimens was not perfectly cubic [37]; (3) The non-parallelism and friction between the faces in contact with the bars were hardly included in the FEM model [37]; (4) The strain rate was not constant in the SHPB, and so on.

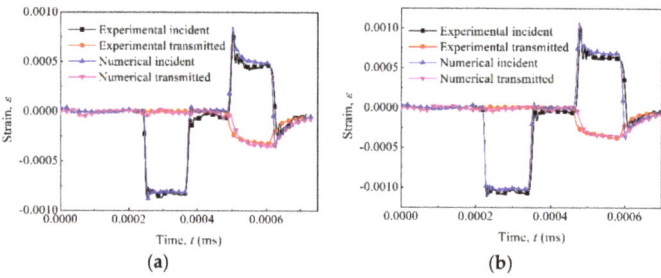

Figure 8. Comparison of strain waves between the simplified Johnson–Cook model and experimental data under strain rates (**a**) 1300 s^{-1}, (**b**) 2200 s^{-1}.

4.3. Prediction of the Crushing Peofrmance of Lattice Structures

As in Section 4.2, to further validate the effectiveness of the developed simplified Johnson–Cook model of the FFRC applied in engineering structure analysis, numerical simulation was also conducted using ABAQUS to simulate lattice structures subjected to quasi-static crushing and drop-hammer impact.

4.3.1. Specimen and FEM Model

Lattice structures were man-made open, porous cellular solids with periodic truss microstructures [38,39], which could meet many stringent requirements of engineering applications, such as blast and ballistic resistance, impact load carrying, and energy absorption. In the present study, FFRCs were used to manufacture pyramidal lattice cores with additional horizontal trusses for structure crashworthiness applications. The manufacturing process is illustrated in Figure 9. Six types of truss strips were cut with a carving machine (3040, Shenzhen Yidiao, Shenzhen, China) with a cutting precision of 0.03 mm from an FFRC sheet and then assembled into a pyramidal lattice structure by strip slot insertion (see Figure 9a–d). To fix the lattice core, two pieces of glass fiber-reinforced composite panels (see Figure 9e) were used to bond both sides of the lattice core forming a sandwich structure (see Figure 9f). The specimen of the lattice structure measured 84 mm in length, 84 mm in width, and 14 mm in height.

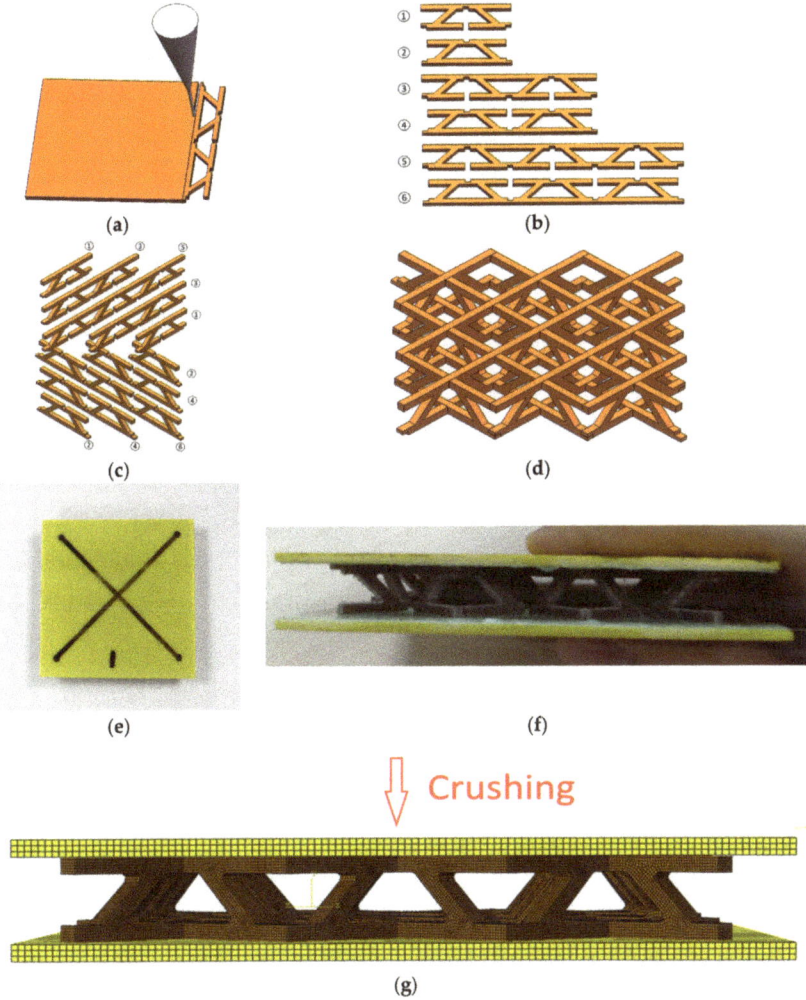

Figure 9. Pyramid lattice manufacturing: (**a**) Cutting truss strips of the lattice core from the fabricated FFRC; (**b**) obtaining six types of truss; (**c**) assembling truss into a pyramidal lattice structure by strip slot insertion; (**d**) assembled lattice structure; (**e**) top view of the specimen; (**f**) left view of the specimen; (**g**) FEM model of the specimen.

Numerical models were then established to predict the crushing performance of the lattice structure under quasi-static and drop-hammer impact conditions in the out-of-plane loading direction. A three-dimensional FEM model with the C3D8R element type was numerically established in ABAQUS, (Figure 9g). The density of the elements was relatively higher in the core, and the element size was 0.3 mm. The FEM model had exactly the same size as the tested specimen with setups. Two plates were considered as rigid bodies to crush the specimen.

4.3.2. Quasi-Static and Impact Experiments

Quasi-static and drop-hammer impact experiments, corresponding to the loading condition the of numerical simulation, were also conducted to validate numerical results to further evaluate the effectiveness of the simplified Johnson–Cook model in typical engineering structures. Quasi-static crushing tests were conducted on INSTRON 8801 (manufactured in Boston, MA, USA), where specimens were placed centrally between the polished moving platen and the stationary platen. To eliminate the influence of glass fiber panels, the platens were bigger than the specimens. All specimens were crushed under the following conditions: stroke distance, 5 mm; loading rate, 2 mm/min; and temperature, 25 °C. The crush-load curve with respect to the moving platen displacement was recorded automatically into a computer. A drop-hammer testing system was used for impact crushing tests to evaluate the effect of strain rate. The drop hammer, with a mass of 6 kg, was lifted by the pulley to different heights and then released through the trip gear to achieve different levels of impact energy, such as 30 and 45 J. The strain rates during impact crushing fell within the range of the SHPB experiments. The specimens were located at the center of the base of the drop-hammer testing system, directly opposite to the center of the mass of the hammer. The size of the hammer was larger than that of the specimen. A force sensor was mounted on the base to measure the impact force. The force response signals could then be recorded using a digital oscilloscope at a sampling frequency of 100,000 Hz and stored on a computer. Thus, the time history of crush load could be obtained from the recorded data. In addition, during the impact tests, a high-speed camera was used to capture the crushing deformation of the specimen with a rate of 2000 frame/s.

4.3.3. Numerical Simulation and Experiment Correlation

Numerical simulation of lattice structures under quasi-static and drop-hammer impact conditions, as well as the corresponding experiments, was conducted. The results are depicted in Figures 10 and 11. At a glance, an acceptable agreement could be observed for both quasi-static and dynamic results, indicating the validity of the simplified Johnson–Cook model. Table 3 summarizes the peak load comparison between the numerical simulation and the experiments.

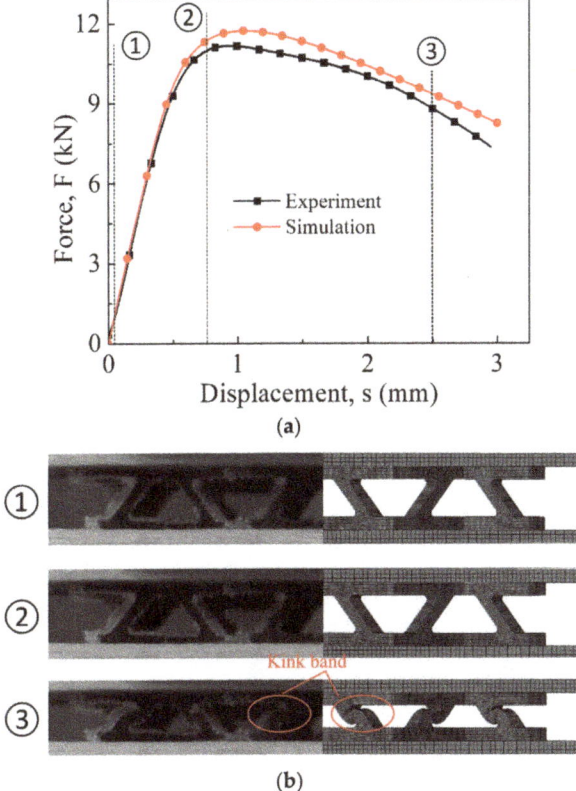

Figure 10. Comparison between the experiment and the numerical simulation: (**a**) Crush load versus displacement curve; (**b**) deformation process.

Under quasi-static conditions, the specimen exhibited typical progressive crushing (Figure 10). When the moving platen came into contact with the specimen, the crushing load was generated and increased linearly until the peak load was reached, corresponding to the critical Euler buckling load of the truss. As compression continued, the crush load was followed by a decrease stage and kink-band formation, resulting in the development of shear stresses, as featured in Figure 10b. In Figure 10a, the crush load curves of both numerical and experimental results exhibited a highly similar trend, except that the former was slightly higher than the latter. The deformation process was also similar with each other; specifically, numerical simulation was able to well predict the kink band formation of the specimen.

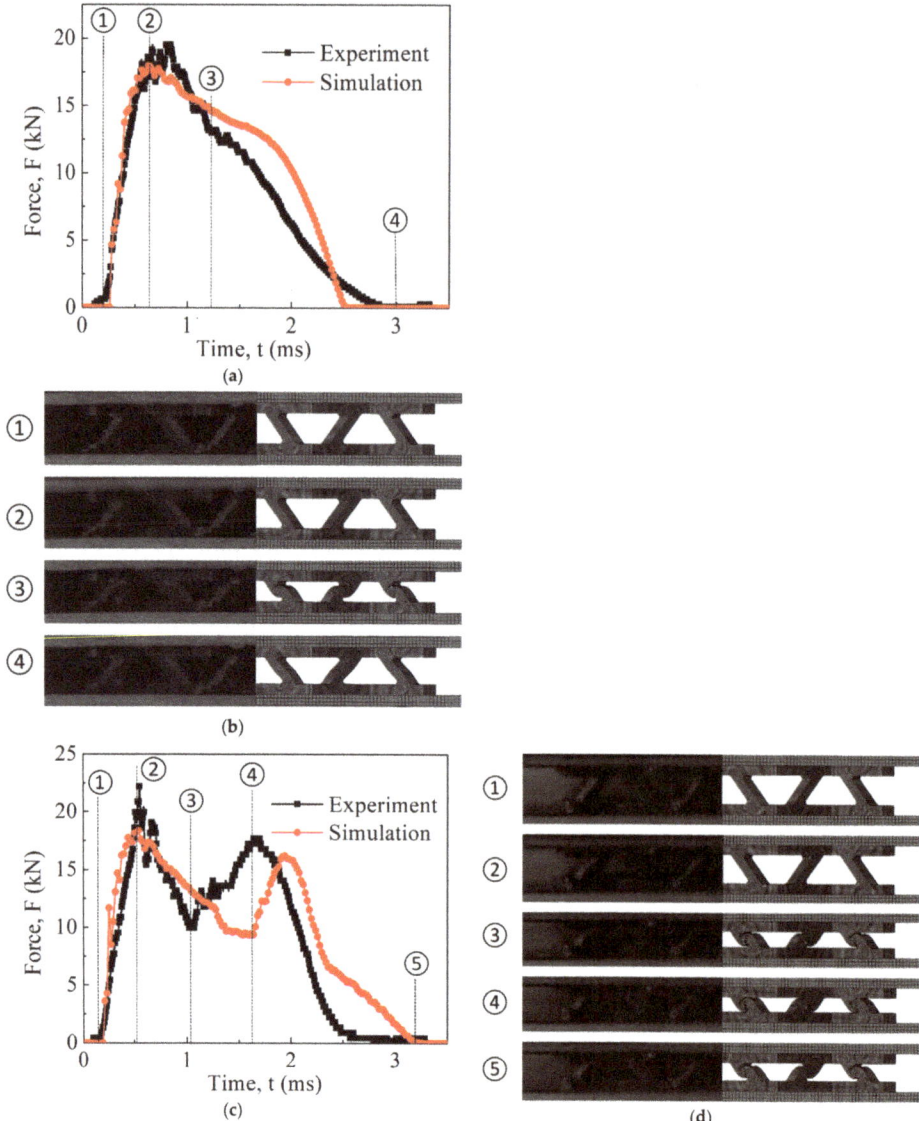

Figure 11. Comparison of the time history of crush load and deformation obtained by FEM calculation with its counterpart from experimental measurement under different impact energy: 30 J (**a**,**b**) and 45 J (**c**,**d**).

Table 3. Comparison of peak load between numerical simulation and experiments.

Impact Energy (J)	Peak Load (kN)					Deviation %
	Experiments				Simulation	
	No. 1	No. 2	No. 3	Mean		
Quasi static	10.89	11.18	11.27	11.11 ± 0.16	11.73	5.58
30 J	15.13	18.74	12.79	15.55 ± 2.45	17.96	15.50
45 J	16.70	14.06	-	15.83 ± 1.32	18.40	16.23

In the drop-hammer impact experiments, dynamic crushing was highly similar to the quasi-static crushing with the same failure mechanism of Euler buckling. Figure 11 presents the time history of crush load of both the numerical and experimental results under impact energy of 30 and 45 J, and corresponding deformation results were also presented. The numerical results were close to the experimental results, with a slight difference. In addition, the peak load was significantly higher than the quasi-static crushing experiment, which illustrates the effect of strain rate. As for the lattice structure subjected to crush loading with impact energy of 30 J, the curves of both the numerical simulation and the experiment were clearly characterized by one peak corresponding to Euler buckling initiation (Figure 11a). The deformation of the numerical simulation was also consistent with that of experiments (Figure 11b). With regard to the lattice structure subjected to crush loading with an impact energy of 45 J, both curves of the numerical simulation and the experiment were clearly characterized by two peaks (Figure 11c). The first peak load also corresponded to Euler buckling initiation. The second peak load was attributed to the impact between the top and bottom panels, when higher impact energy beyond the truss bearing capacity resulted in lattice compaction (Figure 11d). The crush load from the numerical simulation could efficiently predict the experimental results, despite the slight difference with time delay of the second peak between the numerical simulation and the experiment. These differences could be attributed to the following: First, the lattice structure was manually manufactured and assembled, which inevitably introduced geometry errors, including machining errors, assembly errors, size errors, defects, and so on. Euler buckling was known to be extremely sensitive to these geometry errors, which failed to easily perform quantitative evaluation and thus was not considered in the FEM models, leading to an error in crushing response prediction. Second, the damage initiation and evolution model was not considered in the FEM model. The development of an accurate, efficient, and robust damage model remained a challenge. Accordingly, the developed simplified Johnson–Cook model without damage initiation and evolution could not obtain a highly precise failure prediction under complex contact interfaces and stress state.

However, considering the complexity of the actual structure, the accuracy of numerical simulation could be acceptable. Thus, adequate confidence in the present numerical studies with the developed simplified Johnson–Cook model could be extended to engineering structure analysis.

5. Conclusions

Flax fiber-reinforced composites were expected to play an increasingly important role in the design of engineering structures subject to dynamic loadings because of the requirements for lightweight, low cost, recyclability, and excellent mechanical properties. Thus, full understanding of the dynamic material behavior of FFRCs became a priority. In this study, FFRCs were fabricated by VARI. The effects of strain rate on the mechanical properties of the FFRCs were investigated using quasi-static and SHPB experiments. Distinguishing strain-hardening behaviors were observed under both quasi-static and dynamic loading conditions, which revealed that FFRCs exhibited evident strain-rate sensitivity. On the basis of the experimental results, a simplified Johnson–Cook model was obtained and verified by numerical simulation of SHPB experiments. Moreover, the dynamic behavior of the lattice structures composed of FFRCs was numerically simulated and compared with the quasi-static and drop-hammer crushing experiments. The results evidently revealed that the proposed simplified Johnson–Cook model was able to accurately describe the dynamic mechanical behaviors of the FFRC material. Overall, the results of this study could be a solid step to elucidate the dynamic mechanical behaviors of the FFRC material and could provide valuable guidance for future applications in engineering.

Supplementary Materials: The following are available online at http://www.mdpi.com/1996-1944/12/6/854/s1, Figure S1: Tensile stress-strain curves of pure matrix.

Author Contributions: Conceptualization, D.H.; Funding acquisition, D.H. and Z.Z.; Investigation, D.H., L.D., C.Z., and Z.Z.; Methodology, D.H., L.D. and C.Z.; Project administration, D.H.; Resources, D.H.; Validation, Z.Z.; Writing—original draft, D.H., and L.D.; Writing—review and editing, D.H. and Z.Z.

Funding: This research was funded by the National Natural Science Foundation of China [grant number: 11872100]; Joint Funds of the National Natural Science Foundation of China [grant number: U1664250]; Fundamental Research Funds for Central Public Welfare Research Institutes [grant number: 118009001000160001], Defense Industrial Technology Development Program [grant number: JCKY2018601B106], Hebei provincial major S&T Research and Development Projects [grant number: 18277610D],and Fundamental Research Funds for the Central Universities, Beihang University.

Conflicts of Interest: The authors declare no conflict of interest.

References

1. Dittenber, D.B.; Gangarao, H.V.S. Critical review of recent publications on use of natural composites in infrastructure. *Compos. Part A Appl. Sci. Manuf.* **2012**, *43*, 1419–1429. [CrossRef]
2. Fragassa, C.; Pavlovic, A.; Santulli, C. Mechanical and impact characterisation of flax and basalt fibre vinylester composites and their hybrids. *Compos. Part B Eng.* **2018**, *137*, 247–259. [CrossRef]
3. Campana, C.; Leger, R.; Sonnier, R.; Ferry, L.; Ienny, P. Effect of post curing temperature on mechanical properties of a flax fiber reinforced epoxy composite. *Compos. Part A Appl. Sci. Manuf.* **2018**, *107*, 171–179. [CrossRef]
4. López-Alba, E.; Schmeer, S.; Díaz, F. Energy absorption capacity in natural fiber reinforcement composites structures. *Materials* **2018**, *11*, 418. [CrossRef]
5. Claramunt, J.; Ventura, H.; Fernández-Carrasco, L.; Ardanuy, M. Tensile and flexural properties of cement composites reinforced with flax nonwoven fabrics. *Materials* **2017**, *10*, 215. [CrossRef]
6. Amenini, F.; Brocail, J.; Chauvin, M.; Thuillier, S. Dynamical properties of flax fibre reinforced PA11 over a large frequency range. *Compos. Sci. Technol.* **2019**, *171*, 234–243. [CrossRef]
7. Chilali, A.; Assarar, M.; Zouari, W.; Kebir, H.; Ayad, R. Analysis of the hydro-mechanical behaviour of flax fibre-reinforced composites: Assessment of hygroscopic expansion and its impact on internal stress. *Compos. Struct.* **2018**, *206*, 177–184. [CrossRef]
8. Crawford, B.; Pakpour, S.; Kazemian, N.; Klironomos, J.; Stoeffler, K.; Rho, D.; Denault, J.; Milani, A.S. Effect of fungal deterioration on physical and mechanical properties of hemp and flax natural fiber composites. *Materials* **2017**, *10*, 1252. [CrossRef]
9. Ventura, H.; Claramunt, J.; Navarro, A.; Rodriguez-Perez, M.A.; Ardanuy, M. Effects of wet/dry-cycling and plasma treatments on the properties of flax nonwovens intended for composite reinforcing. *Materials* **2016**, *9*, 93. [CrossRef]
10. Joshi, S.V.; Drzal, L.T.; Mohanty, A.K.; Arora, S. Are natural fiber composites environmentally superior to glass fiber reinforced composites? *Compos. Part A Appl. Sci. Manuf.* **2004**, *35*, 371–376. [CrossRef]
11. Koh, R.; Madsen, B. Strength failure criteria analysis for a flax fibre reinforced composite. *Mech. Mater.* **2018**, *124*, 26–32. [CrossRef]
12. Ramesh, M. Flax (*Linum usitatissimum* L.) fibre reinforced polymer composite materials: A review on preparation, properties and prospects. *Prog. Mater. Sci.* **2019**, *102*, 109–166. [CrossRef]
13. Alkbir, M.F.M.; Sapuan, S.M.; Nuraini, A.A.; Ishak, M.R. Fibre properties and crashworthiness parameters of natural fibre-reinforced composite structure: A literature review. *Compos. Struct.* **2016**, *148*, 59–73. [CrossRef]
14. Yan, L.; Chouw, N.; Jayaraman, K. Effect of triggering and polyurethane foam-filler on axial crushing of natural flax/epoxy composite tubes. *Mater. Des.* **2014**, *56*, 528–541. [CrossRef]
15. Yan, L.; Chouw, N. Crashworthiness characteristics of flax fibre reinforced epoxy tubes for energy absorption application. *Mater. Des.* **2013**, *51*, 629–640. [CrossRef]
16. Yan, L.; Kasal, B.; Huang, L. A review of recent research on the use of cellulosic fibres, their fibre fabric reinforced cementitious, geo-polymer and polymer composites in civil engineering. *Compos. Part B Eng.* **2016**, *92*, 94–132. [CrossRef]
17. Ravandi, M.; Teo, W.S.; Tran, L.Q.N.; Yong, M.S.; Tay, T.E. Low velocity impact performance of stitched flax/epoxy composite laminates. *Compos. Part B Eng.* **2017**, *117*, 89–100. [CrossRef]
18. Meredith, J.; Coles, S.R.; Powe, R.; Collings, E.; Cozien-Cazuc, S.; Weager, B.; Müssig, J.; Kirwan, K. On the static and dynamic properties of flax and Cordenka epoxy composites. *Compos. Sci. Technol.* **2013**, *80*, 31–38. [CrossRef]
19. Shishevan, F.A.; Akbulut, H.; Mohtadi-Bonab, M.A. Low velocity impact behavior of basalt fiber-reinforced polymer composites. *J. Mater. Eng. Perform.* **2017**, *26*, 2890–2900. [CrossRef]

20. Rahman, M.Z.; Jayaraman, K.; Mace, B.R. Impact energy absorption of flax fiber-reinforced polypropylene composites. *Polym. Compos.* **2018**, *39*, 4165–4175. [CrossRef]
21. Dhakal, H.N.; Zhang, Z.Y.; Bennett, N.; Reis, P.N.B. Low-velocity impact response of non-woven hemp fibre reinforced unsaturated polyester composites: Influence of impactor geometry and impact velocity. *Compos. Struct.* **2012**, *94*, 2756–2763. [CrossRef]
22. Rajaei, M.; Kim, N.K.; Bhattacharyya, D. Effects of heat-induced damage on impact performance of epoxy laminates with glass and flax fibres. *Compos. Struct.* **2018**, *185*, 515–523. [CrossRef]
23. Shen, Y.; Zhong, J.J.; Cai, S.M.; Ma, H.; Qu, Z.H.; Guo, Y.C.; Li, Y. Effect of temperature and water absorption on low-velocity impact damage of composites with multi-layer structured flax fiber. *Materials* **2019**, *12*, 453. [CrossRef]
24. Al-Hajaj, Z.; Sy, B.L.; Bougherara, H.; Zdero, R. Impact properties of a new hybrid composite material made from woven carbon fibres plus flax fibres in an epoxy matrix. *Compos. Struct.* **2019**, *208*, 346–356. [CrossRef]
25. Hu, D.Y.; Meng, K.P.; Jiang, H.L.; Xu, J.; Liu, R.R. Strain rate dependent constitutive behavior investigation of AerMet 100 steel. *Mater. Des.* **2015**, *87*, 759–772. [CrossRef]
26. Elanchezhian, C.; Ramnath, B.V.; Hemalatha, J. Mechanical behaviour of glass and carbon fibre reinforced composites at varying strain rates and temperatures. *Procedia Mat. Sci.* **2014**, *6*, 1405–1418. [CrossRef]
27. Omar, M.F.; Akil, H.M.; Ahmad, Z.A.; Mazuki, A.A.M.; Yokoyama, T. Dynamic properties of pultruded natural fibre reinforced composites using Split Hopkinson Pressure Bar technique. *Mater. Des.* **2010**, *31*, 4209–4218. [CrossRef]
28. Kim, W.; Argento, A.; Lee, E.; Flanigan, C.; Houston, D.; Harris, A.; Mielewski, D.F. High strain-rate behavior of natural fiber-reinforced polymer composites. *J. Compos. Mater.* **2012**, *46*, 1051–1065. [CrossRef]
29. Rubio-López, A.; Olmedo, A.; Santiuste, C. Modelling impact behaviour of all-cellulose composite plates. *Compos. Struct.* **2015**, *122*, 139–143. [CrossRef]
30. Poilâne, C.; Cherif, Z.E.; Richard, F.; Vivet, A.; Doudou, B.B.; Chen, J. Polymer reinforced by flax fibres as a viscoelastoplastic material. *Compos. Struct.* **2014**, *112*, 100–112. [CrossRef]
31. Rubio-López, A.; Hoang, T.; Santiuste, C. Constitutive model to predict the viscoplastic behaviour of natural fibres based composites. *Compos. Struct.* **2016**, *155*, 8–18. [CrossRef]
32. Jalón, E.; Hoang, T.; Rubio-López, A.; Santiuste, C. Analysis of low-velocity impact on flax/PLA composites using a strain rate sensitive model. *Compos. Struct.* 2018. [CrossRef]
33. Holbery, J.; Dan, H. Natural-fiber-reinforced polymer composites in automotive applications. *JOM* **2006**, *58*, 80–86. [CrossRef]
34. Gama, B.A.; Lopatnikov, S.L.; Gillespie, J.W. Hopkinson bar experimental technique: A critical review. *Appl. Mech. Rev.* **2004**, *57*, 223–250. [CrossRef]
35. Liang, S.X.; Gning, P.-B.; Guillaumat, L. Quasi-static behaviour and damage assessment of flax/epoxy composites. *Mater. Des.* **2015**, *67*, 344–353. [CrossRef]
36. Singh, N.K.; Cadoni, E.; Singha, M.K.; Gupta, N.K. Dynamic tensile behavior of multi phase high yield strength steel. *Mater. Des.* **2011**, *32*, 5091–5098. [CrossRef]
37. Arbaoui, J.; Tarfaoui, M.; El Malki Alaoui, A. Mechanical behavior and damage kinetics of woven E-glass/vinylester laminate composites under high strain rate dynamic compressive loading: Experimental and numerical investigation. *Int. J. Impact Eng.* **2016**, *87*, 44–54. [CrossRef]
38. Berger, J.B.; Wadley, H.N.G.; McMeeking, R.M. Mechanical metamaterials at the theoretical limit of isotropic elastic stiffness. *Nature* **2017**, *543*, 533. [CrossRef]
39. Tankasala, H.C.; Deshpande, V.S.; Fleck, N.A. Tensile response of elastoplastic lattices at finite strain. *J. Mech. Phys. Solids.* **2017**, *109*, 307–330. [CrossRef]

© 2019 by the authors. Licensee MDPI, Basel, Switzerland. This article is an open access article distributed under the terms and conditions of the Creative Commons Attribution (CC BY) license (http://creativecommons.org/licenses/by/4.0/).

Article

Uniaxial Tensile Behavior of Carbon Textile Reinforced Mortar

Fen Zhou [1,2,*], Huanhui Liu [3], Yunxing Du [1,2], Lingling Liu [3], Deju Zhu [1] and Wei Pan [4]

1. Key Laboratory for Green & Advanced Civil Engineering Materials and Application Technology of Hunan Province, College of Civil Engineering, Hunan University, Changsha 410082, China; duyunxing@hnu.edu.cn (Y.D.); dzhu@hnu.edu.cn (D.Z.)
2. Key Laboratory for Damage Diagnosis of Engineering Structures of Hunan Province, Hunan University, Changsha 410082, China
3. College of Civil Engineering, Hunan University, Changsha 410082, China; liuhuanhui@hnu.edu.cn (H.L.); liulingling@hnu.edu.cn (L.L.)
4. Department of Civil Engineering, The University of Hong Kong, Pokfulam, Hong Kong, China; wpan@hku.hk
* Correspondence: zhoufen@hnu.edu.cn; Tel.: +86-158-0263-6298

Received: 27 December 2018; Accepted: 23 January 2019; Published: 25 January 2019

Abstract: This paper investigates the effects of the reinforcement ratio, volume fraction of steel fibers, and prestressing on the uniaxial tensile behavior of carbon textile reinforced mortar (CTRM) through uniaxial tensile tests. The results show that the tensile strength of CTRM specimens increases with the reinforcement ratio, however the textile–matrix bond strength becomes weaker and debonding can occur. Short steel fibers are able to improve the mechanical properties of the entire CTRM composite and provide additional "shear resistant ability" to enhance the textile–matrix bond strength, resulting in finer cracks with smaller spacing and width. Investigations into the fracture surfaces using an optical microscope clarify these inferences. Increases in first-crack stress and tensile strength are also observed in prestressed TRM specimens. In this study, the combination of 1% steel fibers and prestressing at 15% of the ultimate tensile strength of two-layer textiles is found to be the optimum configuration, producing the highest first-crack stress and tensile strength and the most reasonable multi-cracking pattern.

Keywords: carbon textile reinforced mortar; uniaxial tensile tests; debonding failure; steel fibers; prestress; multi-cracking pattern

1. Introduction

Textile reinforced mortar (TRM), also called textile reinforced concrete (TRC) in some cases with slightly coarse aggregates in matrix, refers to an emerging type of cement-based composite material characterized by reinforcing high-performance, fine-grained concrete with high-ductility, alkali-resistant textiles [1–3]. It was born as a result of reinforcing and/or rehabilitating aging masonry and reinforced concrete in conventional structures. The combination of fine-grained concrete and high-strength textiles produces holistic composites that may enable the fabrication of sophisticated and lightweight concrete structures, stay-in-place formwork elements, and prefabricated sandwich panels with extraordinary mechanical performance, very high durability, and enhanced potential for free-form designs in comparison to steel-bar reinforced concrete [4–6]. Moreover, compared with engineered cementitious composites which have a relatively low fiber utilization rate because the short fibers are randomly distributed in the matrix [7–9], TRM has high-performance fibers that are continuously embedded in the mortar matrix along the anticipated principal stress direction which results in a higher fiber utilization rate with the same reinforcement ratio [10]. Therefore, TRM has gradually become an attractive replacement composite material for engineering applications [11]. With recent developments in civil engineering, TRM can also be used in the supporting and connecting components of new structures [12–16].

As TRM is increasingly applied in the construction industry, an in-depth understanding of the fundamental mechanism developed via experimental investigation becomes essential for analysis, modeling, and design. The following paragraphs describe some recent studies on the mechanical behavior of TRM.

A number of studies have shown that the textile reinforcement ratio significantly affects the tensile behavior of TRM. Contamine et al. [17] studied the direct tensile behavior of TRC composites that were produced through the laminating technique. The results showed that composites with high reinforcement ratios were insensitive to defects and, thus, provide reliable test results. Larrinaga et al. [18] observed that specimens reinforced with one layer of basalt textile broke smoothly and an increase in the reinforcement ratio turned the failure mode into a brittle rupture with a sudden load drop during tensile tests, indicating that there exists a critical threshold for positive effects: once a certain ratio has been reached, the potentiation may become weak and inadequate fracture modes may occur. A typical three-stage evolution theory depicting the relationship between the stress and strain of TRC has been unanimously approved. Initially developed in the 2000s at the Technische Universität Dresden [19,20], this theory indicates that TRC materials exhibit distinct strain-hardening behavior. Several researches [21,22] have further refined the tensile behavior of TRC using the classic three-stage stress–strain curve.

With further research, it is gradually becoming clear that the load-bearing capacity of TRM is strongly related to the synergistic effect of the components, which is intensively affected by the bond property between the textile and the matrix [23]. A promising approach for improving the bond is to impregnate textiles with epoxy resin before producing the TRC composite, as reported by Dvorkin et al. [24]. Colombo et al. [25] then investigated the influences of the reinforcement ratio, textile geometry, curing condition, and specimen size on the mechanical properties of AR-glass TRC. The results revealed that the bond strength between textiles and matrix tends to increase with the increasing number of textile layers, weft spacing, and shrinkage caused by different curing conditions; consequently, both the first-crack stress and tensile strength increase. Although the increasing specimen size enhances ductility, no significant changes in tensile strength have been observed.

To further improve the mechanical performance of TRM composites, process modifications and some exterior additions have been developed. In terms of process modifications, pre-tensioning turns out to be an effective approach. Reinhardt et al. [26] showed that the application of prestressing on textiles improved the cracking, tensile strength, and stiffness of cracked sections, with more notable effects occurring in impregnated carbon TRC specimens. As a result, the prestressing process can significantly extend the serviceability of TRC composite materials. With respect to the exterior additions, Barhum and Mechtcherine [27] addressed the influence of short dispersed fibers made of AR glass on the fracture behavior of TRC by uniaxial tests. It is reported that TRC specimens that are reinforced with short dispersed fibers enhance the first-crack stress (by a factor of 2) and form more and finer cracks. Du et al. [28] explored the flexural behavior of basalt textile reinforced concrete (BTRC) with a combination of prestress and chopped steel fibers and found that chopped steel fibers increase the crack number of BTRC specimens; this effect was more obvious at higher prestress levels.

Textiles made from popular fibers, including synthetic groups such as AR (alkali resistant) glass, basalt, carbon, or aramid [29,30] and natural groups such as sisal, hemp, and flax [31,32], have received the most attention from researchers. Carbon textile has been found to provide better supported load capacity and higher strength and Young's modulus when used as traction reinforcement [33,34].

Despite significant efforts to investigate the mechanical properties of TRM composite materials, limited information is available regarding the effects of adding steel fibers with different volume fractions and applying different levels of prestressing force to carbon textile reinforced mortar (CTRM). In particular, no relevant experimental data are available for the uniaxial tensile behavior of prestressed

CTRM composites with the addition of steel fibers, which are known to offer superior tensile strength and have a high elastic modulus [35]. The present research aims to investigate the influence of the textile reinforcement ratio, volume fraction of short steel fibers, and prestressing force on the uniaxial tensile behavior of TRM. In the following sections of this paper, the main materials considered in this study, including carbon textiles, short steel fibers, and fine-grained mortar, are described, and the pre-tensioning of textiles is explained in detail. The experimental profiles and corresponding test results are presented, and the differences in tensile properties (including first-crack stress, tensile strength, crack numbers, and crack spacing) according to the three design variables are discussed. An optical microscope is used to illustrate the distinctions among the failure modes of the test specimens.

2. Materials and Methods

2.1. Materials

2.1.1. Carbon Textile

The carbon textile that was used in this study is manufactured in two orthogonal directions at a nominal spacing of 5 mm. Multifilament yarns consisting of 6000 monofilaments are used to produce the textile. The outer filaments (sleeve filaments) are in direct contact with the matrix and, therefore, have a greater effect on the bond properties than the inner filaments (core filaments). The inner filaments have no direct contact with the matrix because of the low penetration of the matrix. The force on the sleeve filaments is transformed to the core filaments via friction; thus, the untreated textiles exhibit little cooperative bearing ability between the sleeve and core filaments. A large number of experimental studies have shown that epoxy resin can fully penetrate all filaments within the yarns, thereby providing an intact unit reinforcing system [36,37]. Hence, impregnated carbon textiles are used in this study, and the warp yarns (along the length of the textile) are considered to be the reinforcing yarns, as shown in Figure 1a.

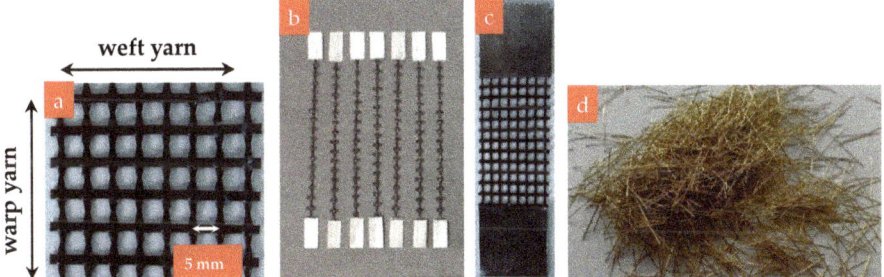

Figure 1. (**a**) Carbon textile impregnated with resin epoxy, (**b**) single yarn samples, (**c**) strip sample, and (**d**) steel fibers.

Details about the physical and mechanical properties of the impregnated carbon yarns (reinforcing direction) are presented in Table 1. The tensile strength, Young's modulus, and elongation of the carbon yarns were determined through direct tensile tests using 100-mm gauge length samples according to the Chinese specification GB/T 3362-2017 (Figure 1b). The cross-sectional area of a single yarn is 0.218 mm^2, calculated as the ratio of tex (the linear density of this material) to its bulk density. Ten 40-mm wide strips of 100-mm gauge length samples were cut from the textiles (Figure 1c) to determine their tensile properties. The average tensile bearing capacity and tensile strength of the 10 test strips were 4 kN and 2293.6 MPa, respectively, as determined via tensile tests.

Table 1. Physical and mechanical properties of the warp carbon yarns (impregnated).

Type	Tensile Strength (MPa)	Young's Modulus (GPa)	Strain Capacity (%)	Density (g/cm^3)	Cross-Sectional Area (mm^2)	Tex (g/km)
6K	2290	230	1	1.8	0.218	390

Note: 6K means that one multifilament yarn consists of 6000 monofilaments.

2.1.2. Short Steel Fibers

In view of the limitations of the mesh size of carbon textiles and the mold height (20 mm), copper-coated steel fibers of length 12–15 mm were used, as shown in Figure 1d. The mechanical properties of the steel fibers are listed in Table 2. The density was obtained by dividing the mass by the volume, and the volume was measured using the drainage method.

Table 2. Mechanical properties and geometric parameters of steel fibers.

Diameter (mm)	Length (mm)	Density (g/cm^3)	Tensile Strength (MPa)	Young's Modulus (GPa)
0.18–0.23	12–15	8.5	2850	200

2.1.3. High-Performed Fine-Grained Mortar

The matrix of TRM specimens ought to have super fluidity and self-compactness, so the maximum particle size of the aggregates should be less than 2 mm. Given that textiles in some specimens were pre-tensioned, the matrix should also possess high early strength in order to reduce the loss of prestress. Therefore, the mix proportion of high-performance fine-grained mortar designed by Du et al. [35] was used (see Table 3). The fly ash improves the fluidity and the silica fume and slag ash are designed to improve the early strength. Samples with dimensions of 40 mm × 40 mm × 160 mm were produced to study the mechanical properties of the matrix, especially the early mechanical properties. The flexural strength of the matrix was measured through three-point bending tests, and the values after 7 and 28 days were 11.5 MPa and 12.3 MPa, respectively. The compressive strength of the matrix was determined on the two broken parts that were obtained from the three-point bending tests, and the values after 7 and 28 days were 62.5 MPa and 76.7 MPa, respectively, indicating that the mortar matrix prepared according to the designed mix proportion satisfied the high early strength requirement.

Table 3. Composition of TRM (textile reinforced mortar) matrix.

Materials	Cement Type II 52.5	Fly ash	Silica Fume	Slag	Fine Sand	Super-Plasticizer	Water
Contents (kg/m^3)	800	100	50	50	1200	2.0	286

2.2. Experimental Program

2.2.1. Testing Series

In this study, the tensile performance of carbon TRM was examined with respect to the reinforcement ratio of the textile, the addition of steel fibers at different volume fractions, and the prestressing force applied to the carbon textiles. The specific test schemes included two primary specimen series in which TRM composites that were reinforced with one or two textile layers were produced; control specimens with no textile layers were also produced.

The reinforcement ratios of the specimens that were produced with one and two layers of textile are 0.4% and 0.8%, respectively. The reinforcement ratio can be expressed as follows:

$$\rho_f = \frac{A_f}{A_c} \quad (1)$$

where A_f and A_c are the cross-sectional areas of the carbon yarns and TRM specimens, respectively.

The effects of steel fibers with different volume fractions (0.5%, 1%, and 2%) were studied in each series, and the effects of prestress on the mechanical behavior of the TRM composites were investigated by applying different prestressing forces to the one-layer (10% and 20% of the ultimate tensile strength of one-layer textile) and two-layer (15% of the ultimate tensile strength of two-layer textiles) specimens. Moreover, prestressed thin plates with 1% steel fibers added were researched to determine their effects on the mechanical behavior. The test specimens were named as follows: for example, P15C2S1 represents specimens with a prestress of 15% of the ultimate tensile strength of two-layer textiles (P15), a two-layer carbon textile arrangement (C2), and the addition of 1% (by volume) steel fibers (S1).

2.2.2. Tensioning System

Figure 2 shows the device that was used to apply pre-tensioning to the carbon textiles. Considering that the pre-tensioning will decrease slightly over time, the pre-tensioning of the carbon textiles should be completed before casting the mortar matrix. The device is employed as follows:

1. Firstly, the two free ends of the carbon textiles, which are not impregnated with the epoxy resin within the range of around 80 mm, are anchored at either end of the device using the self-locking principle (see Figure 2). During the process, the roller beneath the chute slides to adjust the position of the chute. Note that warp fibers in the upper and lower layers (when there are two textile layers) must be strictly aligned and parallel to the edge of the textile groove.
2. Stretch the carbon textiles by tightening the nut at one end. Note that the loading process should be uniform and slow so that the fiber bundles are evenly and cooperatively stressed. The pre-tensioning force is measured by the load cell at the other end, output to the data acquisition system, and is finally displayed on the digital terminal.
3. When the target pre-tension is reached, the loading is paused. After 5–10 minutes, the loss of pre-tension is measured and recorded. The pre-tension force is then re-applied to reach the target value. This operation calls additional tensioning, which should be repeated afterwards. To maintain a stable pre-tension level, the textile should stay stretched for 24 h, during which time additional tensioning should be implemented every 8 h; in other words, the additional tensioning is repeated three times in total.
4. Finally, the mortar matrix is cast on the textile.

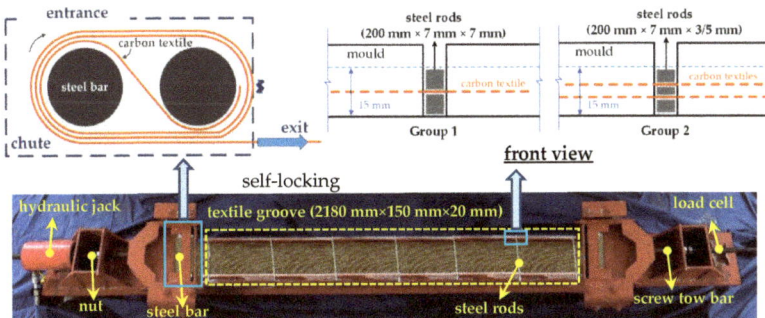

Figure 2. Illustration of the tensioning system.

2.2.3. Preparation of the Composite Specimens

The manufacturing process started with all textiles tautly positioned on the groove (2180 mm × 150 mm × 20 mm) of the tensioning system (Figure 2), i.e., one layer of textile was positioned at the middle height of the specimen or two layers of textiles were placed uniformly along the height of the specimen. Two groups of overlapping steel rods (thicknesses of 3, 5, 7 mm) were used to determine the location of the textile layers and the thickness of the specimens (Figure 2). Note that the global thickness of all specimens was set to 15 mm.

Firstly, the carbon textiles were fixed in the groove and slightly stretched. As for prestressed specimens, the textiles must be tensioned as described in Section 2.2.2. The fresh mortar matrix was then poured into the groove and a flat vibrator was used to eliminate pores; the top surface was later smoothed by a roller (Figure 3a). After the surface of the mortar had hardened, the plate was covered with wet towels and cured at room temperature. The prestressed plates were cured for four days before being released and demolded, whereas the other specimens were only cured for one day before demolding. The prestressing force was released by cutting off the textiles in a tensioned state, and then the plates were removed from the molds and stored in a climate-controlled room at 20 °C and 90% relative humidity until the age of 28 days. The TRM plates were then cut into specimens with dimensions of 240 mm × 40 mm (length × width) using a water-cooled cutting machine. For each test condition, there were at least six valid specimens. As soon as the specimens had dried, a thin layer of white paint was applied to the surface to aid the observation of the crack pattern.

Figure 3b,c show the distributions of steel fibers. To facilitate more uniform distributions of steel fibers in the matrix and effective stress transmission between the textile and matrix, some of the steel fibers were inserted vertically or obliquely into the textile grids, and the remainder were thoroughly mixed up with the mortar matrix and then poured into the mold.

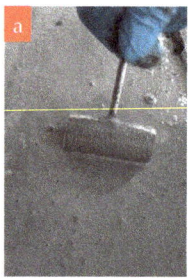  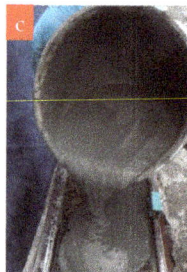

Figure 3. (a) Surface treatment of TRM composite, (b) steel fibers inserted into the grids of the textile, (c) flowable fresh mortar matrix with steel fibers.

2.2.4. Uniaxial Tensile Test Setup

Uniaxial tensile tests were performed using an MTS (MTS System Corporation, Shenzhen, China) load frame (C43.304) with a load capacity of 30 kN and a maximum sampling rate of 1000 Hz (Figure 4). In this study, the tests were carried out by displacement control (0.5 mm/min) with a sampling rate of 20 Hz. The deformations were measured using an extensometer with a gauge length of 100 mm positioned in the central area of the specimens. Both the loads and deformations were recorded simultaneously through a computer that was connected to the testing machine. The stress was calculated by dividing the load by the cross-section of the specimen. A simple and cost-effective device was used to avoid the negative effects of possible eccentricity and misalignment, as shown in Figure 4. Five-ball joints were installed at both the top and bottom of the machine grips as universal joints, and 2-mm-thick aluminum plates with 14.5-mm diameter pinholes were attached to both ends of the specimens (Figure 5); the specimens were connected to the universal joints with a 14.5 mm diameter pin (Figure 5). Using this method, the applied axial load could be transformed from the machine grips to the specimens without causing bending effects [17,38]. Notably, the pin and pinhole must fit perfectly to prevent any rotation of the specimen at the beginning of the test. The specimens had a preload of 15 N applied. To observe the crack development, a digital camera was used to record the cracks during the loading process at intervals of 15 s.

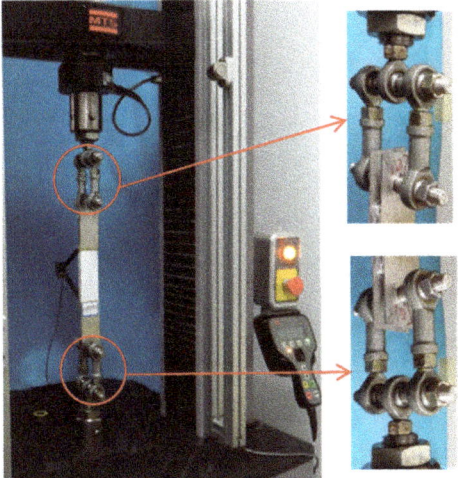

Figure 4. Uniaxial tensile test setup.

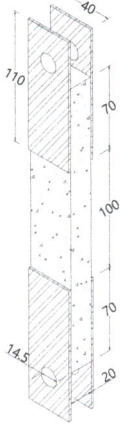

Figure 5. Dimensions of the TRM specimens (units: mm).

3. Results and Discussion

Representative curves from each testing protocol, showing the relationship between the stress and strain of the test specimens, are displayed in Figure 6. They are in good agreement with previous three-stage TRM stress–strain curves identified under tensile testing [18,39]. In stage I, the stiffness and volume proportion of both the matrix and textiles determine the slope of the curve, and the matrix plays a decisive role. In this stage, TRM exhibits nearly linear-elastic behavior until the point at which the stress reaches the tensile strength of the matrix, leading to the formation of the first crack. Stage II corresponds to a multi-cracking stage: the length and slope of this part of the stress–strain curve depend on the quality of the textile–matrix bond properties. Stage III is regarded as a strain-hardening stage and is characterized by high tensile strength and high strain capacity. In this stage, the existing cracks continuously widen and few further cracks appear. Furthermore, TRM exhibits linear behavior in this region, with the textiles carrying the whole load until the composite fails. In several tests where the transition from the second to third stages is not apparent, some extra cracks may also develop in stage III.

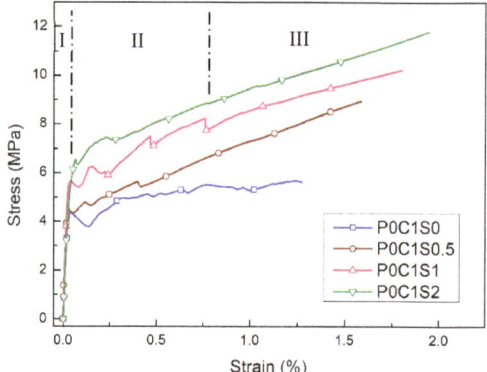

Figure 6. Typical tensile stress—strain curves of TRM specimens.

The results of uniaxial tensile tests are presented in Table 4. An effective factor (EF) is used to highlight the bond properties. This is determined by dividing the peak load of CTRM specimens by that of corresponding carbon textile strips during the uniaxial test [25]. EF < 1 corresponds to a weak bond property, whereas EF > 1 indicates the existence of strain-hardening, namely, stage III in the stress–strain curve of the TRM specimens (Figure 6).

Table 4. Mechanical properties of uniaxial tensile tests.

Specimen	First-crack Stress (MPa)	Tensile Strength (MPa)	Strain Capacity (%)	Crack Number (/)	Crack Spacing (mm)	EF (/)
P0C1S0	4.66 (0.30)	6.04 (0.49)	1.22 (0.13)	6 (2.10)	17.56 (1.86)	0.60 (0.20)
P0C1S0.5	4.88 (0.10)	8.71 (0.32)	1.61 (0.04)	4.8 (0.84)	19.03 (2.67)	0.87 (0.13)
P0C1S1	5.61 (0.11)	9.71 (0.51)	1.74 (0.07)	7 (1.58)	15.51 (2.55)	0.97 (0.20)
P0C1S2	6.43 (0.11)	11.99 (0.28)	1.89 (0.05)	8 (1.58)	13.37 (3.03)	1.20 (0.11)
P0C2S0	4.95 (0.36)	9.88 (0.12)	1.03 (0.13)	9.8 (2.39)	10.92 (3.46)	0.49 (0.05)
P0C2S0.5	5.43 (0.40)	10.84 (0.71)	1.25 (0.07)	9.2 (2.59)	9.98 (2.25)	0.54 (0.28)
P0C2S1	6.65 (0.23)	12.72 (0.68)	1.27 (0.18)	12.25 (2.36)	8.68 (2.49)	0.64 (0.27)
P0C2S2	9.69 (0.75)	22.63 (0.96)	1.69 (0.11)	13 (2.64)	7.66 (2.12)	1.13 (0.38)
P10C1S0	4.80 (0.002)	6.61 (0.44)	1.45 (0.10)	4.6 (1.14)	23.84 (2.89)	0.66 (0.18)
P20C1S0	5.86 (0.36)	7.88 (0.20)	1.58 (0.25)	9.33 (2.08)	10.9 (3.41)	0.79 (0.08)
P10C1S1	5.06 (0.13)	11.15 (0.72)	2.04 (0.07)	7.4 (1.52)	12.39 (1.79)	1.12 (0.29)
P20C1S1	7.06 (0.25)	12.92 (0.51)	1.92 (0.15)	9.8 (1.92)	10.57 (2.56)	1.29 (0.20)
P15C2S0	7.17 (0.30)	14.94 (0.37)	1.18 (0.14)	8.5 (1.29)	11.31 (2.39)	0.75 (0.15)
P15C2S1	10.01 (0.32)	22.12 (0.25)	1.60 (0.18)	11.5 (3.70)	9.18 (2.320)	1.11 (0.10)

Note: Values in parentheses denote the standard deviations.

3.1. Influence of Reinforcement Ratio on the TRM Tensile Behavior

Figure 7 shows the stress–strain curves that were obtained from the tensile tests for specimens P0C0S0, P0C1S0, and P0C2S0. The stress–strain behavior of P0C0S0 is linear-elastic before reaching the ultimate stress. The stress then drops to zero after the brittle failure, indicating that only one crack forms in this specimen. Nevertheless, for P0C1S0 and P0C2S0, the stress increases approximately linearly with the strain until the first crack appears in the TRM specimens. An apparent drop of tensile stress occurs after reaching the first-crack stress in both P0C1S0 and P0C2S0 and then the stress continues to increase until dropping again upon the formation of a new crack. This procedure is repeated until the ultimate stress is reached, at which point the specimens fail completely.

Generally, the tensile strength of both P0C1S0 and P0C2S0 increases with the reinforcement ratio. The average tensile strengths of P0C1S0 and P0C2S0 composites were 6.04 MPa and 9.88 MPa, respectively, representing increases of 0.47% and 1.41% over the unreinforced specimen (P0C0S0). The EF value of P0C1S0 was ~0.60, and that of P0C2S0 was ~0.49. Consistent with the specimen failure modes observed in Figure 8, the textiles did not break when the specimens failed, however became separated from the matrix, resulting in debonding failure and low utilization rate of carbon textiles. Figure 8 further reveals that the longitudinal debonding along the yarns parallel to the load direction occurred gradually as the load increased. Moreover, the tensile response of P0C1S0 and P0C2S0 exhibits a bilinear behavior (Figure 7). Stage III, typically observed in TRMs, did not occur in P0C1S0 and P0C2S0 because of the debonding failure mode. The lack of hardening in stage III is caused by the poor interfacial properties of TRM. Thus, practical measures should be taken to enhance the bond strength between the textiles and the matrix.

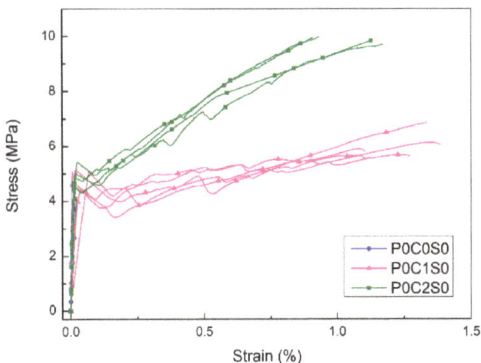

Figure 7. Stress–strain curves of the test specimens: P0C0S0, P0C1S0, and P0C2S0.

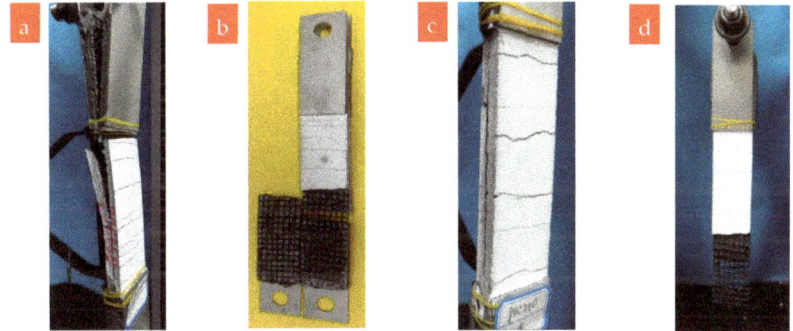

Figure 8. Debonding failure of TRM specimens: (**a**,**b**) P0C1S0 and (**c**,**d**) P0C2S0.

The cracking patterns of P0C1S0 and P0C2S0 after uniaxial tensile tests are shown in Figure 9. Compared with the plain matrix, Figure 9 indicates that the TRM specimens possessed a uniform distribution of fine cracks. The increase in reinforcement ratios also affected the crack patterns. As shown in Figure 10 and Table 4, the number of cracks increased from 6 to 9.8 as the reinforcement ratio increased from 0.4% to 0.8%, accompanied by a reduction in the distances between cracks and the crack widths. Based on Figures 7 and 8, the cracking mechanisms of P0C1S0 and P0C2S0 under tensile loading can be described as follows. (i) The first crack formed in the TRM composites when the tensile stress of the specimens reached the tensile strength of the cementitious matrix. (ii) The load originally carried by the matrix was transferred to the carbon textiles located in the crack. (iii) The bond strength between the matrix and textile allowed the textiles to transfer the load to the uncracked matrix located at either side of the crack. (iv) A new crack formed in the TRM when the stress in the uncracked matrix reached its tensile strength. (v) The stress constantly transferred between the textiles and the uncracked matrix and a multi-cracking pattern formed. (vi) No sequent cracks formed in the specimens when the interfacial bond property was so poor that the stress could not be transferred. (vii) Finally, the specimens suffered debonding failure.

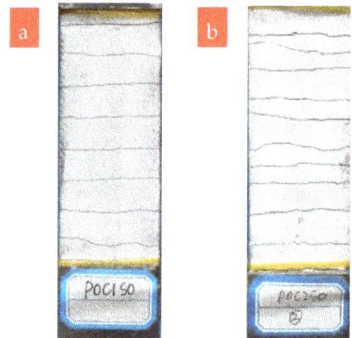

Figure 9. Cracking in TRM specimens: (**a**) P0C1S0 and (**b**) P0C2S0.

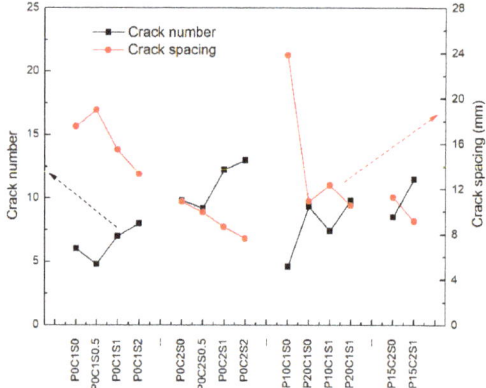

Figure 10. Crack number and spacing of TRM specimens.

3.2. Effect of Steel Fibers

Figure 11 shows the experimental stress–strain responses of the TRM composites (with one or two layers of textile reinforcement) without and with short steel fibers in proportions of 0.5%, 1%, and 2% by volume. The first-crack stress and tensile strength of the TRM specimens without and with steel fibers are depicted in Figure 12. The numerical values, including the first-crack stress, uniaxial tensile strength, ultimate strain capacity, crack number, crack spacing, and EF values, are summarized in Table 4. Moreover, the stress–strain curves of the TRM with steel fibers are generally above those of the TRM without steel fibers, as shown in Figure 11. Therefore, the bearing capacity of TRM increases noticeably through the addition of steel fibers. Within the scope of this test, the improvements in tensile mechanical behavior were significantly correlated with the proportion of steel fibers.

From the experimental results, it can be inferred that short steel fibers distributed randomly in the grids of the textiles as secondary reinforcement improve the bond strength between the textiles and the matrix. The excellent bond strength is mainly attributable to the "shear resistant ability" of the steel fibers inserted vertically or obliquely into the grids of the textiles. Adding steel fibers to TRM could improve the cracking resistance of the matrix and further enhance the bearing capacity of the composites. These improved mechanical properties are characterized by a higher first-crack stress, smaller reduction in stiffness after cracking, smaller fluctuations in the stress–strain curves, and higher ultimate tensile strength compared with the specimens without steel fibers (Figure 11).

With respect to the TRM reinforced with a single layer of textile and with 0.5%, 1%, and 2% steel fibers by volume, the ultimate tensile strength increased by 0.44%, 0.61%, and 0.98%, respectively, compared with the TRM specimens without steel fibers (Figure 12, Table 4). The strain capacity increased remarkably as a result of the strong bridging action of the steel fibers in cracks, and the maximum strain capacity of the TRM reinforced with one layer of textile was observed to increase by 1.89%.

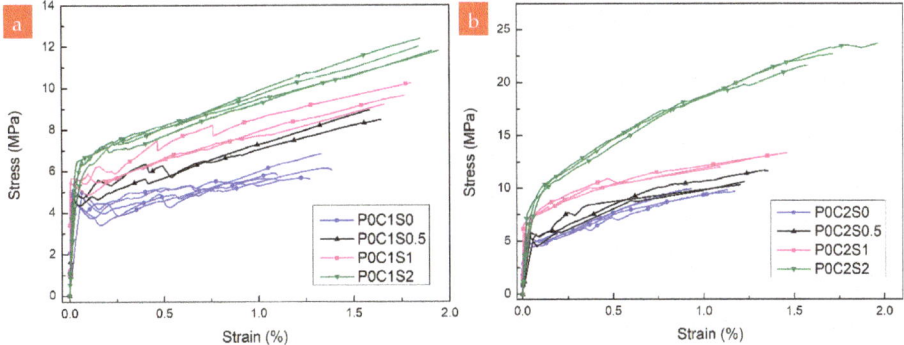

Figure 11. Stress–strain curves of the TRM specimens with varying volume fractions of short steel fibers: (**a**) one-layer and (**b**) two-layer textiles.

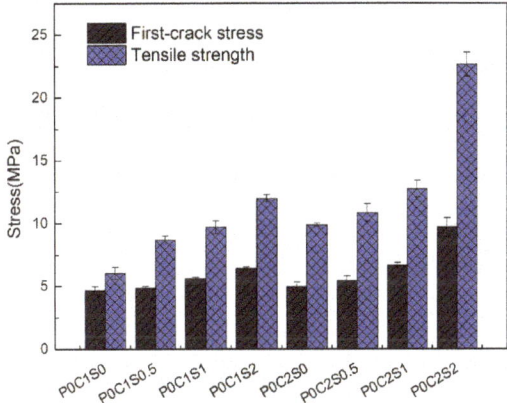

Figure 12. First-crack stress and tensile strength of the TRM specimens.

The positive effects of steel fibers on the mechanical performance of TRM reinforced with two-layer textiles are clearly noticeable in Figure 11b. Significant improvements occurred in all mechanical properties of P0C2S2 compared with those of P0C2S0, with a 129% increase in ultimate tensile strength, 95.8% increase in first-crack stress, and 64.1% increase in strain capacity. However, P0C2S0.5 exhibited only a moderate increase in tensile strength. A slight increase in the first-crack stress of P0C2S0.5 compared with that of P0C2S1 and P0C2S2 was also observed. These findings can be attributed to the addition of steel fibers, enabling the textile to bond with the matrix. Clearly, adding higher proportions of steel fibers results in better bond properties. Improved mechanical properties are limited by the distribution and orientation of the short fibers. The number and extent of fluctuations in the curves decreased with the increasing proportion of steel fibers, indicating that the bond between the textile and the matrix was better for TRM specimens with a higher proportion of steel fibers.

Figure 13 compares the cracking patterns of P0C2S0, P0C2S0.5, P0C2S1, and P0C2S2, and clearly shows the differences resulting from varying steel fiber proportions. The visual surface inspection of the TRM specimens found a large number of micro-cracks in TRM specimens with steel fibers. The cracking patterns were also transformed from relatively straight and flat continuous cracks to irregular, short cracks. The cracks propagated along a more complex path, growing not only along the width of the specimen, however also along the length. The steel fibers, distributed randomly in the grids of the textiles, enabled some resistance to micro- and macro-crack propagation and changed the direction of the development of the cracks as well as the cracking patterns.

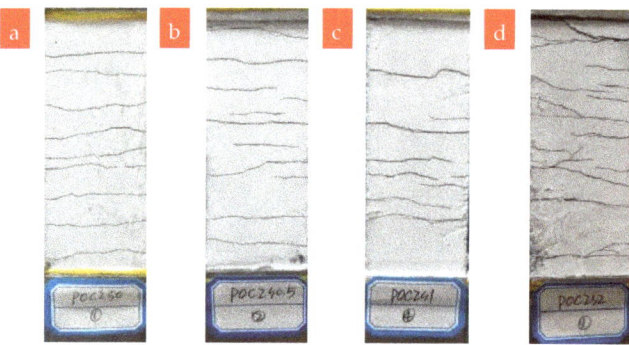

Figure 13. Cracking in TRM specimens with different volume fractions of steel fibers: (**a**) P0C2S0, (**b**) P0C2S0.5, (**c**) P0C2S1, and (**d**) P0C2S2.

The fracture surfaces of TRM specimens with steel fibers are shown in Figures 14 and 15. For both P0C1S1 and P0C2S1, only a small number of specimens failed because of the complete fracture of carbon textiles. Most of the specimens still exhibited debonding failure. With the increase in the steel fiber proportions in the TRM, the failure mode of both P0C1S2 and P0C2S2 transformed into a complete fracture of carbon textiles, i.e., a high utilization rate of carbon textiles was achieved. Adding 2% steel fibers to the TRM composites can be regarded as an effective means of enhancing the textile–matrix bond property. Additionally, it can be seen in Figure 10 that P0C2S2 had the maximum crack number as well as the minimum crack spacing among all the specimens. P0C2S2 had an average of 13 cracks and an average crack spacing of 7.66 mm.

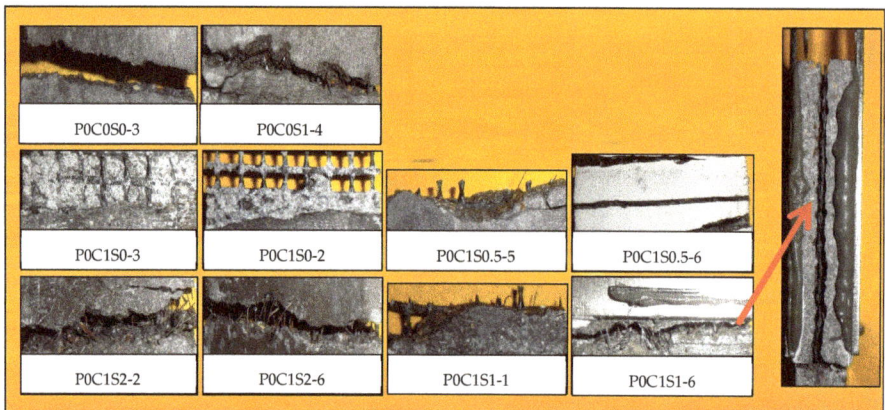

Figure 14. Fracture behavior of plain matrix, matrix reinforced with 1% steel fibers, and TRM (one-layer) specimens with steel fibers.

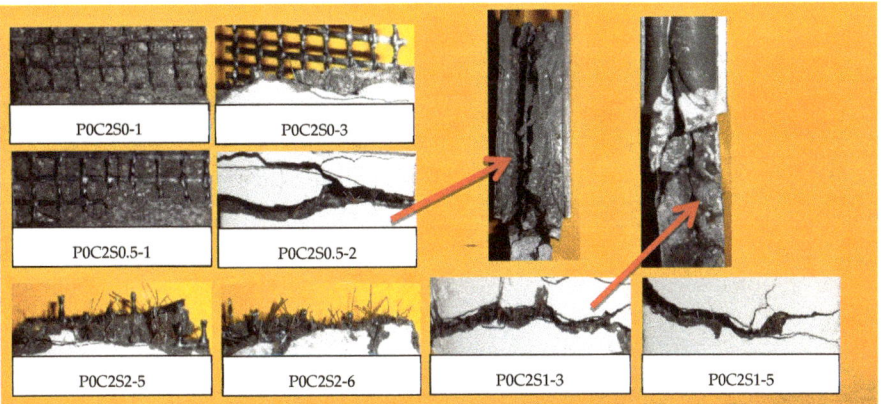

Figure 15. Fracture morphology of TRM specimens with steel fibers and two-layer textiles.

The improvements in strength and failure behavior of TRM resulting from the addition of steel fibers may be ascribed to the following mechanisms:

1. As shown in Figure 16a, the steel fibers are evenly distributed throughout the cross-section of the specimen. Well-distributed steel fibers form a good bond with their surrounding matrix and further reduce the shrinkage of the matrix, thus reducing the resulting internal defects in the matrix. Moreover, steel fibers also inhibit the formation of micro-cracks, which develop because of shrinkage.
2. Steel fibers play a bridge role in micro-cracks and delay the formation of the first macro-crack. Thus, higher stress is needed to cause the transition from micro-cracks to macro-crack. The bridging of steel fibers in micro-cracks contributes to the improved first-crack stress of TRM specimens with steel fibers. In addition, the bridging action of steel fibers in macro-cracks causes additional stress transfer over the cracks (Figure 16b,c); thus, a new crack can be formed at a smaller distance from an existing crack. Thus, fine multiple cracking can develop and lead to pronounced ductility.
3. The elastic modulus of carbon textile is greater than that of mortar matrix, so their deformation after tensioning is different, causing a relative sliding trend to occur. For TRM composites without the addition of short fibers, the bond properties between the textile and the cementitious matrix depend on the friction and adhesive caused by the matrix hydration products. For TRM composites with the addition of steel fibers, the bonding properties are enhanced. Steel fibers inserted into the textile grids can mitigate the trend, enabling a better cooperation of the carbon textile and mortar. As shown in Figure 17, steel fibers there can resist shearing force. Hence, it can be summarized that steel fibers in the textile grids provide additional "shear resistant ability" between the carbon fabric and the mortar, thus improving the bonding properties between carbon fabric and inorganic mortar in CTRM. Investigation of the fracture surfaces of TRM with steel fibers using an optical microscope has helped to explain this new link (Figure 16d).

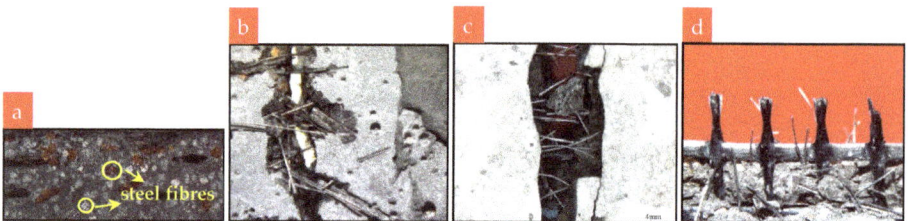

Figure 16. (a) Well-distributed steel fibers; bridging capacity of steel fibers in cracks: (b) P0C1S1, (c) P0C2S1; (d) fracture surface of TRM with steel fibers.

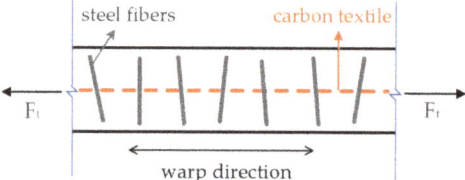

Figure 17. Side view of steel fiber distribution in the CTRM (carbon textile reinforced mortar).

3.3. Effect of Prestress

Figure 18 presents the stress–strain curves of the prestressed TRM specimens. The results, including a summary of the tensile strength, ultimate strain, first-crack stress, crack number, crack spacing, and EF values are listed in Table 4. The first-crack stress and tensile strength of the prestressed TRM specimens are depicted in Figure 19. According to the results shown in Figure 18a, pre-tension to 10% and 20% of the ultimate tensile capacity of the one-layer textile increased the specimens' first-crack stress. Although both the first-crack stress and tensile strength improved with increases in the prestress level, the improvements are not obvious (Figure 19). For example, the first-crack stress and ultimate tensile strength of P20C1S0 only increased by 25.7% and 30.5%, respectively, compared with those in P0C1S0. Figure 18b shows that adding 1% steel fibers to the prestressed TRM specimens improves their mechanical properties. The first-crack stress, ultimate tensile strength, and strain capacity of P20C1S1 increased by 51.5%, 114%, and 57.4%, respectively, compared with those in the reference specimen P0C1S0.

According to the results presented in Figures 18c and 19, TRM specimens pre-tensioned to 15% of the ultimate tensile capacity of the two-layer textiles (P15C2S0, P15C2S1) exhibit higher first-crack stress and tensile strength than the control TRM specimen (P0C2S0). Compared with P0C2S0, the first-crack stress of P15C2S0 and P15C2S1 increased by 44.8% and 102%, and the tensile strength increased by 51.2% and 124%, respectively. The average crack number and crack spacing of P15C2S1 increased from 9.8 and 11.92 mm to 11.5 and 9.18 mm in comparison with P0C2S0 (Figure 10 and Table 4). It can be concluded that adding 1% steel fibers by volume to prestressed TRM specimens (P15C2S1) is an effective method of improving the specimens' mechanical performance.

The effect of prestress can be explained by referring to Figure 20. Point O is the origin of the tension force N with respect to displacement Δl. The curve includes three distinctive stages, i.e., elastic, multiple cracking, and post-cracking. In this case, N_{cr} indicates the critical tensile load at which the mortar matrix first cracks, corresponding to the moment that the tensile load is mostly transferred to the textiles, whereas N_u is the ultimate tensile load, beyond which the specimen loses its load capacity. Whether the slope of the post-cracking stage is steeper or shallower than that in stage I depends on the stiffness of the reinforcing textiles. Once the textiles in the TRM specimen are pre-tensioned, the origin O shifts to point O', permitting the previous short uncracked stage to extend to an ideal duration [26]. With respect to this particular curve, a prestressing force is exerted on the textiles at point O' and released at point N_p. Thus, an initial compressive stress on the concrete matrix was achieved, leading to an increase in the first-crack stress [35]. Moreover, after releasing the prestress on the mortar matrix, the bond strength between the textile and matrix is considerably improved, so the ultimate tensile strength of the TRM specimens also increases (Figures 18 and 19). The development process after releasing the prestress in TRM specimens runs along the same path as the un-prestressed specimens until the final failure. As a result, exerting a prestressing force on the textiles extends the serviceability limit states of TRM and produces more reliable workability.

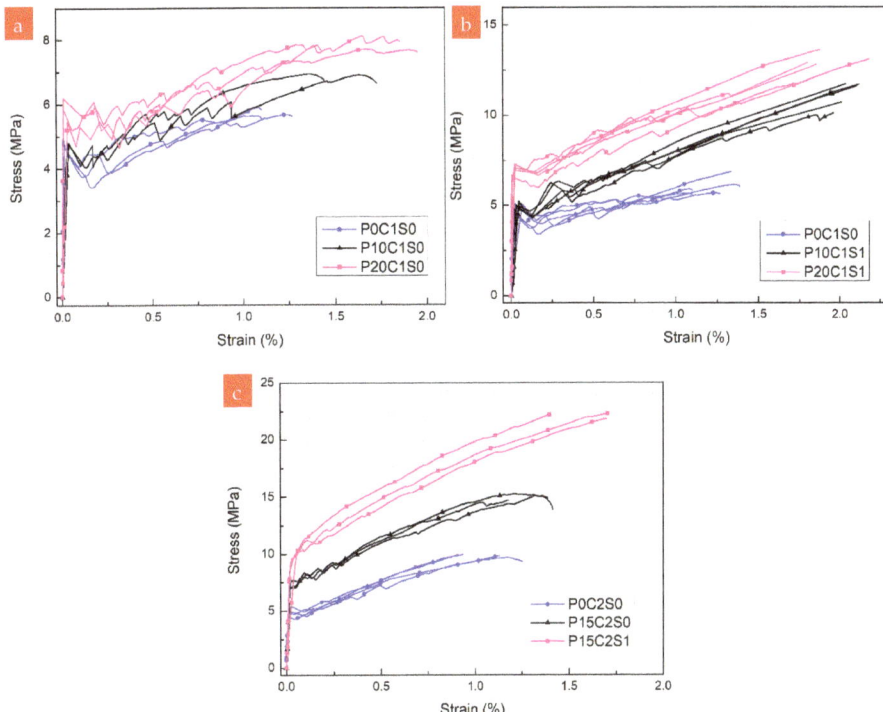

Figure 18. Stress–strain curves of the prestressed TRM specimens: (**a**) reinforced with one-layer textile and without steel fibers, (**b**) reinforced with one-layer textile and 1.0% steel fibers, and (**c**) reinforced with two-layer textiles.

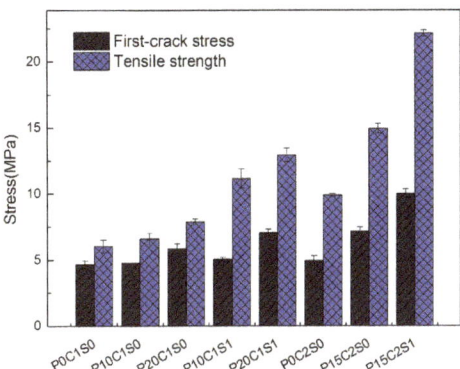

Figure 19. First-crack stress and tensile strength of the prestressed TRM specimens.

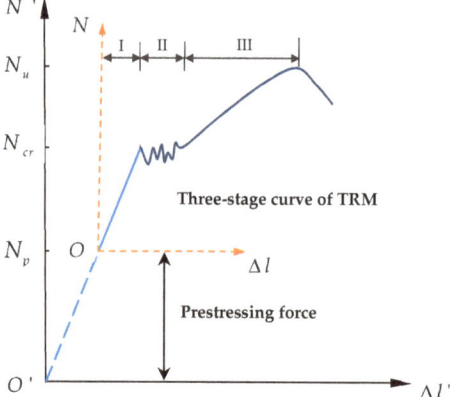

Figure 20. Influence of prestress on TRM.

Although prestressing the textiles improves the cooperative bearing ability between the textile and the matrix to a certain extent, specimens P10C1S0, P20C1S0, and P15C2S0 exhibited debonding failure. The failure modes of P10C1S1, P20C1S1, and P15C2S1 changed from debonding to the complete fracture of carbon textiles; thus, adding 1% steel fibers by volume to prestressed TRM specimens could significantly improve the textile–matrix bond properties. The fracture morphologies of prestressed TRM specimens are shown in Figures 21 and 22.

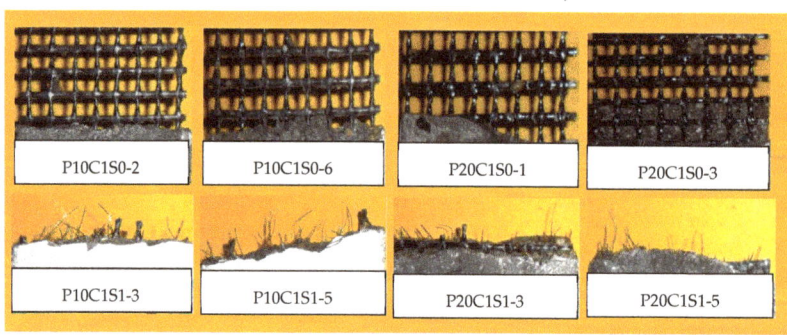

Figure 21. Fracture morphology of prestressed TRM specimens with one-layer textile.

Figure 22. Fracture morphology of prestressed TRM specimens with two-layer textiles.

4. Conclusions

In this research, the effects of the reinforcement ratio, short steel fibers as additional reinforcements, and prestressing have been explored with regard to the strength and failure behavior of TRM subjected to tensile loading. The following conclusions can be drawn:

(1) Generally, the tensile strength of both P0C1S0 and P0C2S0 increased with an increase in the reinforcement ratio. However, the EF values of specimens P0C1S0 and P0C2S0 decreased as the reinforcement ratio increased, indicating weakened textile–matrix bond strength. The textiles did not snap when P0C1S0 and P0C2S0 failed, however slid from the matrix, thereby resulting in debonding failure and low utilization rate of the carbon textiles. In contrast, the utilization rate of carbon textiles increased remarkably when short dispersed steel fibers of 2% volume fraction were inserted into the grids of the textile.

(2) Short steel fibers are able to improve the mechanical properties of the mortar and the entire composite in experiments. Moreover, steel fibers improve the textile–matrix bond strength, which can be attributed to the "shear resistant ability" of steel fibers inserted into the grids of the textile. Increases in tensile strength were clearly observed in all specimens with added steel fibers. An investigation of the fracture surfaces using an optical microscope further revealed that short steel fibers added to TRM cause finer cracks with smaller spacings and widths. Furthermore, within the scope of this test, the improvements in tensile mechanical behavior were highly correlated with the increase in the steel fiber proportion. Compared with the results that were obtained for the reference TRM plates, the tensile strength increased by approximately 100% following the addition of 2% steel fibers by volume.

(3) Increases in first-crack stress and tensile strength were also observed in prestressed TRM specimens. The enhanced first-crack stress was attributed to the extension (caused by pre-compression of the mortar matrix after being released) of stage I, i.e., the uncracked state corresponding to the related mortar matrix. The tensile strength increased as the bond behavior improved, a result of the strengthened interaction effect between the surface of the textile and the matrix activated by prestressing. Therefore, the serviceability limit states of TRM composites can be extended by exerting a prestressing force on the textiles.

(4) Adding steel fibers at 1% volume to prestressed TRM specimens is an effective method of improving the specimens' mechanical performance, dramatically enhancing the bond strength between matrix and textiles. As a result, the failure mode changes from debonding to the complete fracture of the carbon textiles. In this study, the combination of 1% steel fibers and prestress calculated at 15% of the ultimate tensile strength of the two-layer textiles was found to be the optimum configuration, producing the highest first-crack stress and tensile strength and the most reasonable multi-cracking pattern.

Author Contributions: F.Z. designed the experiments; H.L. and Y.D. analyzed the data; L.L. performed the experiments; D.Z. and W.P. revised the paper; all the authors reviewed and approved the paper.

Funding: This work was supported by the funds from the Natural Science Foundation of Hunan Province (Grant No. 2018JJ2043), the Major Project of Sci-Tech Plan of Changsha City (Grant No. kq1804002, kq1703002), and the Project of Sci-Tech Plan of Changsha City (Grant No. kq1701032).

Acknowledgments: The authors gratefully acknowledge Hunan Good Bond Construction Technic Development Co., Ltd (Changsha, China) for supplying the cementitious materials, super-plasticizer, sand, and epoxy resin.

Conflicts of Interest: The authors declare no conflict of interest.

References

1. Tran, M.T.; Vu, X.H.; Ferrier, E. Experimental and analytical analysis of the effect of fibre treatment on the thermomechanical behaviour of continuous carbon textile subjected to simultaneous elevated temperature and uniaxial tensile loadings. *Constr. Build. Mater.* **2018**, *183*, 32–45. [CrossRef]

2. Schneider, K.; Lieboldt, M.; Liebscher, M.; Fröhlich, M.; Hempel, S.; Butler, M.; Schröfl, C.; Mechtcherine, V. Mineral-based coating of plasma-treated carbon fibre rovings for carbon concrete composites with enhanced mechanical performance. *Materials* **2017**, *10*, 360. [CrossRef] [PubMed]
3. Mechtcherine, V. Novel cement-based composites for the strengthening and repair of concrete structures. *Constr. Build. Mater.* **2013**, *41*, 365–373. [CrossRef]
4. Mechtcherine, V. Towards a durability framework for structural elements and structures made of or strengthened with high-performance fibre-reinforced composites. *Constr. Build. Mater.* **2012**, *31*, 94–104. [CrossRef]
5. Butler, M.; Lieboldt, M.; Mechtcherine, V. Application of textile-reinforced concrete (TRC) for structural strengthening and in prefabrication. In Proceedings of the International Conference on Advanced Concrete Materials (ACM), Stellenbosch, South Africa, 2010; pp. 125–134.
6. Papanicolaou, C.G. Applications of textile-reinforced concrete in the precast industry. In *Textile Fibre Composites in Civil Engineering*; Triantafillou, T., Ed.; Woodhead Publishing: Patras, Greece, 2016; pp. 227–244. ISBN 9781782424467. (online).
7. Arundhathy, S.; Vasugi, V. Engineered cementitious composites for sustainable construction. *Key Eng. Mater.* **2016**, *692*, 17–26. [CrossRef]
8. Zhang, R.; Matsumoto, K.; Hirata, T.; Ishizeki, Y.; Niwa, J. Application of PP-ECC in beam-column joint connections of rigid-framed railway bridges to reduce transverse reinforcements. *Eng. Struct.* **2015**, *86*, 146–156. [CrossRef]
9. Qudah, S.; Maalej, M. Application of Engineered Cementitious Composites (ECC) in interior beam-column connections for enhanced seismic resistance. *Eng. Struct.* **2014**, *69*, 235–245. [CrossRef]
10. Hegger, J.; Will, N.; Bruckermann, O.; Voss, S. Load-bearing behaviour and simulation of textile reinforced concrete. *Mater. Struct. Constr.* **2006**, *39*, 765–776. [CrossRef]
11. Si Larbi, A.; Contamine, R.; Ferrier, E.; Hamelin, P. Shear strengthening of RC beams with textile reinforced concrete (TRC) plate. *Constr. Build. Mater.* **2010**, *24*, 1928–1936. [CrossRef]
12. Hegger, J.; Horstmann, M.; Zell, M. Textile Reinforced Concrete–Realization in applications. *Tailor Made Concr. Struct.* **2008**, 357–362.
13. Raupach, M.; Cruz, C.M. Textile-reinforced concrete: Selected case studies. In *Textile Fibre Composites in Civil Engineering*; Triantafillou, T., Ed.; Woodhead Publishing: Patras, Greece, 2016; pp. 275–299, ISBN 9781782424467.
14. Hegger, J.; Horstmann, M.; Feldmann, M.; Pyschny, D.; Raupach, M.; Büttner, T.; Feger, C. Sandwich panels made of TRC and discrete and continuous connectors. *Int. RILEM Conf. Mater. Sci.* **2010**, *I*, 381–392.
15. Williams Portal, N.; Zandi, K.; Malaga, K.; Wlasak, L. GFRP connectors in textile reinforced concrete sandwich elements. In Proceedings of the 19th IABSE Congress Stockholm, IABSE c/o ETH Hönggerberg, Sweden, 2016; pp. 1331–1338.
16. Brückner, A.; Ortlepp, R.; Curbach, M. Textile reinforced concrete for strengthening in bending and shear. *Mater. Struct. Constr.* **2006**, *39*, 741–748. [CrossRef]
17. Contamine, R.; Si Larbi, A.; Hamelin, P. Contribution to direct tensile testing of textile reinforced concrete (TRC) composites. *Mater. Sci. Eng. A* **2011**, *528*, 8589–8598. [CrossRef]
18. Larrinaga, P.; Chastre, C.; Biscaia, H.C.; San-José, J.T. Experimental and numerical modeling of basalt textile reinforced mortar behavior under uniaxial tensile stress. *Mater. Des.* **2014**, *55*, 66–74. [CrossRef]
19. Frank, J. Tragverhalten von Filamentgarnen in zementgebundener Matrix. Ph.D. Thesis, Technische Universitaet Dresden, Dresden, Germany, 2004.
20. Barhum, R.; Mechtcherine, V. Effect of short, dispersed glass and carbon fibres on the behaviour of textile-reinforced concrete under tensile loading. *Eng. Fract. Mech.* **2012**, *92*, 56–71. [CrossRef]
21. Contamine, R.; Junes, A.; Si Larbi, A. Tensile and in-plane shear behaviour of textile reinforced concrete: Analysis of a new multiscale reinforcement. *Constr. Build. Mater.* **2014**, *51*, 405–413. [CrossRef]
22. Rambo, D.A.S.; de Andrade Silva, F.; Toledo Filho, R.D.; da Fonseca Martins Gomes, O. Effect of elevated temperatures on the mechanical behavior of basalt textile reinforced refractory concrete. *Mater. Des.* **2015**, *65*, 24–33. [CrossRef]
23. Xu, S.; Krüger, M.; Reinhardt, H.-W.; Ožbolt, J. Bond Characteristics of Carbon, Alkali Resistant Glass, and Aramid Textiles in Mortar. *J. Mater. Civ. Eng.* **2004**, *16*, 356–364. [CrossRef]

24. Dvorkin, D.; Poursaee, A.; Peled, A.; Weiss, W.J. Influence of bundle coating on the tensile behavior, bonding, cracking and fluid transport of fabric cement-based composites. *Cem. Concr. Compos.* **2013**, *42*, 9–19. [CrossRef]
25. Colombo, I.G.; Magri, A.; Zani, G.; Colombo, M.; Di Prisco, M. Textile Reinforced Concrete: Experimental investigation on design parameters. *Mater. Struct. Constr.* **2013**, *46*, 1933–1951. [CrossRef]
26. Reinhardt, H.W.; Krüger, M.; Große, C.U. Concrete Prestressed with Textile Fabric. *J. Adv. Concr. Technol.* **2003**, *1*, 231–239. [CrossRef]
27. Barhum, R.; Mechtcherine, V. Influence of short dispersed and short integral glass fibres on the mechanical behaviour of textile-reinforced concrete. *Mater. Struct. Constr.* **2013**, *46*, 557–572. [CrossRef]
28. Du, Y.; Zhang, X.; Zhou, F.; Zhu, D.; Zhang, M.; Pan, W. Flexural behavior of basalt textile-reinforced concrete. *Constr. Build. Mater.* **2018**, *183*, 7–21. [CrossRef]
29. Ou, Y.; Zhu, D. Tensile behavior of glass fiber reinforced composite at different strain rates and temperatures. *Constr. Build. Mater.* **2015**, *96*, 648–656. [CrossRef]
30. Zhu, D.; Peled, A.; Mobasher, B. Dynamic tensile testing of fabric-cement composites. *Constr. Build. Mater.* **2011**, *25*, 385–395. [CrossRef]
31. Yan, L.; Kasal, B.; Huang, L. A review of recent research on the use of cellulosic fibres, their fibre fabric reinforced cementitious, geo-polymer and polymer composites in civil engineering. *Compos. Part B Eng.* **2016**, *92*, 94–132. [CrossRef]
32. Tan, B.K.; Ching, Y.C.; Poh, S.C.; Abdullah, L.C.; Gan, S.N. A review of natural fiber reinforced poly(vinyl alcohol) based composites: Application and opportunity. *Polymers* **2015**, *7*, 2205–2222. [CrossRef]
33. Kazim, A. Carbon fibers. In *Fiber Technology for Fiber-Reinforced Composites*; Seydibeyoğlu, M.Ö., Mohanty, A.K., Misra, M., Eds.; Woodhead Publishing: Patras, Greece, 2017; pp. 123–151, ISBN 9780081018712.
34. Newcomb, B.A. Processing, structure, and properties of carbon fibers. *Compos. Part A Appl. Sci. Manuf.* **2016**, *91*, 262–282. [CrossRef]
35. Du, Y.; Zhang, X.; Liu, L.; Zhou, F.; Zhu, D.; Pan, W. Flexural Behaviour of Carbon Textile-Reinforced Concrete with Prestress and Steel Fibres. *Polymers* **2018**, *10*, 98. [CrossRef]
36. Donnini, J.; Corinaldesi, V.; Nanni, A. Mechanical properties of FRCM using carbon fabrics with different coating treatments. *Compos. Part B Eng.* **2016**, *88*, 220–228. [CrossRef]
37. Yin, S.; Xu, S.; Li, H. Improved mechanical properties of textile reinforced concrete thin plate. *J. Wuhan Univ. Technol. Mater. Sci. Ed.* **2013**, *28*, 92–98. [CrossRef]
38. Soe, K.T.; Zhang, Y.X.; Zhang, L.C. Material properties of a new hybrid fibre-reinforced engineered cementitious composite. *Constr. Build. Mater.* **2013**, *43*, 399–407. [CrossRef]
39. Du, Y.; Zhang, M.; Zhou, F.; Zhu, D. Experimental study on basalt textile reinforced concrete under uniaxial tensile loading. *Constr. Build. Mater.* **2017**, *138*, 88–100. [CrossRef]

 © 2019 by the authors. Licensee MDPI, Basel, Switzerland. This article is an open access article distributed under the terms and conditions of the Creative Commons Attribution (CC BY) license (http://creativecommons.org/licenses/by/4.0/).

Article

Investigation of Shear-Induced Deformation of Reinforcing Textiles by Optical Measurement Devices

Stefan Rothe [1,*], Ellen Wendt [1], Sybille Krzywinski [1], Marianna Halász [2], Peter Bakonyi [2], Peter Tamás [3] and Attila Bojtos [3]

1. Institute of Textile Machinery and High Performance Material Technology (ITM), TU Dresden, 01062 Dresden, Germany; ellen.wendt@tu-dresden.de (E.W.); sybille.krzywinski@tu-dresden.de (S.K.)
2. Department of Polymer Engineering, Budapest University of Technology and Economics, 1111 Budapest, Hungary; halaszm@pt.bme.hu (M.H.); bakonyi@pt.bme.hu (P.B.)
3. Department of Mechatronics, Optics and Engineering Informatics, Budapest University of Technology and Economics, 1111 Budapest, Hungary; tamas@mogi.bme.hu (P.T.); bojtos@mogi.bme.hu (A.B.)
* Correspondence: stefan.rothe3@tu-dresden.de; Tel.: +49-351-463-39739

Received: 4 March 2019; Accepted: 25 March 2019; Published: 28 March 2019

Abstract: When fiber-reinforced plastic (FRP) components are designed, it is very important to ensure that textiles are formed into complex 3D geometries without folds, and that the reinforcing structure is oriented appropriately. Most research in this context is focused on finite element (FE) forming simulations and the required characterization of textile reinforcements. However, the early stage of the design of FRPs, where kinematic draping simulations are used, is barely considered. In particular, the need for a critical shear angle for the execution and evaluation of kinematic draping simulations is often neglected. This paper presents an extended picture frame test stand with an optical device recording shear-induced deformations with the help of a laser line emitter. Associated hardware and software for detecting and quantifying the fold formation during a picture frame test were developed. With the additional recorded information, a material-specific critical shear angle can be determined, material behaviors can be compared, and FE-based simulation methods can be evaluated. This innovative test stand and the associated software tools will help engineers to decide on suitable materials and improve transparency in the early stages of the design process.

Keywords: textiles; composite preforming; mechanical properties; shear behavior; surface analysis; picture frame test; kinematic draping simulation

1. Introduction

The importance of energy and resource efficiency is constantly growing, consequently, the significance of innovative and interdisciplinary lightweight technologies rises as well. This applies, in particular, to passenger and freight transport, and mechanical and plant engineering [1]. For this reason, high-strength fiber-reinforced plastics (FRPs) are increasingly used, in addition to traditional construction materials, such as aluminum or steel. Due to their fibrous structure, high formability and adjustable directional properties, textile reinforcements (e.g., woven, knitted, braided or laid fabrics) are commonly used in the manufacturing of FRPs [2].

Such FRPs are lightweight, but expensive and often require a high effort of manual labor. Nevertheless, a double-digit growth of carbon fiber-reinforced plastics is expected by 2020 [3]. Therefore, it is crucial to significantly reduce costs by shortening product development cycles and making manufacturing resource-efficient. Computer-aided methods can greatly help with this as they are particularly suitable for complex and even double-curved FRP products. Especially in the early stages of designing an FRP, when various models are considered, costs can be avoided with kinematic draping simulations. In contrast to FE-based simulations, which are suitable for the subsequent planning of automated processes, kinematic draping simulations are faster, easier to handle and less sophisticated with regard to material characterization. Nevertheless, the manufacturing process is taken into account by predefined handling methods. Two important aspects in evaluating kinematic draping results and FE-based forming simulations are the fold-free forming of reinforcing textiles and a load-adapted orientation of reinforcing fibers. In particular, folds occurring during manufacturing must be strictly avoided as they lead to a significant reduction in strength or impaired damage behavior of the final FRP product [4,5].

While numerous investigations focus on FE-based simulation methods and the related characterization of materials [6–10], kinematic draping simulations and their requirements are investigated sparsely. However, only one parameter that limits the formability is needed for the assessment of a kinematic draping simulation. In available software solutions and in the industry, this material-specific parameter is often named "critical shear angle", which is not the same as the "locking angle" as used in the FE-based simulations literature. The "critical shear angle" can be determined by shear behavior testing, although there is no standardized method yet. The investigations and the developed test stand presented in this paper aim to determine the critical shear angle reproducibly and comprehensibly for reliable kinematic draping simulations, and to give an assessment in the product development cycle.

1.1. Determination of Shear Deformation Behavior

Besides tensile and flexural behavior, shear behavior is essential for assessing the overall deformation behavior of reinforcing textiles [11–13]. While the former two are evaluated according to standardized test methods [1], no standard for the determination of shear behavior exists yet, in spite of numerous investigations regarding 2D and 3D characterization [14]. This obviously shows the difficulty of determining textile shear behavior. In developing a shear test method, three assumptions are made for textiles based on less extensible fibers, e.g., glass or carbon: The fibers are inextensible, no slippage occurs between fibers and the flexural stiffness of fibers is negligible [15]. These assumptions are the same as the assumptions in kinematic draping simulations [16–18]. Therefore, shear behavior tests with these assumptions are suitable for determining the "critical shear angle". This angle is defined here as the shear angle at which a specific fold height occurs during shear testing.

The shear process of textiles is divided into two main shear principles: pure shear and simple shear [19]. For simple shear, the distance between the two clamping lines remains constant during the shear process. This leads to an elongation of the unclamped sample edges. It not only changes the angle between the yarns, but also causes tensile strain due to yarn torsion at the crossing points. In contrast, pure shear only changes the angle between the yarns without yarn elongation, because the distance between the clamping lines does not remain constant. Based on these assumptions, shear processes inducing both pure and constant shear on textiles are preferred, e.g., the bias extension test or picture frame test [12].

1.1.1. Bias Extension Test

The bias extension test (BET) is widely used because of its simplicity. A 45° oriented textile sample is fixed into a tensile testing device and loaded with tension [20–22]. Additionally, the BET can be performed as a biaxial tensile test with a modified sample geometry and clamping in the second direction [23]. In a standard clamping arrangement, there are different shear zones, Figure 1a. In zone A, pure and constant shear is induced, due to the free edges of the relevant yarns. In zone B, only half the shear of zone A is induced due to the partly clamped yarns. Finally, in zone C, no shear occurs due to the completely clamped yarns. During testing, the force/displacement diagram is recorded. However, the measured force reflects the whole effort of sample shearing. As this effort depends on the sample size, a normalized shear force was introduced [12]. Calculating the normalized shear force requires a comprehensive knowledge of engineering mechanics.

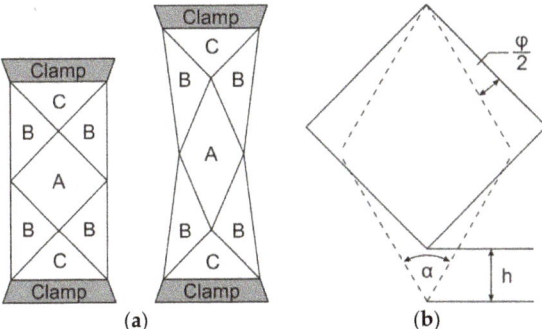

Figure 1. Principle of shear deformation in the bias extension test (**a**) and in the picture frame test (**b**) [1].

As mentioned above, the BET is widely used and well-investigated. Most of the research only considers woven fabrics as to their "pin-joint" arrangement. In several studies, a material-specific locking angle was determined, describing the shear angle when "locking" occurs. The locking phenomenon is observed when warp and weft yarns cannot rotate further without inducing non-negligible in-plane compressions on their neighboring yarns. As it is not possible to measure the moment of locking directly, it is derived from the shear force/shear angle diagram or by additional optical measurements. Theoretically, the shear force/shear angle diagram shows a steep rise at the moment of locking, so it can be derived quite easily. However, it was shown in [24] that folding was observed before locking. This additionally supports the distinction in the terminology used. In [25], it is shown that folding depends on textile properties, whole strain and stress distribution and on boundary conditions. Folding is mentioned as a multiscale problem. Thus, the determination of the locking angle does not seem to be a reliable criterion for deriving the critical shear angle needed for kinematic draping simulations.

In addition, there are a few studies also investigating multi- and uni-directional non-crimp textiles (NCF and UD-NCF, respectively) [26–29]. For NCFs, slippage was observed starting at a shear angle of 40° [27]. This is a major disadvantage of BET and leads to a minor deviation between the mathematically calculated shear angle and the optically measured shear angle. Consequently, the applicability of BET is questionable for NCFs, because of the missing pin joint arrangement and the spuriously made assumption regarding slippage. Nevertheless, such effects are also encountered in the practice of manufacturing FRP products and need to be considered.

Optical measurements of the out-of-plane deformation behavior during BET are rare. In [23], a multistep biaxial bias extension test was performed to measure the force required to unfold an already sheared woven fabric sample. At each step, the resulting deformation was 3D scanned. Therefore, the sample and its surroundings needed to be additionally prepared, as compared to the standard bias extension test. The scans were edited to achieve smooth evaluable surfaces and the dimensions of the fold were derived. This procedure was not aimed at determining a critical shear angle as needed for kinematic draping simulations, but it can be used for the comparison of materials. It is not a continuous measurement of folding, i.e., the beginning of folding cannot be determined. In addition, sample preparation, the test procedure, data preparation, and evaluation are time consuming and cost intensive.

1.1.2. Picture Frame Test

In the picture frame test (PFT) [12,21,29,30], pure and constant shear is achieved by fixing the edges of a square sample (initial state $\alpha = 90°$) in a hinged frame. Tensile loading at two of its opposite corners causes the sample to deform into a rhomboid shape depending on transversal displacement h (Figure 1b). In contrast to BET, the induced shear is uniform. The force and displacement are recorded analogously to BET. The sample shear angle φ can be calculated with the use of the transversal displacement h and the picture frame edge length a, as in Equation (1):

$$\varphi = 90° - 2 \times \cos^{-1}\left(\frac{1}{\sqrt{2}} + \frac{h}{2 \times a}\right) \tag{1}$$

The shear force F_{sh} can be calculated directly for standard hinged picture frames depending on the shear angle, with Equations (2) and (3) [30,31]:

$$F_{sh} = \frac{F}{2 \times \cos \alpha} \tag{2}$$

$$\alpha = 90° - \varphi \tag{3}$$

A group of academic and industrial researchers conducting suitable picture frame test methods as well as BET methods was set up for benchmarking. Five picture frame devices were investigated and compared [12]. All presented picture frames were based on rigid clamps, but the dimensions differed and consequently lead to different sample sizes. Three normalization methods were compared. Finally, a normalization method with an energy approach considering both the side length of the fabric and of the picture frame [30] was recommended [12]. Besides different dimensions, two picture frames use a lever mechanism, for which the amplification needs to be considered in the calculation of shear force. A main disadvantage of all presented picture frames were the induced in-plane tensions due to rigid clamping as they significantly influence the beginning of folding [25,32]. Additionally, the frame itself and each sample require elaborate preparation. In particular, precise alignment of the sample while being clamped on the frame is essential, or else additional tensions are generated, which also cause folding to begin earlier.

The applicability of the PFT has been investigated for several textiles, mainly for woven fabrics, but also for UD-NCF [26] and pre-consolidated textiles [33]. Furthermore, BET and PFT have been compared extensively [20–22,34]. Cao et al. [12] investigated balanced twill and plain woven fabrics, as well as unbalanced twill woven fabrics with both tests. It was shown that the PFT is feasible for balanced woven fabrics, but not for unbalanced woven fabrics. The obtained shear force/shear angle diagrams were comparable to the results of BET after normalization. Additionally, different preparation methods have also been investigated, leading to the conclusion that mechanical conditioning (e.g., shearing the sample a few times by hand) increases reproducibility. However, it was considered that such a procedure is not feasible in industry. Another conclusion was that it is necessary to remove transverse yarns from the clamping arms to prevent the initiation of folding in these zones, due to the rigid clamps [35]. A comparison of woven fabrics and UD-NCFs with adapted clamping revealed that shear tests are not suitable for UD-NCFs, as shearing is not an intrinsic behavior either in PFT or BET [26]. For pre-consolidated textiles, alternative sample preparation was needed to allow PFT, otherwise clamping would have prevented shearing due to the consolidated state of the material. The authors mentioned that the locking angle cannot be determined by PFT for pre-consolidated textiles. Furthermore, the definition of locking angle needs to be reconsidered. In [36], the theoretical determination of the locking angle based on textile parameters for one specific woven fabric is described and experimentally investigated by PFT and BET. Due to the test procedure of PFT, where intra-plane yarn shear was induced, it was found that locking theoretically occurs at higher angles than in BET. This was confirmed with the experiments for PFT and BET. Additionally, the authors assumed that folding would appear after locking. Even if this was confirmed by their experiments, other investigations showed diverging tendencies, i.e., this observation cannot be generalized.

Other investigations of the last decade have mainly focused on FE-based simulation methods. Therefore, the relationship between tension and shear behavior was examined extensively to improve simulation models and results of forming simulations. In addition, there are several studies dealing with the purposeful generation of pretensions in shear testing and in manufacturing to prevent folding or delay the onset of folding. These studies are necessary for FE-based simulations and contribute to a better understanding of textile deformation behavior in general. However, they are less essential for kinematic draping simulations as they do not determine the critical shear angle or consider the global phenomenon of folding.

Nevertheless, the achieved shear force/shear angle data can be used for determining the critical shear angle by further data analysis. Souter [37] introduced a method to identify the critical shear angle by intersecting two fitting linear regressions in the first and last five degrees of the shear force/shear angle data (Figure 2a). Another method is described in [38,39], where a linear regression is fitted in the shearing phase at the very beginning of the shear force/shear angle data. Once the measured data exceed a deviation of 5% or more as compared to linear regression, the critical shear angle is obtained (Figure 2b).

However, both methods are questionable for several reasons. First of all, these shear force/shear angle diagrams correspond to ideal shear behavior. For realistic data sets, it is more challenging to define different linear zones (Figure 2c). Also, the subjective definition of linear zones leads to minor deviations. Additionally, in Souter's method, the definition of the posterior linear zone is strongly dependent on the procedure parameter of maximum transverse distance, and consequently, influences the intersection point. As can be seen, the post-determination of critical shear angle from shear force/shear angle data is not reliable and prone to error.

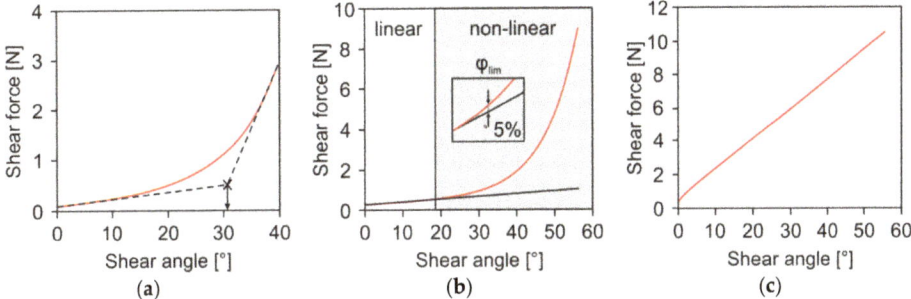

Figure 2. Evaluation of a shear force/shear angle diagram. Determination of the critical shear angle (**a**) according to [37] and (**b**), according to [38,39], (**c**) shear force/shear angle diagram, biaxial non-crimp fabric ±45°, sewing thread (ST) 0°.

Of course, this fact is known and has led to several approaches to determine the critical shear angle in a shear test. Non-destructive optical measurements are suitable for this purpose as the critical shear angle depends on folding. A widely used optical method is DIC (Digital Image Correlation), which can be used to capture local elongations. The distribution of elongations over the sheared samples is examined and evaluated in the matter to determine the critical shear angle. This procedure is very elaborate in regards to data processing. In [39], the distribution of elongation on a PFT sample was averaged at each transversal distance step. As folding occurred, the mean value of elongation rose significantly and was visible in the overall diagram. However, no information of the folding dimensions was obtained.

Another approach is to identify folding by means of a gray-scale image. However, subsequent image analysis is extremely time-consuming and the results are often inaccurate. Commercial 3D deformation detection systems are provided by GOM. However, this technique is quite costly and laborious in terms of sample preparation (PONTOS system) and the evaluation of the recorded data (ARAMIS system). Our own experience has shown that both solutions are prone to error as data points were missing during the progressive test process.

The lack of reproducibility and traceability as well as the discontinuity and high effort of existing solutions (for both PFT and BET) for determining the critical shear angle by means of fold dimensions were the main drivers for the developed test stand presented in this study. This initial situation motivated the joint research project of the Institute of Textile Machinery and High Performance Material Technology of TU Dresden, and the Department of Polymer Engineering of TU Budapest. The objective was to exploit the detection and quantification of folds during the picture frame test and, subsequently, to achieve a cost-efficient solution. Therefore, we investigated image processing methods in terms of suitability and added necessary equipment to the picture frame test stand.

2. Materials and Methods

The investigations were based on the picture frame shear test, described above in Section 1.1.2. A picture frame modified at the ITM was used [19]. In contrast to the commonly used picture frames, with a clamping mechanism holding the samples at their edges, the samples are held by needle bars. This allows the twisting of the reinforcing threads and the force to be transferred from the picture frame into the sample (Figure 3). This picture frame test stand was completed by an option for the continuous optical detection of shear deformation. During the picture frame test, fold deformation is most pronounced at the center of the sample; it is therefore sufficient to measure the central deflection of the sample. In this respect, our device is different from commercially available 3D deformation measuring systems offered by the company GOM or by a recent study [23], which includes the deformation of the entire surface.

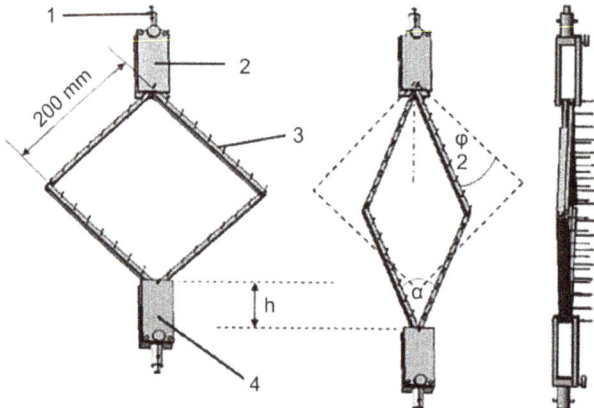

Figure 3. Modified picture frame (1: adapter; 2: upper frame holder; 3: needle bar; 4: lower frame holder).

The test conditions were as follows:

Test device:	Monoaxial tensile test machine Z 2.5, ZWICK, Ulm, Germany
Force sensor:	500 N
Sample size:	300 mm × 300 mm
Picture frame edge length:	200 mm × 200 mm
Deformation distance:	60 mm (corresponds to a shear angle of $\varphi \approx 28°$)
Test speed:	100 mm/min
Test conditions:	Standard climate (DIN EN ISO 139), $T = 20\,°C$, rel. humidity = 65%
Recorded data:	Force F at increments of $\Delta h = 0.1$ mm in transverse direction Displacement

The extended test setup consists of a device including a laser line emitter, a camera and a computer with specific data processing software. The equipment is attached to the tensile testing machine in addition to the picture frame. The attachment is above the force sensor so that it does not influence the capture of the shear force. The laser line emitter, which is fixed at a defined distance, projects a horizontal beam orthogonally onto the sample surface (constant over 150 mm width). A Basler acA 14 µm camera (Basler AG, Ahrensburg, Germany) for the optical recording of the laser beam is placed at a defined angle and distance to the sample plane (Figures 4 and 5). Due to its kinematic design with fixed and guided joints, the equipment ensures a constant position of the projected laser beam in the middle of the sample (Figure 4b). The distance of laser line emitter and camera remains constant over testing.

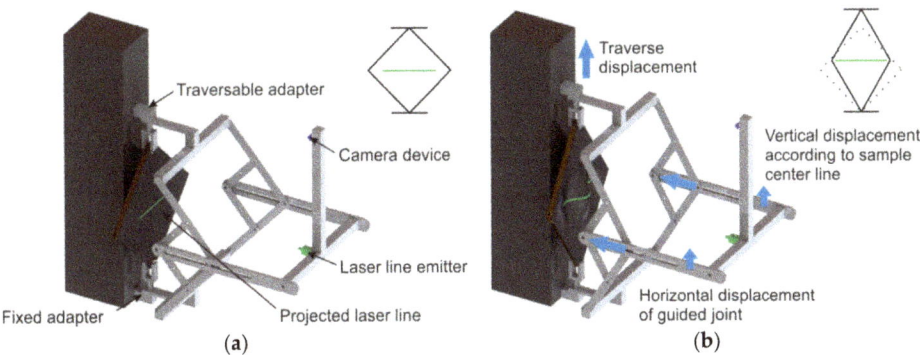

Figure 4. Extended picture frame test stand. Initial (**a**) and deflected (**b**) state.

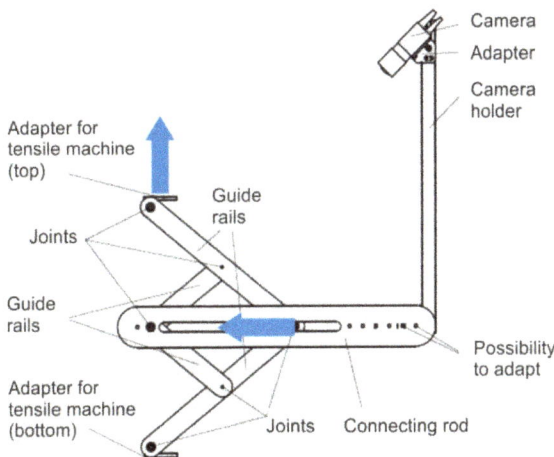

Figure 5. Setup of the extended picture frame test stand.

During the shear test, the geometrical change of the laser beam caused by the shear deformation of the sample was continuously recorded with a Software Development Kit (SDK) from Basler AG. In the initial state, the distance between the left and right joints of the shear frame was 282.8 mm. With a shear deformation of 28° (corresponding to a deflection of 60 mm), the distance was 206 mm. The laser beam was captured over a constant width of approx. 150 mm during the picture frame test. Image processing software supported by OpenCV calculated the deflection curves of the laser beam for every single time step. After calibration (see Section 3.1) and interpolation, the geometric data of the deflection curves were determined as a function of time (Figure 6) and were available for further analysis.

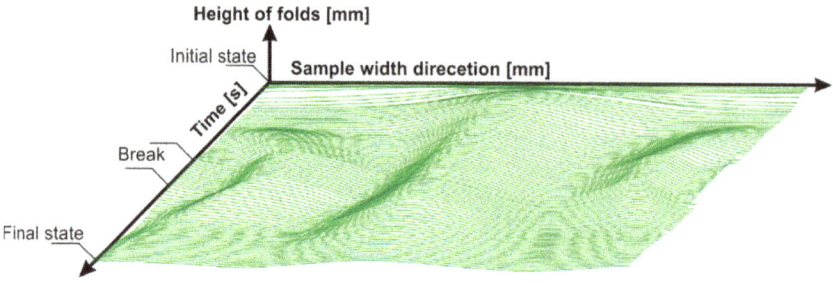

Figure 6. Measured height profile of the laser beam as a function of time (SDK software, Basler) over a constant width of 150 mm—carbon fiber (CF) biaxial non-crimp fabric ±45°, ST 0°; Hysteresis with 10 seconds break at maximum deflection.

The error of the measurement system resulting from an incompletely level position of the sample is eliminated by comparing the geometry data of each time step with the first time step (initial state). The force/displacement diagram and the shear force/shear angle diagram are recorded simultaneously in accordance with the optical detection of the laser beam. This allows a distinct force and a corresponding shear angle to be related to the respective shear deformation for each time step (Figure 7).

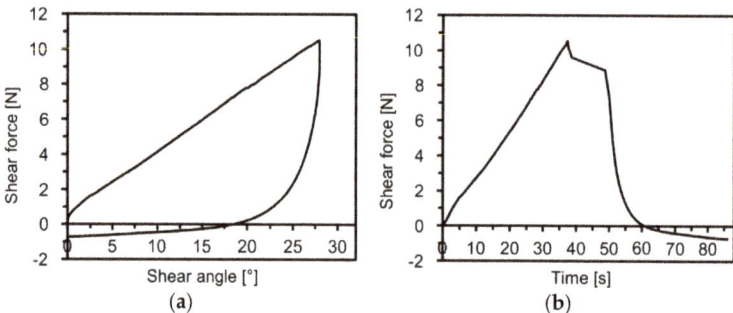

Figure 7. Shear force/shear angle diagram (**a**). Shear force/time diagram (**b**) for a CF-biaxial non-crimp fabric.

Device Calibration

It is necessary to calibrate the data recorded by the camera for accuracy. Data evaluation is based on the theory of 3D scanning (triangulation).

The laser beam projects a planar curve onto the material to be measured. Taking a picture of the projected curve on the material surface is a plane-to-plane perspective transformation as a bijection. A perspective transformation by homogenous coordinates is a linear transformation [40] projecting a quadrangle to a quadrangle. The transformation matrix has eight independent coordinates ($p_0, p_1, \ldots, p_7$), as shown in Equation (4):

$$\underline{\underline{P}} = \begin{bmatrix} p_0 & p_1 & p_2 \\ p_3 & p_4 & p_5 \\ p_6 & p_7 & 1 \end{bmatrix} \quad (4)$$

Corners of rectangular calibration equipment are appropriate to define the matrix coordinates (Figure 8).

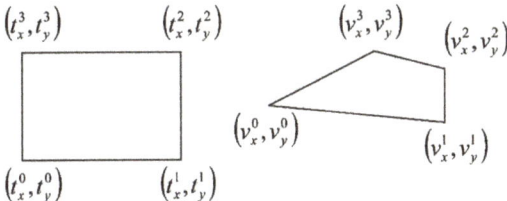

Figure 8. Planar perspective projection.

The corners of the calibration equipment are $\left(t_x^i, t_y^i\right)$, and the corners of its picture are $\left(v_x^i, v_y^i\right)$ (i = 0, 1, 2, 3); based on this, the transformation is shown in Equation (5):

$$\begin{bmatrix} v_x^i \\ v_y^i \\ 1 \end{bmatrix} = \begin{bmatrix} p_0 & p_1 & p_2 \\ p_3 & p_4 & p_5 \\ p_6 & p_7 & 1 \end{bmatrix} \cdot \begin{bmatrix} t_x^i \\ t_y^i \\ 1 \end{bmatrix} \quad (5)$$

There are eight unknown coordinates and eight equations, Equation (6):

$$\begin{bmatrix} v_x^i \\ v_y^i \\ 1 \end{bmatrix} = \begin{bmatrix} p_0 & p_1 & p_2 \\ p_3 & p_4 & p_5 \\ p_6 & p_7 & 1 \end{bmatrix} \cdot \begin{bmatrix} t_x^i \\ t_y^i \\ 1 \end{bmatrix} \quad i = 0, 1, 2, 3 \quad (6)$$

Knowing the real positions of four points in the lighted plane (t^i) allows the real coordinates of other points on the lighted plane to be computed.

The system is calibrated with a rectangular device with an LED in each corner (Figure 9a). During the calibration process, the rectangle is positioned in the laser lighted plane. This means that the device is aligned in the picture frame in such a way that it is positioned centrally and planarly in all planes, sagittal, transverse and frontal, with the frontal plane corresponding to the position of the shear specimen in the initial state. Figure 9b shows an image of the calibration tool. With the measured picture coordinates and the real coordinates of the LEDs, the eight coordinates of the homogenous transformation matrix can be calculated.

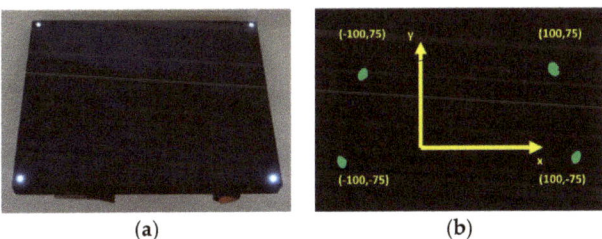

Figure 9. The calibration device (a) and its image during calibration (b) with its real dimensional coordinates.

In order to validate the hardware and software for recording and evaluating the shear deformation that occurs during the picture frame test, a reference geometry with a defined height profile (Figure 10a) was designed and manufactured. The aim of validation is to ensure that the height profile is measured correctly. Since shear deformation is only measured centrally, based on the deflection of the projected laser beam, the dimensions of the reference geometry do not need to correspond to the sample geometry, only allow the projection of the laser beam. For this purpose, the reference geometry was positioned on the lateral pivot point of the square picture frame and the laser beam is projected. The optical measurement of this height profile and its comparison to nominal dimensions is made stationary state and requires all measuring system components to be properly aligned and a validated calibration.

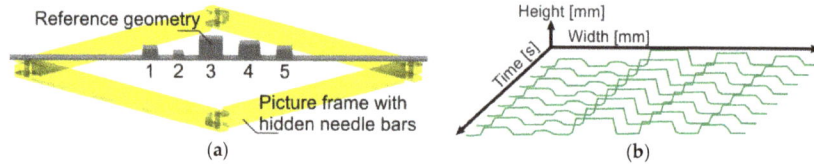

Figure 10. Reference geometry with defined peaks (1 to 5) adapted to the picture frame (**a**) and its measured height profile (**b**).

For experimental investigations, different reinforcing textiles which are widely used, e.g., in automotive and wind power plant engineering, were selected, Table 1. They vary in terms of fabric construction and/or fiber material, e.g., CF, glass fiber (GF), polyamide (PA) or polyester (PES), thus representing the group of commonly investigated materials.

Table 1. Material specifications.

Identification	Appearance	Mass Per Unit Area	Warp Yarn	Weft Yarn	Sewing Thread		
					Material/Orientation	Linear Density	Stitch Length
		(g/m^2)	(tex)	(tex)	(°)	(dtex)	(mm)
CF biaxial non-crimp fabric ±45°		300	1600	1600	PES/0	36	2.9
CF biaxial non-crimp fabric 0°/90°		349	800	400	PES/0	180	2.5
CF monoaxial non-crimp fabric 0°		330	1600	-	PES/0 PES/90	820 340	2.5
CF woven fabric, twill 2/2		400	800	800	-	-	-
GF/PA woven fabric, plain		625	1800	1800	-	-	-

[1] For non-crimp fabrics the front and back and for woven fabrics, only the fronts are shown

3. Results

To determine the accuracy of the extended picture frame test stand, the reference geometry was measured after calibrating. The deviations between measured and nominal height were mostly less than 5%, Table 2. Only at peak 4 (nominal height = 15 mm) the deviation was marginally higher than 5%. This could be caused by the averaging over relevant data points of this plateau of peak. Defining those relevant data points is quite challenging, because from the gained data, it is not clear whether a data point belongs to the plateau of the peak or to the very beginning of the inclined flank. In the visualization of data set points the inclined flanks can be seen (Figure 10b). Minor deviations in horizontal direction result from the limited number of data set points, which are only linearly connected for visualization. This effect is less relevant for plain and sheared textiles compared to the smooth surfaces without any hard transitions.

Table 2. Evaluation of reference geometry within the extended picture frame test stand.

Criteria	Peak 1	Peak 2	Peak 3	Peak 4	Peak 5
Nominal value (mm)	10.00	5.00	20.00	15.00	10.00
Measured value (mm)	9.59	4.83	19.72	14.21	10.23
Absolute Deviation (%)	4.12	3.44	1.42	5.28	2.35

The selected materials were investigated (Figure 11), the force/displacement diagram was obtained and the respective deflection of the laser beam was measured optically over time. Due to its fabric construction, the ± 45° CF biaxial non-crimp fabric was tested in two directions to investigate the influence of the sewing thread (in test direction/90° relative to test direction) on shear behavior. In order to avoid errors in the optical detection due to reflections of the tested reinforcing textiles (glass/carbon), the samples were treated with a digitizing spray in the area of the laser beam. This had no effect on the shear result. Subsequently, the shear force, the shear angle (Figure 12), and the geometrical data of shear deformation in the area of the laser beam (Figure 13) were calculated.

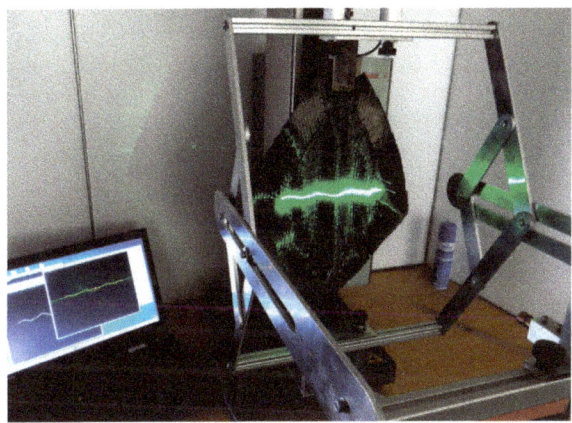

Figure 11. Extended picture frame test stand with CF biaxial non-crimp ± 45°, ST 0° at maximum deflection.

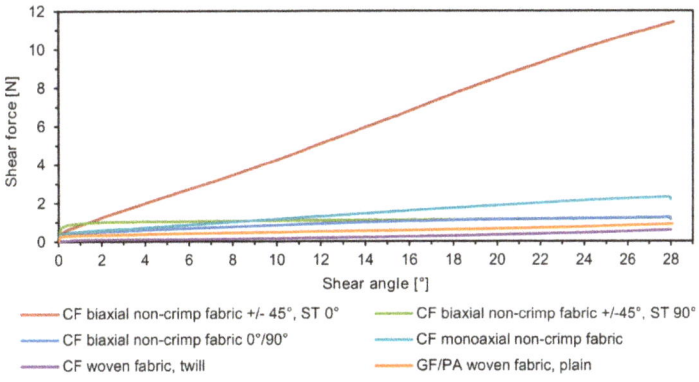

Figure 12. Shear force/shear angle diagram of investigated materials (average curve from nine samples for each material).

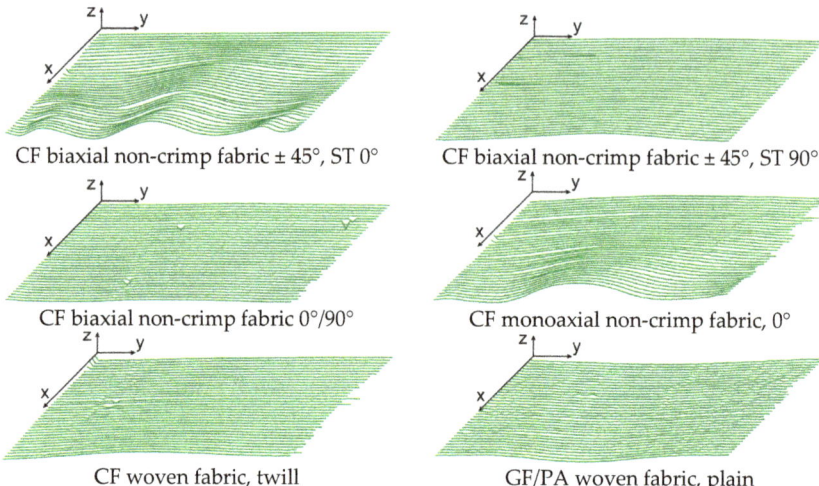

Figure 13. Measured height profiles of the laser beam as a function of time (one sample per material is shown); x-axis: time; y-axis: width; z-axis: height (also see Figure 6).

To evaluate the reproducibility of the results, nine samples were tested for each material and material direction. Figures 14 and 15 show the shear force/shear angle curves as well as the recorded surface deformations for the ± 45° biaxial non-crimp fabric with the sewing thread in the tensile direction. The deviations of about 20% in the shear force/shear angle curve result from the properties of the semi-finished product caused by fabric construction. For non-crimp fabric the yarn systems are not crossed but held by a sewing thread to ensure good drapeability. The missing crossing lead to a nearly perfectly stretched out situation of the reinforcing fiber. Thus, high mechanical characteristics in the reinforced directions (e.g., mono-, bi- tri- or quadriaxial) can be realized in FRP products. But its low resistance to displacement, restricts its handleability, making it almost impossible to achieve exactly reproducible sample preparation in the shear test.

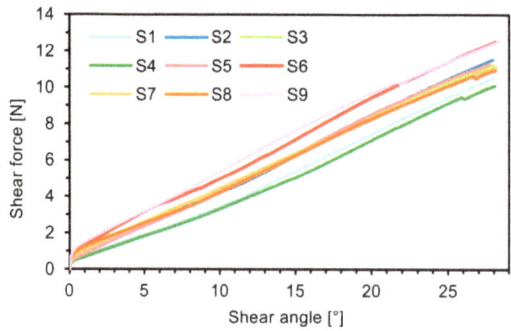

Figure 14. Shear force/shear angle diagram—CF biaxial non-crimp fabric ± 45°, ST 0°.

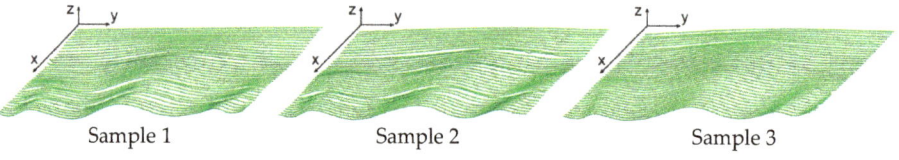

Figure 15. Cont.

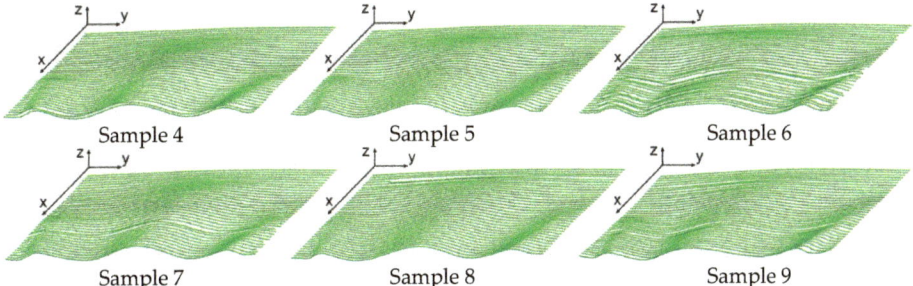

Figure 15. Measured height profiles of the laser beam as a function of time—CF biaxial non-crimp fabric ± 45°, ST 0°; x-axis: time; y-axis: width; z-axis: height (also see Figure 6)

The recorded surface deformations show generally similar results (Figure 15). Minor differences in the characteristics are the result of the above-mentioned issues in sample preparation. Moreover, it is obvious that similar shear force/shear angle curves do not lead to similar deformation behavior. To determine the exact beginning of fold formation and the geometric fold characteristics under shear stress, further analysis of the measured surface deformation is needed.

3.1. Additional Options for the Analysis of Optically Measured Shear Deformation

The geometry data of the time-dependent shear deformation can be imported into a software environment for statistical calculations and graphical output. The software used in this application case was "R" [41]. As described in Section 3.1, all data of a test cycle are leveled based on a first time step (initial state of the shear test), in order to eliminate the intrinsic error of the measurement system. The cleansed data can be used in a further process to visualize the deflection of the laser beam over time as a 3D area. Figure 16 shows the characteristics of the folds including their minima and maxima over a constant sample width. A color gradation allows the assignment of numerical values to different fold heights. Moreover, the amplitude maxima and minima of each time step can be illustrated both in a separate diagram (Figure 17a) and as a sum of their amount (i.e., as peak-to-peak value, Figure 17b) as a function of time and the shear angle.

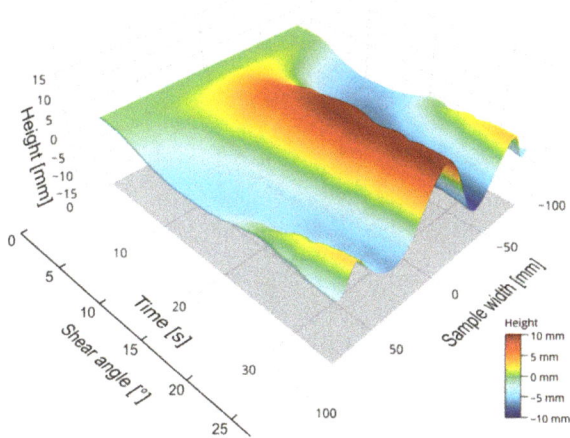

Figure 16. Deflection of the laser beam over time of CF biaxial non-crimp fabric ± 45°, ST 0°.

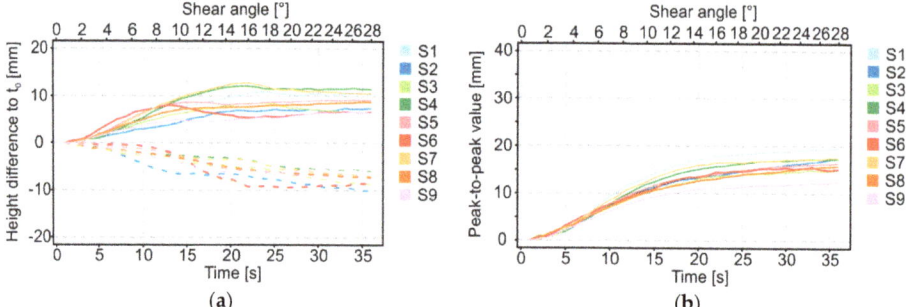

Figure 17. Amplitudes of maxima and minima (**a**). Peak-to-peak value (**b**). CF biaxial non-crimp fabric ± 45°, ST 0°, nine samples.

The amplitudes of maxima and minima revealed qualitatively different curve progressions. As an example, a maximum is initially formed in the case of sample 6, which decreased after about 12 s, as the minimum increased from this point forward. In contrast, the amplitude progression of sample 4 shows a continuous maximum, whereas the minimum remained relatively low. The analysis of the amplitude curve explains the differences of samples of a material variant in shear behavior (cf. Figure 15). The evaluation of the peak-to-peak value, however, clearly indicates qualitatively comparable curve progressions and thus the reproducibility of the developed shear deformation. The deviations that occur (Figure 17b) result from inaccuracies in sample preparation as described above and cannot be avoided even if the test is performed with great accuracy. Figure 18 shows the averaged results of the peak-to-peak values and the standard deviations over time for each material and material direction. Besides the absolute measured heights, which strongly differ, a closer view on reproducibility is possible. It is obvious that, for the CF-based woven and biaxial non-crimp fabrics, the standard deviations at the beginning of picture frame testing were lower than at higher deflections. Furthermore, the increase in standard deviation seems to be correlated with the beginning of folding. For the CF-based monoaxial non-crimp fabric and the GF/PA-based woven fabric, the standard deviations were nearly constant over time, but significantly higher than for the other materials. Especially for the GF/PA-based woven fabric, reflections occurred during testing, influencing the base data set and leading to the obtained high standard deviations. Furthermore, the observed results for CF monoaxial non-crimp fabric need to be considered critically as earlier investigations [26] revealed that shearing is not an intrinsic behavior of monoaxial non-crimp fabrics. Additionally, sample preparation is much more challenging, given by the loose construction. Overall, the high standard deviations are caused by material-specific behavior, not by the accuracy of the extended picture frame test. Hence, the additional optical evaluation of fold height is more reliable than an objective evaluation of folding.

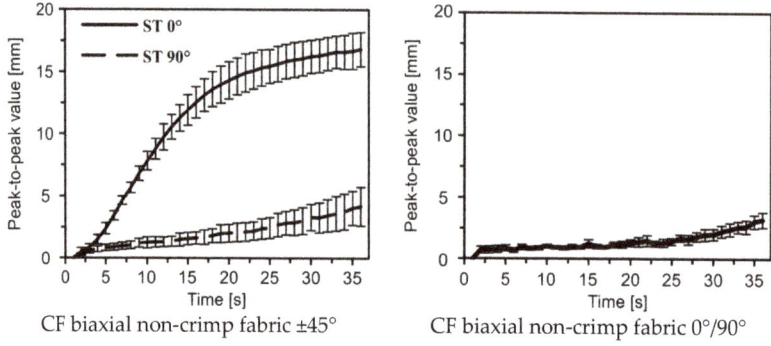

Figure 18. *Cont.*

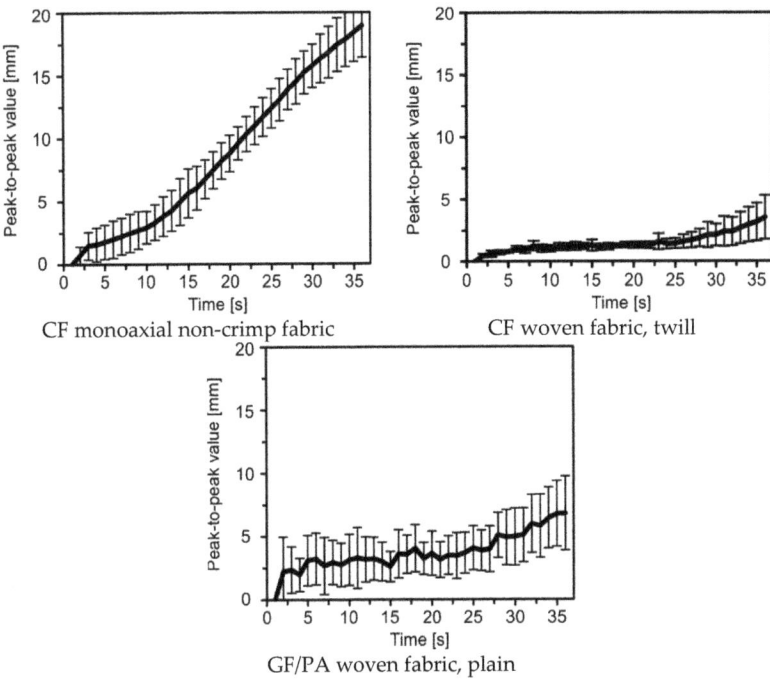

Figure 18. Averaged peak-to-peak value with standard deviations of each material over time.

This provides the user with a tool for automatically detecting the exact point of time of fold formation in a reproducible manner and for relating it to a critical shear angle. Depending on the geometry of the construction part to be produced and on the specific requirements, a limit value can be defined for a critical fold height. If this limit value is exceeded, the critical shear angle can be assigned, based on the time step. Consequently, the shear angle serves as a reliable decision criterion in kinematic draping simulations.

4. Conclusions

One of the most important requirements for the application of fiber-reinforced composites in large-scale productions is the reproducible production of FRP components with a constant quality. Reducing the time and resources necessary for the development of prototypes and to enhance virtual component development, appropriate methods for material tests and forming simulations are required. These methods, in turn, demand an application-oriented and, even more importantly, conclusive test technology to determine the characteristic material properties.

For kinematic draping simulations, a critical shear angle needs to be determined. Practical experiments have revealed that simulation parameters cannot be clearly defined for all material constructions based on the shear force/shear angle curve of the picture frame test. Instead, additional information is necessary to comprehensively characterize shear behavior.

The hardware and software developed for the recording and graphical evaluation of shear deformations by means of a laser beam enables fold formations to be automatically detected and quantified, and to be related to shear force/shear angle curves. Hence, a tool offering a wide variety of additional information for analyzing the shear behavior of reinforcing textiles is available. In the future, this tool may help users in practice to decide on suitable materials and to improve transparency in the design and construction process.

The extended test setup can be integrated into the currently established picture frame test without significant efforts. Moreover, no additional sample preparation is required, which results in considerable time savings as compared to currently available 3D deformation measuring systems. Therefore, this method is a cost-efficient solution tailored to small and medium-sized companies. Moreover, the accuracy of the extended picture frame test stand has been demonstrated for several high performance textile fabrics. Additionally, further knowledge was gained regarding the influence of fabric construction, e.g., the linear density of sewing thread, on deformation behavior under shear stress. This is a very interesting inspiration for more detailed investigations in future, because this affects not only the users of such materials, but also the producers. Another interesting application is the possibility to evaluate FE-based simulation by comparison of the attained surface deformations.

Author Contributions: Conceptualization, E.W., S.R. and S.K.; methodology, E.W., M.H., S.R. and P.B.; software, P.T., A.B. and S.R.; validation, E.W. and S.R.; formal analysis, P.T. and S.R.; investigation, P.B., E.W. and S.R.; data curation, S.R. and P.T.; writing—original draft preparation, E.W., S.R. and P.T.; writing—review and editing, S.K., M.H., P.T, A.B., E.W. and S.R.; visualization, S.R., P.T. and E.W.; supervision, S.K. and M.H.; project administration, S.K. and M.H.; funding acquisition, S.K. and M.H.

Funding: This research was funded by Bundesministerium für Bildung und Forschung (BMBF, Germany, project executing organization DLR, grant number 01DS14023) and the National Research, Development and Innovation Office (NKFIH, Hungary, grant number TÉT-12-DE-1-2013-0006, K100949 and PD116122). The APC was funded by the SLUB/TU, Dresden.

Conflicts of Interest: The authors declare no conflict of interest. The funders had no role in the design of the study; in the collection, analyses, or interpretation of data; in the writing of the manuscript, or in the decision to publish the results.

References

1. Cherif, C. (Ed.) *Textile Materials for Lightweight Constructions: Technologies–Methods–Materials–Properties*, 1st ed.; Springer: Berlin, Germany, 2016.
2. Kenkare, N.; May-Plumlee, T. Evaluation of drape characteristics in fabrics. *Int. J. Cloth. Sci. Technol.* **2005**, *17*, 109–123. [CrossRef]
3. Sauer, M.; Kühnel, M. Composite Market Report 2017. Carbon Composite e.V., Ed. Available online: https://www.carbon-composites.eu/media/2996/ccev-avk-marktbericht-2017.pdf (accessed on 28 February 2019).
4. Lemanski, S.L.; Wang, J.; Sutcliffe, M.P.F.; Potter, K.D.; Wisnom, M.R. Modelling failure of composite specimens with defects under compression loading. *Compos. Part A Appl. Sci. Manuf.* **2013**, *48*, 26–36. [CrossRef]
5. Mukhopadhyay, S.; Jones, M.I.; Hallett, S.R. Compressive failure of laminates containing an embedded wrinkle; experimental and numerical study. *Compos. Part A Appl. Sci. Manuf.* **2015**, *73*, 132–142. [CrossRef]
6. Yu, W.R.; Harrison, P.; Long, A. Finite element forming simulation for non-crimp fabrics using a non-orthogonal constitutive equation (vol 36, pg 1079, 2005). *Compos. Part A Appl. Sci. Manuf.* **2005**, *36*, 1600. [CrossRef]
7. Hamila, N.; Boisse, P.; Sabourin, F.; Brunet, M. A semi-discrete shell finite element for textile composite reinforcement forming simulation. *Int. J. Numer. Methods Eng.* **2009**, *79*, 1443–1466. [CrossRef]
8. Haanappel, S.P.; Ten Thije, R.H.; Sachs, U.; Rietman, B.; Akkerman, R. Formability analyses of uni-directional and textile reinforced thermoplastics. *Compos. Part A Appl. Sci. Manuf.* **2014**, *56*, 80–92. [CrossRef]
9. Smith, J.R.; Vaidya, U.K.; Johnstone, J.K. Analytical modeling of deformed plain woven thermoplastic composites. *Int. J. Mater. Form.* **2014**, *7*, 379–393. [CrossRef]
10. Huebner, M.; Rocher, J.-E.; Allaoui, S.; Hivet, G.; Gereke, T.; Cherif, C. Simulation-based investigations on the drape behavior of 3D woven fabrics made of commingled yarns. *Int. J. Mater. Form.* **2016**, *9*, 591–599. [CrossRef]
11. Krzywinski, S. *Verbindung von Design und Konstruktion in der textilen Konfektion unter Anwendung von CAE*; TUDpress Verlag der Wissenschaften GmbH, Habilitation; Technische Universität Dresden: Dresden, Germany, 2005.

12. Cao, J.; Akkerman, R.; Boisse, P.; Chen, J.; Cheng, H.S.; de Graaf, E.F.; Gorczyca, J.L.; Harrison, P.; Hivet, G.; Launay, J.; et al. Characterization of mechanical behavior of woven fabrics: Experimental methods and benchmark results. *Compos. Part A Appl. Sci. Manuf.* **2008**, *39*, 1037–1053. [CrossRef]
13. Mohammed, U.; Lekakou, C.; Dong, L.; Bader, M.G. Shear deformation and micromechanics of woven fabrics. *Compos. Part A Appl. Sci. Manuf.* **2000**, *31*, 299–308. [CrossRef]
14. Zhu, B.; Yu, T.X.; Tao, X.M. An experimental study of in-plane large shear deformation of woven fabric composite. *Compos. Sci. Technol.* **2007**, *67*, 252–261. [CrossRef]
15. Boisse, P.; Hamila, N.; Guzman-Maldonado, E.; Madeo, A.; Hivet, G.; dell'Isola, F. The bias-extension test for the analysis of in-plane shear properties of textile composite reinforcements and prepregs: A review. *Int. J. Mater. Form.* **2017**, *10*, 473–492. [CrossRef]
16. Mack, C.; Taylor, H.M. 39—The fitting of woven cloth to surfaces. *J. Text. Inst. Trans.* **1956**, *47*, T477–T488. [CrossRef]
17. Van Der Weeën, F. Algorithms for draping fabrics on doubly-curved surfaces. *Int. J. Numer. Methods Eng.* **1991**, *31*, 1415–1426. [CrossRef]
18. Cherouat, A.; Borouchaki, H.; Billoeet, J.-L. Geometrical and mechanical draping of composite fabric. *Eur. J. Comput. Mech.* **2005**, *14*, 693–707. [CrossRef]
19. Orawattanasrikul, S. *Experimentelle Analyse der Scherdeformation biaxial Verstärkter Mehrlagengestricke*; Dissertation: Dresden, Germany, 2006.
20. Hivet, G.; Duong, A.V. A contribution to the analysis of the intrinsic shear behavior of fabrics. *J. Compos. Mater.* **2011**, *45*, 695–716. [CrossRef]
21. Harrison, P.; Clifford, M.J.; Long, A.C. Shear characterisation of viscous woven textile composites: A comparison between picture frame and bias extension experiments. *Compos. Sci. Technol.* **2004**, *64*, 1453–1465. [CrossRef]
22. Launay, J.; Hivet, G.; Duong, A.V.; Boisse, P. Experimental analysis of the influence of tensions on in plane shear behaviour of woven composite reinforcements. *Compos. Sci. Technol.* **2008**, *68*, 506–515. [CrossRef]
23. Rashidi, A.; Milani, A.S. A multi-step biaxial bias extension test for wrinkling/de-wrinkling characterization of woven fabrics: Towards optimum forming design guidelines. *Mater. Des.* **2018**, *146*, 273–285. [CrossRef]
24. Zhu, B.; Yu, T.X.; Tao, X.M. Large deformation and slippage mechanism of plain woven composite in bias extension. *Compos. Part A Appl. Sci. Manuf.* **2007**, *38*, 1821–1828. [CrossRef]
25. Boisse, P.; Hamila, N.; Vidal-Salle, E.; Dumont, F. Simulation of wrinkling during textile composite reinforcement forming. Influence of tensile, in-plane shear and bending stiffnesses. *Compos. Sci. Technol.* **2011**, *71*, 683–692. [CrossRef]
26. Schirmaier, F.J.; Weidenmann, K.A.; Kaerger, L.; Henning, F. Characterisation of the draping behaviour of unidirectional non-crimp fabrics (UD-NCF). *Compos. Part A Appl. Sci. Manuf.* **2016**, *80*, 28–38. [CrossRef]
27. Bel, S.; Boisse, P.; Dumont, F. Analyses of the deformation mechanisms of non-crimp fabric composite reinforcements during preforming. *Appl. Compos. Mater.* **2012**, *19*, 513–528. [CrossRef]
28. Bel, S.; Hamila, N.; Boisse, P.; Dumont, F. Finite element model for NCF composite reinforcement preforming: Importance of inter-ply sliding. *Compos. Part A Appl. Sci. Manuf.* **2012**, *43*, 2269–2277. [CrossRef]
29. Lomov, S.V.; Barburski, M.; Stoilova, T.; Verpoest, I.; Akkerman, R.; Loendersloot, R.; Ten Thije, R.H. Carbon composites based on multiaxial multiply stitched preforms. Part 3: Biaxial tension, picture frame and compression tests of the preforms. *Compos. Part A Appl. Sci. Manuf.* **2005**, *36*, 1188–1206. [CrossRef]
30. Peng, X.Q.; Cao, J.; Chen, J.; Xue, P.; Lussier, D.S.; Liu, L. Experimental and numerical analysis on normalization of picture frame tests for composite materials. *Compos. Sci. Technol.* **2004**, *64*, 11–21. [CrossRef]
31. Lomov, S.V.; Boisse, P.; Deluycker, E.; Morestin, F.; Vanclooster, K.; Vandepitte, D.; Verpoest, I.; Willems, A. Full-field strain measurements in textile deformability studies. *Compos. Part A Appl. Sci. Manuf.* **2008**, *39*, 1232–1244. [CrossRef]
32. Harrison, P.; Abdiwi, F.; Guo, Z.; Potluri, P.; Yu, W.R. Characterising the shear-tension coupling and wrinkling behaviour of woven engineering fabrics. *Compos. Part A Appl. Sci. Manuf.* **2012**, *43*, 903–914. [CrossRef]
33. Dangora, L.M.; Hansen, C.J.; Mitchell, C.J.; Sherwood, J.A.; Parker, J.C. Challenges associated with shear characterization of a cross-ply thermoplastic lamina using picture frame tests. *Compos. Part A Appl. Sci. Manuf.* **2015**, *78*, 181–190. [CrossRef]

34. Lebrun, G.; Bureau, M.N.; Denault, J. Evaluation of bias-extension and picture-frame test methods for the measurement of intraply shear properties of PP/glass commingled fabrics. *Compos. Struct.* **2003**, *61*, 341–352. [CrossRef]
35. Li, L.; Zhao, Y.; Vuong, H.g.; Chen, Y.; Yang, J.; Duan, Y. In-plane shear investigation of biaxial carbon non-crimp fabrics with experimental tests and finite element modeling. *Mater. Des.* **2014**, *63*, 757–765. [CrossRef]
36. Hosseini, A.; Kashani, M.H.; Sassani, F.; Milani, A.S.; Ko, F.K. Identifying the distinct shear wrinkling behavior of woven composite preforms under bias extension and picture frame tests. *Compos. Struct.* **2018**, *185*, 764–773. [CrossRef]
37. Souter, B.J. *Effect of Fiber Architecture on Formability of Textile Preforms*; Dissertation: Nottingham, UK, 2001.
38. Zhu, B.; Yu, T.X.; Teng, J.; Tao, X.M. Theoretical Modeling of Large Shear Deformation and Wrinkling of Plain Woven Composite. *J. Compos. Mater.* **2009**, *43*, 125–138. [CrossRef]
39. Cherif, C. *Drapierbarkeitssimulation von Verstärkungstextilien für den Einsatz in Faserverbundwerkstoffen mit der Finite-Element-Methode*; Dissertation: Aachen, Germany, 1999.
40. Kim, D.K.; Jang, B.T.; Hwang, C.J. A Planar Perspective Image Matching using Point Correspondes and Rectangle-to-Quadrilateral Mapping. In Proceedings of the Fifth IEE Southwest Symposium on Image Analysis and Interpretation, Santa Fe, NM, USA, 7–9 April 2002; pp. 1532–1537. [CrossRef]
41. R Core Team. *R: A Language and Environment for Statistical Computing*; R Foundation for Statistical Computing: Vienna, Austria, 2016. Available online: URLhttps://www.R-project.org/ (accessed on 28 February 2019).

© 2019 by the authors. Licensee MDPI, Basel, Switzerland. This article is an open access article distributed under the terms and conditions of the Creative Commons Attribution (CC BY) license (http://creativecommons.org/licenses/by/4.0/).

Article

Piezo-Sensitive Fabrics from Carbon Black Containing Conductive Cellulose Fibres for Flexible Pressure Sensors

Julia Ullrich [1], Martin Eisenreich [1], Yvonne Zimmermann [1], Dominik Mayer [2], Nina Koehne [2], Jacqueline F. Tschannett [3,†], Amalid Mahmud-Ali [3,†] and Thomas Bechtold [3,*,†]

- [1] Textilforschungsinstitut Thüringen-Vogtland e.V., Zeulenrodaer Straße 42, D-07973 Greiz, Germany; j.ullrich@titv-greiz.de (J.U.); m.eisenreich@titv-greiz.de (M.E.); y.zimmermann@titv-greiz.de (Y.Z.)
- [2] Kelheim Fibres GmbH, Regensburger Straße 109, D-93309 Kelheim, Germany; Dominik.Mayer@kelheim-fibres.com (D.M.); Nina.Koehne@kelheim-fibres.com (N.K.)
- [3] Research Institute of Textile Chemistry and Textile Physics, Leopold-Franzens-University of Innsbruck, Hoechsterstraße 73, A-6850 Dornbirn, Austria; Jacqueline.Tschannett@uibk.ac.at (J.F.T.); Amalid.Mahmud-Ali@uibk.ac.at (A.M.-A.)
- * Correspondence: thomas.bechtold@uibk.ac.at; Tel.: +43-(0)-5572-28753
- † Research Institute of Textile Chemistry and Textile Physics: Member of EPNOE–European Polysaccharide Network of Excellence, www.epnoe.eu.

Received: 18 October 2020; Accepted: 13 November 2020; Published: 16 November 2020

Abstract: The design of flexible sensors which can be incorporated in textile structures is of decisive importance for the future development of wearables. In addition to their technical functionality, the materials chosen to construct the sensor should be nontoxic, affordable, and compatible with future recycling. Conductive fibres were produced by incorporation of carbon black into regenerated cellulose fibres. By incorporation of 23 wt.% and 27 wt.% carbon black, the surface resistance of the fibres reduced from 1.3×10^{10} Ω·cm for standard viscose fibres to 2.7×10^3 and 475 Ω·cm, respectively. Fibre tenacity reduced to 30–50% of a standard viscose; however, it was sufficient to allow processing of the material in standard textile operations. A fibre blend of the conductive viscose fibres with polyester fibres was used to produce a needle-punched nonwoven material with piezo-electric properties, which was used as a pressure sensor in the very low pressure range of 400–1000 Pa. The durability of the sensor was demonstrated in repetitive load/relaxation cycles. As a regenerated cellulose fibre, the carbon-black-incorporated cellulose fibre is compatible with standard textile processing operations and, thus, will be of high interest as a functional element in future wearables.

Keywords: conductive fibres; cellulose fibres; pressure sensor; smart textiles; viscose fibres; carbon black

1. Introduction

The Scientific and Technology Options Assessment Panel of the European Parliament (STOA) identified wearables as one of the 10 technologies which will change our lives with very promising market prospects for wearables, forecasted to increase to USD 150 billion by 2026 [1]. The introduction of sensors and electronic devices into textile products allows integration of additional functionalities and opens access to the market of intelligent products for new applications [2].

The integration of electrical devices into textiles requires development of flexible conductive structures and the availability of sensor systems to translate external physical stimuli into electrical signals [3,4]. The intended application of a product determines selection of a certain sensor principle;

thus, dependent on the technical requirements, a wide number of different sensor concepts have been developed for a given physical parameter [5].

A number of principles have been reported in the literature for measurement of pressure, i.e., force per area, e.g., via measurement of electrical capacity [6], resistivity [7,8], or optical effects [9]. Changes in inductance and the corresponding resonant frequency of circuits were also proposed as a principle to operate a pressure sensor [10].

The combination of conductive films and deformable insulating layers permits construction of capacitors with pressure-sensitive capacity [11–13]. Similarly, two conductive lines, woven into a fabric or wrapped in ply structure and separated by an elastic material, have been proposed as thread-like capacitive pressure sensors [14–16].

In another approach, the piezo-resistive properties of compressible structures were used to build flexible textile-based pressure sensors [17,18]. Space-resolved pressure-sensitive layers have been assembled via a combination of piezo-resistive nonwoven material and conductive elements with the aim to develop pressure sensors in shoes [19]. Other approaches integrated elastic piezo-sensitive layers in sensor pads [20], shoes [21], seats [22,23], sensor mats [24], and carpets [25–27].

In many cases, the sensor element in such devices is built from a compressible conductive material which then changes its electrical resistance upon compression through external forces [28]. The conductive structure is often a web fabric containing conductive fibres [29]. Furthermore, carbon-nanotube-coated three-dimensional (3D) spacer textiles or conductive silicones have been proposed as pressure-sensitive flexible structures [30–32].

Higher electrical conductivity is required for the electrical connections between sensor and data processor. Thus, metal-based coatings of threads, braided wires, and wrapped yarns with the use of thin metal films are used [3]. These lines should exhibit low electrical resistance; thus, contribution of the electrical connection to the total resistance of a device remains in the dimension of a few ohms.

Different types of conductive fibre-based structures can be applied to build a pressure-sensitive structure, e.g., use of conductive polymers [33], fibre coatings with carbon-based layers [34], or integration of conductive material into fibres to obtain intrinsically conductive fibres. Moreover, formation of carbon foam through carbonisation of melamine foam has been reported as a route to prepare flexible conductive structures [35].

Conductive cellulose fibres can be obtained using a number of techniques, e.g., electroless deposition of metal layers on the fibre surface or incorporation of a conductive material, e.g., graphite, carbon black (CB), or carbon nanotubes, into the fibre matrix during fibre production [36,37]. In the viscose process, cellulose at first is steeped in concentrated NaOH solution. The formed alkali cellulose is then reacted with carbon disulphide to form the alkali-soluble cellulose xanthogenate. The aqueous alkaline solution of the cellulose xanthogenate is then spun into a coagulation bath containing a mixture of Na_2SO_4 and H_2SO_4. In the acidic coagulation bath, at first, coagulation of the xanthogenate occurs, then hydrolysis of the xanthogenate into cellulose and carbon disulphide takes place. The chemical inertness of CB makes this material favourable for incorporation during the viscose fibre formation, as the rather harsh chemical conditions applied during the viscose fibre process cause surface corrosion of metal particles such as Ag and Cu.

In addition to chemical inertness during the process of viscose fibre formation, the conductive additives must not disturb the process of fibre spinning through agglomeration and formation of larger particle structures, which then clog the bores of the spinneret [37].

CB is technically used as a pigment for printing and paint formulation. This nontoxic material is available in bulk amounts and at low costs. Thus, research on the incorporation of CB in viscose fibres for production of conductive cellulose is of high interest to elaborate the fundamentals of a scalable and commercially viable technical process. In this study, the formation of conductive viscose fibres through incorporation of CB was studied as a function of added CB. The fibres were characterised by scanning electron microscopy, conductivity measurement, and determination of fibre strength. The conductive viscose fibres were processed into fibre webs to obtain plane piezo-sensitive layers, which were characterised in static and cyclic load experiments for their functionality as pressure sensors.

2. Experimental

2.1. Preparation and Characterisation of Conductive Viscose Fibres

As a first step, an aqueous dispersion of 20 wt.% carbon black (CB, low structure, regular colour furnace (RCG), average particle size 27 nm; Printex 300, Orion Engineered Carbons, Luxemburg) was prepared with use of an anionic surfactant as dispersant (sodium lignosulphonate). The dispersion was then added to the standard spinning dope (10 wt.% cellulose) to obtain spinning dopes with 3.2, 10, and 30 wt.% CB. Laboratory spinning devices and a pilot-scale spinning unit (both Kelheim Fibres, Kelheim, Germany) were used for viscose fibre production. The viscose dope was filtered and spun to viscose fibres with fineness of 1.7 or 3.3 dtex.

The actual content of CB in the viscose fibres was determined by photometry (double-beam spectrophotometer, Perkin Elmer Lambda 25, Rodgau, Germany).

Laser scanning microscopy of CB-incorporated viscose fibres was undertaken with a laser scanning 3D microscope (VK-X100 series LSM 3D Profile Measurement, KEYENCE, Tokyo, Japan).

Scanning electron microscope photos (SEM) of the fibres were taken with use of a DSM 940A electron microscope (Zeiss, Oberkochen, Germany).

Fibre tenacity and elongation to break were measured using a single-fibre tensile strength tester on the basis of DIN EN ISO 5079 (Fafegraph HR in combination with Vibromat ME, Textechno, Mönchengladbach, Germany). The fineness of the fibres was determined using the vibrational method on the basis of DIN EN ISO 1973 (Vibromat ME, Textechno, Mönchengladbach, Germany).

To characterise fibres in an oriented arrangement, so-called rotor rings were prepared. A mass of 10 g fibres was processed into a parallelised fibre band (Device built by ITV-Denkendorf, Denkendorf, Germany). The surface resistance along the fibre band was measured with a multimeter (Figure S1, Supplementary Materials; Fluke 1587, Glottertal, Germany). To determine the volume resistivity, the fibre rotor ring was packed between two copper plates with 1 cm distance, and the electrical resistance was measured. The fibre conductivity was calculated as the reciprocal value of the measured electrical resistance.

Additionally, the specific surface resistance and the volume resistance of rotor rings were measured using a ring electrode according to DIN EN 1149-1:2006-09. The measurements were undertaken at three different climate conditions (20 °C, 65% relative humidity (RH), 23 °C, 50% RH, and 23 °C, 40% RH). The samples were conditioned in the respective climate for 24 h before measurement. Rotor rings were cut into pieces of 20 mm length, and a mass of 5 g was placed in the ring electrode. A ring electrode with 50.4 mm diameter and a mass of 460 g was used to compress the fibres (Textilelektrode TE 50, H.-P. Fischer Elektronik GmbH&Co, Industrie und Labortechnik KG, Mittenwalde, Germany). The surface and volume resistances were determined with a tera-ohm meter according to DIN EN 1149 (Figure S2, Supplementary Materials; Milli-Tera-Ohmmeter Milli-TO 3, H.-P. Fischer Elektronik GmbH&Co, Industrie und Labortechnik KG).

Fibre samples were processed into yarn using a lab spinning unit (Kelheim fibres, Germany). The determination of the yarn resistance for a length of 10 cm was performed using a textile stripe electrode and a tera-ohm meter (Figure S3, Supplementary Materials; Textilstreifenelektrode TSE 1, H.-P. Fischer Elektronik GmbH&Co, Industrie und Labortechnik KG). Readings were taken after 60 s of equilibration. Results are given as the mean value of five repetitions.

Nonwoven fabrics (mass per area 250 g/m^2) were prepared through needle punching to obtain a piezo-resistive structure.

2.2. Preparation and Characterisation of Piezo-Resistive Nonwovens

A nonwoven material made from 100% viscose fibres (No. 7, 3.3 dtex, 40 mm, 23 wt.% CB) was used to study the piezo-electric behaviour of the material. The plane samples then were placed between two conductive metal foils. The sandwich was placed on a balance, and compression of the web was achieved by stepwise increase of the weight placed on top of the assembly (area 56.8 cm^2). A photograph of the set-up is given in Figure S4 (Supplementary Materials). The resistance between the top foil and the bottom foil was measured by means of a laboratory multimeter. Three repetitive cycles of a stepwise increase in pressure, followed by stepwise relaxation of pressure, were performed to analyse the recovery after compression. Results are given as the mean value and standard deviation of three independent experiments.

In a next approach, the conductive viscose fibres were blended with synthetic fibres to improve elastic recovery of the nonwoven fabrics. Two different types of nonwoven fabric were prepared with use of a 3.3 dtex viscose fibre (No. 8, 27 wt.% CB content):

Material A contained 50 wt.% CB-incorporated viscose (3.3 dtex, fibre length 40 mm) and 50 wt.% polyester fibre (3.3 dtex, fibre length 60 mm).

Material B contained 65 wt.% CB-incorporated viscose (3.3 dtex, fibre length 40 mm), 30 wt.% polyester fibre (3.3 dtex, fibre length 60 mm), and 5 wt.% polyester bi-component fibre (2.2 dtex, fibre length 51 mm).

For the repetitive load/relaxation cycles, a modified tensile testing unit was used (Zwick Roell Z010). A sandwich structure (100 mm × 100 mm) of two copper plates with a conductive fibre web as the middle structure was mounted in the testing device (Figure S5, Supplementary Materials). A series of 50 load/relaxation cycles was performed, and the change in conductivity as a function of applied pressure was recorded. The cycling was performed within pressure limits of 500 Pa (5 N/100 cm^2) and 2300 Pa (230 N/100 cm^2).

3. Results and Discussion

3.1. Fibre Characterisation

The CB was added to the spinning dope in the form of a CB dispersion, which was stabilised by addition of an anionic or nonionic dispersant. The concentration of the conductive material and the state of dispersion of the CB in the fibre determined the final CB content in the fibre and the measured electrical conductivity. The percolation threshold, as well as the fibre conductivity, depended on the amount of conductive additive, the particle size, and distribution in the fibre. Through dispersion with a high shear rate and addition of a dispersing agent, the particle size of the dispersion could be kept below 10 µm, which is a prerequisite to avoid blocking of the spinneret. Immediately before fibre spinning, any larger aggregates of CB were removed by filtration. At the stage of fibre regeneration, weakly bound CB was washed out into the coagulation bath. Thus, the analytically determined amount of CB incorporated in the fibres was lower than the theoretical amount of CB calculated from the addition of CB to the spin dope.

The conductivity of the fibres increased with increasing content of CB; however, as a result of the presence of dispersed CB, a reduction in mechanical properties of the fibres was observed in parallel. The increase in fibre conductivity and the decrease in tenacity and elongation with increasing content of CB are shown in Table 1. A substantial reduction in tenacity was observed at a CB content above 10 wt.%. The threshold for an increase in conductivity was reached at 15 wt.% CB incorporation, which indicates the lower limit for percolation of the CB particles inside the fibre structure (Figure 1).

Table 1. Mechanical and electrical properties of viscose fibres as a function of carbon black (CB) content added to the spin dope (fibre length 40 mm).

No.	Fibre Fineness	CB Added	CB Incorporated	CB Loss	Tenacity	Elongation	Surface Resistance	Conductivity
	dtex	%	%	%	cN/tex	%	$\Omega \cdot cm$	S/m
1	3.3	0	0	0.00	20.5	24	1.3×10^{10}	7.7×10^{-9}
2	1.7	0	0	0.00	26	20	1.3×10^{10}	7.7×10^{-9}
3	1.7	3.2	3.2	−0.3	21.2	24.0	1.3×10^{10}	7.7×10^{-9}
4	1.7	10	7.3	26.9	12.7	17.4	1.3×10^{10}	7.7×10^{-9}
5	3.3	20	16.4	18.0	7.9	15.6	1.1×10^{9}	9.4×10^{-8}
6	3.3	25	19.8	20.6	7.1	14.5	11.4×10^{6}	8.8×10^{-6}
7	3.3	30	23.1	23.2	6.9	15.2	2.7×10^{3}	0.044
8	3.3	36.7	27.2	25.9	9.53	30.35	475	0.210

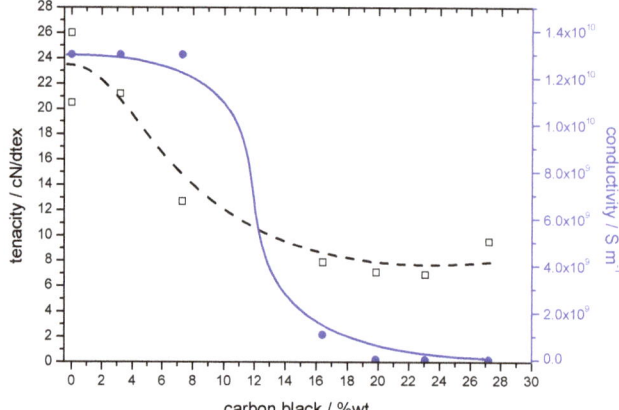

Figure 1. Tenacity and conductivity of viscose fibres as a function of the analytically determined CB content.

Fibres with different content in CB and a standard viscose fibre without addition of CB were analysed by laser scanning microscopy. Fibres with 23.1 wt.% CB content were then selected for scanning electron microscopy, to visualise the state of dispersion of the CB in the fibres using a higher resolution. Representative examples are shown in Figures 2 and 3.

When compared to the surface of a standard viscose fibre (Figure 2e) the incorporation of CB led to an increasingly rougher surface (Figure 2a–d). In particular, at high concentration of CB (Figures 2d and 3a), the presence of CB particles could be observed in the photomicrographs. Bigger agglomerates also led to the appearance of bulges at the fibre surface. In viscose fibre spinning, the cellulose fibre was regenerated from a diluted alkaline solution, which contained approximately 10 wt.% cellulose. During fibre regeneration, solid cellulose forms and a substantial shrinkage in cellulose structure occurred. As an estimate, the area of the cross-section was reduced to 10–15% of the initially extruded viscose solution, and the diameter was reduced to one-third. The dimensions of the CB agglomerates in the dope remained constant; thus, visible bulges began to appear at the fibre surface at higher CB content and larger agglomerates appeared at the fibre surface (Figure 2d,e). The diameter of these agglomerates was still substantially smaller than the diameter of the regenerated fibre, as larger agglomerates were already filtered off before the spinning dope passed the spinneret. The presence of a few larger agglomerates of CB can also be observed in the SEM photomicrographs taken with fibres containing 27.2 wt.% CB. Most of the CB, however, was present inside the fibre in highly dispersed form, which was the condition to achieve percolation and electrical conductivity.

The increase in conductivity with CB content and the moisture dependence of the volume resistivity were also studied in measurements of the volume resistance at rotor rings (Figure 4). Volume resistance measured with the rotor rings was used to characterise the conductivity of the fibres as bulk material with fibres packed in low oriented state. Measurements at different relative humidity demonstrated the influence of the ambient conditions on the electrical resistivity. Fibres with low conductivity exhibited a substantial reduction in resistance with increasing relative humidity, while fibres with 23.1 wt.% CB exhibited an increase in resistance at 65% RH.

The electrical resistance of yarn samples was measured to characterise the conductive behaviour of the fibres in a longitudinally oriented arrangement. The lower amount of conductive material in the cross-section of a yarn and the longer distance between the contact points led to high electrical resistance compared to the tests with rotor rings. Again, the resistance of the samples was reduced with increasing relative humidity due to the contribution of absorbed water to the overall conductivity (Figure 5).

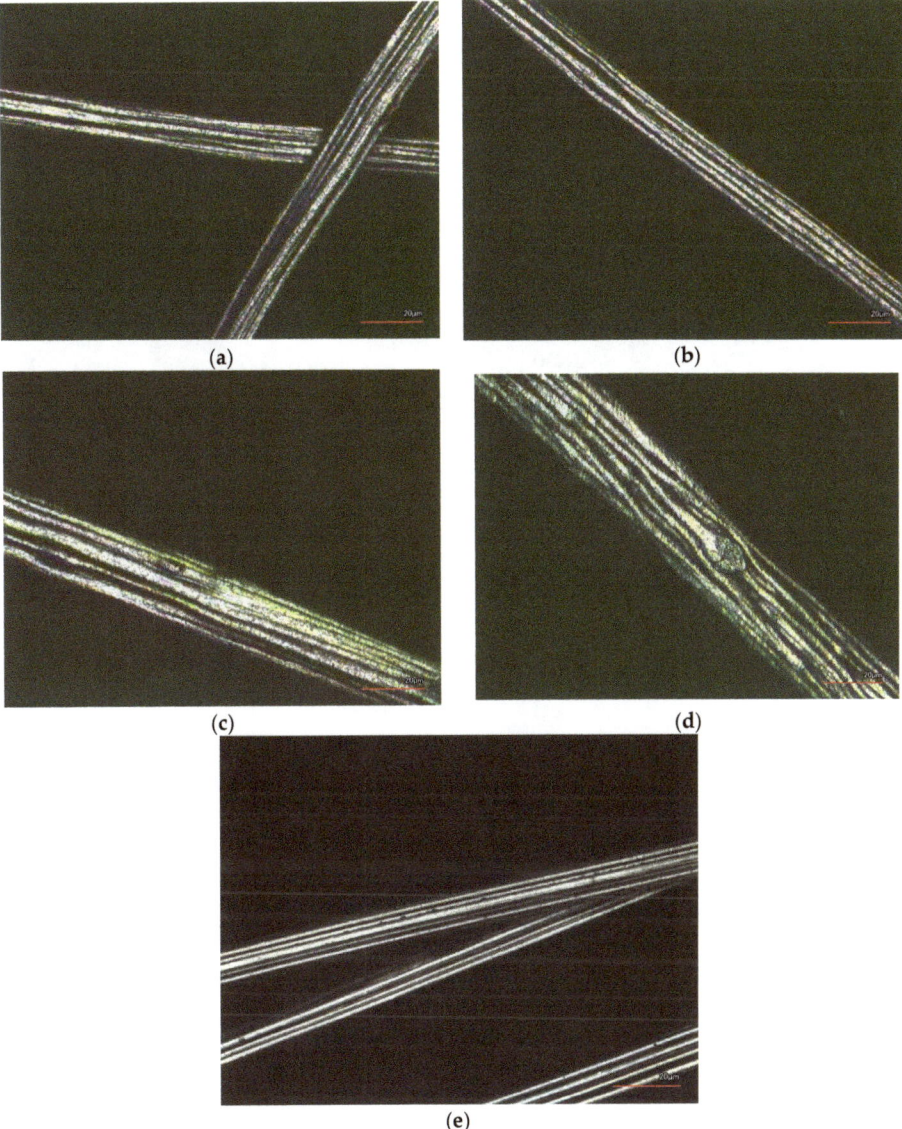

Figure 2. Laser scanning microscopy of CB incorporated viscose fibres: (**a**) sample (4) 1.7 dtex, 7.3 wt.% CB; (**b**) sample (5) 1.7 dtex, 16.4 wt.% CB; (**c**) sample (7) 3.3 dtex, 23.1 wt.% CB; (**d**) sample (8) 3.3 dtex, 27.2 wt.% CB; (**e**) standard viscose fibres 1.3 dtex.

All samples exhibited a distinct change in electrical conductivity with a change in relative humidity. The absorption of water into the cellulose structure and on the fibre surface created two effects:

- Water adsorbed in the cellulose structure and on the fibre surface contributed to the overall conductivity; thus, a reduction in resistivity was observed. The contribution to the conductivity was, however, low; thus, this effect was observed only in fibre assemblies which exhibited a relatively low conductivity. Thus, in the case of fibres assemblies with a volume resistance on the magnitude of gigaohms, the uptake of moisture contributed to the relatively low conductivity.

- The adsorption of water molecules also led to changes in fibre dimensions and to the formation of molecular layers of water on the fibre surface. These effects could lead to a reduction in conductivity, which was observed only in the case of fibre assemblies with volume conductivity in the dimension of several kiloohms. Here, the uptake of moisture reduced the level of percolation, e.g., through hygral fibre expansion, thus leading to a reduction in conductivity with increasing moisture content (Figure 5).

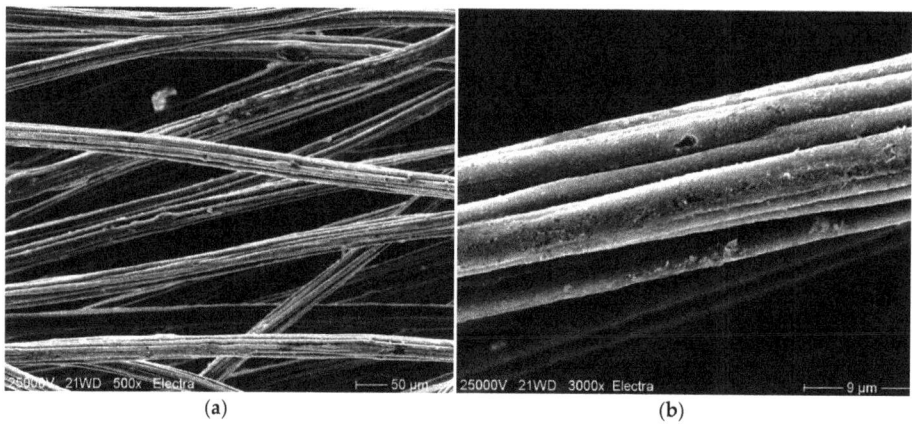

Figure 3. Scanning electron microscopy of CB-incorporated viscose fibres (8) (3.3 dtex, 27.2 wt.% CB); (a) magnification 500×, (b) magnification 3000×.

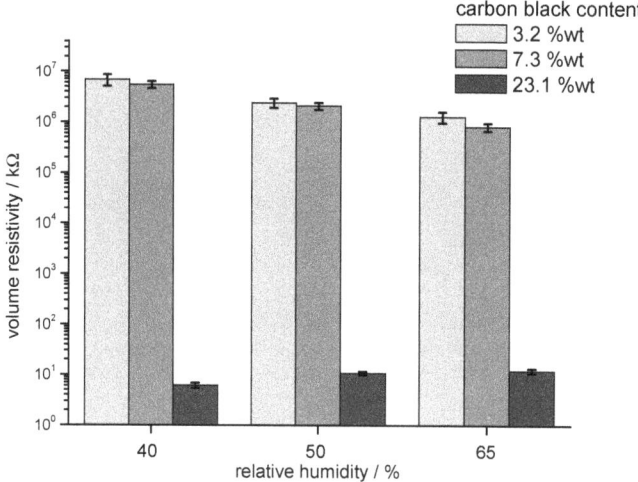

Figure 4. Volume resistivity of rotor rings measured at 40% RH, 50% RH, and 65% RH as a function of incorporated CB: sample (3) 3.2 wt.% CB, sample (4) 7.3 wt.% CB, and sample (7) 23.1 wt.% CB.

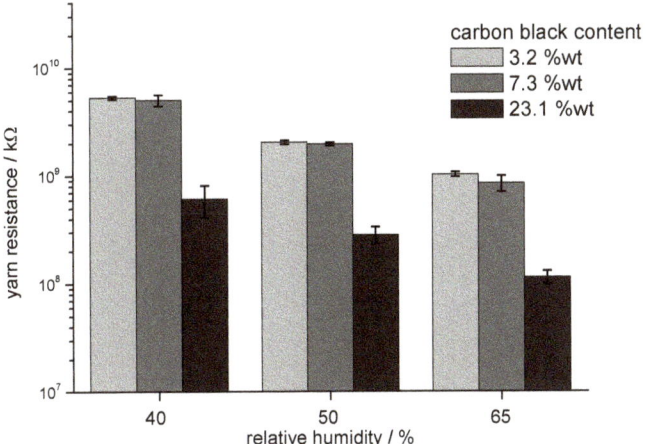

Figure 5. Yarn resistance (measured over a length of 10 cm) measured at 40% RH, 50% RH, and 65% RH as a function of incorporated CB: sample (3) 3.2 wt.% CB, sample (4) 7.3 wt.% CB, and sample (7) 23.1 wt.% CB.

3.2. Needle-Punched Nonwoven Material as a Pressure Sensor

A needle-punched nonwoven material consisting of 100% conductive viscose fibre (sample (7), 3.3 dtex, fibre length 40 mm, 23.1 wt.% CB) was used as a piezo-sensitive layer. A representative example for the compression/relaxation behaviour during three repetitive load/relaxation cycles is given in Figure 6.

The results in Figure 6 demonstrate the pressure sensitivity of the electrical resistance of a needle-punched fibre nonwoven material in the low-pressure region between 200 and 1000 Pa. At a load below 400 Pa, hysteresis between pressure increase and relaxation appeared. During the expansion of the nonwoven material, a higher number of contact points in the nonwoven region and a higher resistance were observed during the phase of relaxation. During the first cycle, fibres in the nonwoven realigned into a more stable structure; thus, the resistance measured during the following load/relaxation cycles stabilised. In the pressure range between 400 and 1000 Pa, a stable relationship and minimal hysteresis between applied pressure and electrical resistance of the nonwoven material were observed.

To improve the load/relaxation behaviour of the nonwoven material, the conductive viscose fibres were blended with more elastic polyester fibres and polyester bicomponent fibres. These fibres contributed to the recovery of the compressed fibre nonwoven material during the relaxation. Two types of samples were studied:

- Material A (50 wt.% CB-incorporated viscose, 3.3 dtex, fibre length 40 mm and 50 wt.% polyester fibre, 3.3 dtex, fibre length 60 mm)
- Material B (65 wt.% CB-incorporated viscose, 3.3 dtex, fibre length 40 mm, 30 wt.% polyester fibre, 3.3 dtex, fibre length 60 mm, and 5 wt.% polyester bicomponent fibre, 2.2 dtex, fibre length 51 mm).

The pressure sensitivity of the electrical resistance of the two different nonwoven materials was studied in three repetitive load/relaxation cycles. Results for Material A and Material B are given in Figure 7.

The repeatability and durability of the nonwoven materials under a high number of repetitive load/relaxation cycles were tested using a modified tensile testing unit.

Representative examples for the resistance change during load/relaxation cycles are shown in Figure 8.

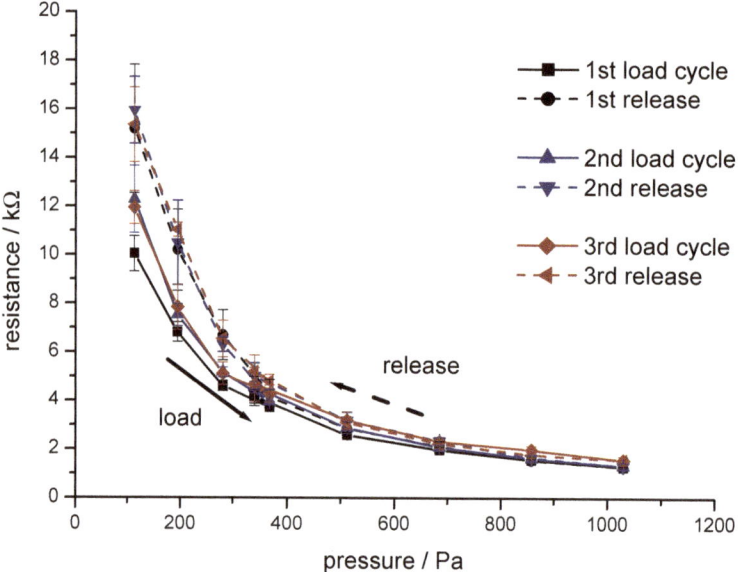

Figure 6. Electrical resistance of a fibre web made of 100% conductive viscose during three repetitive load relaxation cycles (sample (7), 3.3 dtex, fibre length 40 mm, 23.1 wt.% CB, mass per area 250 g/m², test sample area 56.8 cm²).

Due to the higher content in conductive fibres, a lower resistance was measured with sample B. The presence of crimped bicomponent fibres in sample B also supported the recovery of the compressed structure and, thus, contributed to a higher signal stability in the cyclic tests.

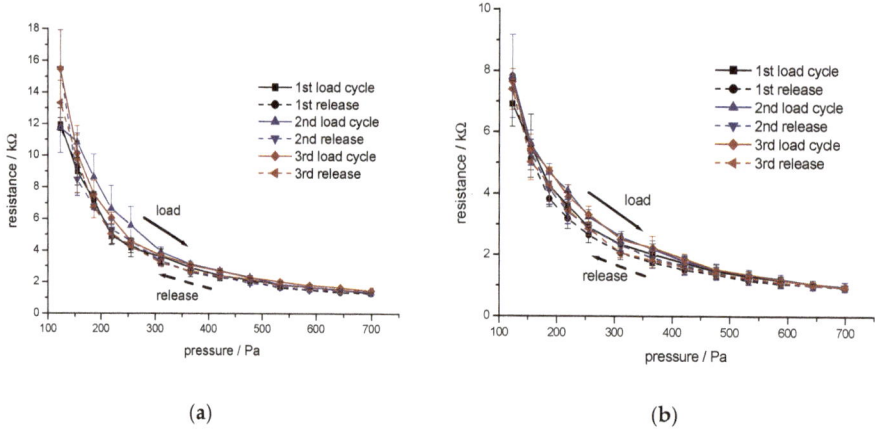

Figure 7. Electrical resistance of a fibre web made of 100% conductive viscose during three repetitive load/relaxation cycles: (**a**) Material A (50 wt.% CB-incorporated viscose, 3.3 dtex, fibre length 40 mm and 50 wt.% polyester fibre, 3.3 dtex, fibre length 60 mm); (**b**) Material B (65 wt.% CB-incorporated viscose, 3.3 dtex, fibre length 40 mm, 30 wt.% polyester fibre, 3.3 dtex, fibre length 60 mm, and 5 wt.% polyester bicomponent fibre, 2.2 dtex, fibre length 51 mm).

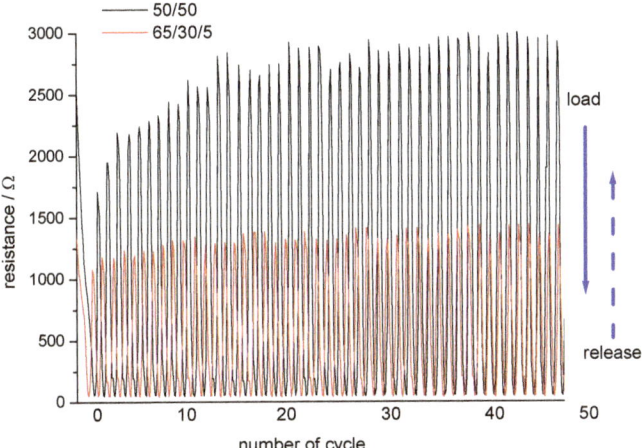

Figure 8. Repeatability and durability of the pressure sensing nonwovens. Resistance change of Material A (50 wt.% CB-incorporated viscose, 3.3 dtex, fibre length 40 mm and 50 wt.% polyester fibre, 3.3 dtex, fibre length 60 mm) shown in black and Material B (65 wt.% CB-incorporated viscose, 3.3 dtex, fibre length 40 mm, 30 wt.% polyester fibre, 3.3 dtex, fibre length 60 mm, and 5 wt.% polyester bicomponent fibre, 2.2 dtex, fibre length 51 mm) shown in red during 50 repetitive load/relaxation cycles.

With increasing load, the electrical resistance of samples A and B was reduced to 50 Ω and then increased during the relaxation to the maximum value of 1250 Ω for the Sample B, while, for sample A, higher resistance of 2500–3000 Ω was observed. The improved mechanical stability of sample B was demonstrated with higher reproducibility of the resistance observed at low pressure. Sample A showed a continuous increase in resistance at low pressure during the first 30 cycles. This was an indication of a prolonged phase of fibre reorganisation in the nonwoven structure in a relaxed state, thus leading to a lower number of contact points available for current transport.

The mechanical stability of Material A and Material B led to a rapid recovery of the electrical resistance when the pressure was reduced to the lower limit (Figure 8). The repeatability and durability of the electrical signal over 50 load/relaxation cycles proved the rapid and reproducible recovery of the conductive structure in unloaded stage. Rupture of brittle fibre segments, breakage of conductive fibres, and loss in the number of electrical contacts in the fibre web would lead to a continuous increase in resistance at low pressure. Both nonwoven materials exhibited very stable signals at the upper pressure limit. A more stable electrical resistance at low pressure was measured with Material B.

The sensor pads exhibited a pressure-sensitive resistance on the magnitude of several kiloohms; thus, measurement of the pressure-dependent signal would be possible with the use of standard electronic devices for data processing. The influence of the electrical resistance of the connections on the signal-evaluating device would then be negligible. A higher sensor resistance, e.g., in the dimension of several megaohms would increase the risk of shunt currents in the electrical connections to the sensor, particularly at high humidity or in a wet state. Piezo-resistive structures with very high resistance would, thus, make the experimental extraction of the pressure dependent signal more difficult.

Changes in relative humidity in the ambient atmosphere would be of substantial influence on the electrical resistance; thus, for practical applications, the nonwoven structure should be covered by either coating or wrapping, to avoid any disturbing influence due to climate-dependent moisture sorption or desorption.

4. Conclusions

Regenerated cellulose fibres can be modified through incorporation of CB to implement electrical conductivity. Addition of CB into the viscose dope requires formation of stable dispersions and rather high concentrations of CB to achieve percolation. Through the addition of 20 wt.% CB to the viscose dope, a regenerated cellulose fibre with content of 16.4 wt.% CB could be obtained. The conductivity of a standard viscose fibre of 7.7×10^{-9} increased to 9.4×10^{-8} S/m for the modified fibre. A further increase in CB content to 19.8 wt.% and 23 wt.% increased the conductivity substantially, to 8.8×10^{-6} and 0.044 S/m, respectively. The presence of particulate matter in the fibre structure, however, reduced the tenacity of the fibres to 30–50% of the value of a standard fibre.

In experiments to measure conductivity of rotor rings and yarns, a remarkable influence of relative humidity present in ambient air was observed. Sorption of water into fibre assemblies with high electrical resistance, e.g., 10^9–10^{10} Ω, led to a decrease in resistance. In the case of rotor rings manufactured from fibres with 23.1 wt.% CB, however, the initially low volume resistivity of 10 kΩ increased slightly, most probably due to hygral fibre expansion and adsorption of water onto the fibre surface.

Through a combination of the more rigid CB-containing viscose fibres with elastic polyester fibres, a piezo-sensitive nonwoven fabric was manufactured, which demonstrated pressure sensitivity in the range of very low pressure of 400–1000 Pa. Repetitive load/relaxation cycles demonstrated the repeatability and durability of the sensor mat and the stability of the signal.

The results highlight the potential of CB-incorporated viscose fibres as a cheap functional material for pressure sensor production in smart textile applications. The incorporation of carbon black into the viscose fibres led to a black colour, which limits their application in the visible parts of a garment. However, their use for pressure sensing inside a garment and therapeutic compression textiles, e.g., bandages, in the form of pressure-sensing pads, could be potential applications of the material. The material is of particular interest for sensor design, as the CB-incorporated cellulose fibres are nontoxic, compatible with future recycling, and able to be produced at affordable costs.

The conductive cellulose fibres exhibit high potential for the substitution of nonbiodegradable synthetic material used in other applications and, thus, could become a greener alternative to existing materials used in smart textiles.

Supplementary Materials: The following are available online at http://www.mdpi.com/1996-1944/13/22/5150/s1. Figure S1. Rotor ring and measurement of resistance measurement with multimeter; Figure S2. Experimental set-up for the measurement of volume resistance; Figure S3. Experimental set-up for the measurement of yarn resistance; Figure S4. Experimental set-up for the measurement of resistance as function of pressure; Figure S5. Experimental set-up for the cyclic load/release experiments in a tensile testing unit.

Author Contributions: Conceptualization, Y.Z., D.M., and T.B.; investigation, M.E., Y.Z., D.M., N.K., J.F.T., A.M.-A., and T.B.; methodology, J.U., Y.Z., and T.B.; project administration, D.M.; writing—original draft, J.U., Y.Z., D.M., and T.B.; writing—review and editing, T.B. All authors have read and agreed to the published version of the manuscript.

Funding: This research was funded by Austrian research promotion agency (FFG) K-Project tccv (860474) Textile Competence Centre Vorarlberg and the FFG-talente program.

Acknowledgments: The authors thank Anna-Lena Moosbrugger for technical support in the experimental work.

Data Statement: The datasets generated and/or analysed during the current study are available from the corresponding author on reasonable request.

Conflicts of Interest: The authors declare no conflict of interest.

References

1. European Commission. *Smart Wearables Reflection and Orientation Paper Including Feedback from Stakeholders December 2017*; European Commission: Bruxelles, Belgium, 2017.
2. Castano, L.M.; Flatau, A.B. Smart Fabric Sensors and E-Textile Technologies: A Review. *Smart Mater. Struct.* **2014**, *23*, 053001. [CrossRef]

3. Pham, T.; Bechtold, T. Conductive Fibres. In *Handbook of Fibres as Advanced Materials*; Hu, J.H., Kumar, B., Lu, J., Eds.; Wiley-VCH Verlag: Weinheim, Germany, 2020.
4. Islam, G.M.N.; Ali, A.; Collie, S. Textile Sensors for Wearable Applications: A Comprehensive Review. *Cellulose* **2020**, *27*, 6103–6131. [CrossRef]
5. Gonçalves, C.; da Silva, A.F.; Gomes, J.; Simoes, R. Wearable E-Textile Technologies: A Review on Sensors, Actuators and Control Elements. *Inventions* **2018**, *3*, 14. [CrossRef]
6. Meyer, J.; Lukowicz, P.; Tröster, G. Textile Pressure Sensor for Muscle Activity and Motion Detection. In Proceedings of the 2006 10th IEEE International Symposium on Wearable Computers, Montreux, Switzerland, 11–14 October 2006; pp. 69–74. [CrossRef]
7. Root, W.; Wright, T.; Caven, B.; Bechtold, T.; Pham, T. Flexible Textile Strain Sensor Based on Copper-Coated Lyocell Type Cellulose Fabric. *Polymers* **2019**, *11*, 784. [CrossRef] [PubMed]
8. Pizarro, F.; Villavicencio, P.; Yunge, D.; Rodríguez, M.; Hermosilla, G.; Leiva, A. Easy-to-Build Textile Pressure Sensor. *Sensors* **2018**, *18*, 1190. [CrossRef] [PubMed]
9. Rothmaier, M.; Luong, M.P.; Clemens, F. Textile Pressure Sensor Made of Flexible Plastic Optical Fibers. *Sensors* **2008**, *8*, 4318–4329. [CrossRef]
10. Nie, B.; Huang, R.; Yao, T.; Zhang, Y.; Miao, Y.; Liu, C.; Liu, J.; Chen, X. Textile-Based Wireless Pressure Sensor Array for Human-Interactive Sensing. *Adv. Funct. Mater.* **2019**, *29*, 1–10. [CrossRef]
11. Enokibori, Y.; Suzuki, A.; Mizuno, H.; Shimakami, Y.; Mase, K. E-Textile Pressure Sensor Based on Conductive Fiber and Its Structure. In Proceedings of the 2013 ACM Conference on Pervasive and Ubiquitous Computing Adjunct Publication—UbiComp'13 Adjunct, New York, NY, USA, 8–12 September 2013; pp. 207–210. [CrossRef]
12. Holleczek, T.; Rüegg, A.; Harms, H.; Tröster, G. Textile Pressure Sensors for Sports Applications. In Proceedings of the 2010 IEEE Sensors, Kona, HI, USA, 1–4 November 2010; pp. 732–737. [CrossRef]
13. Tessarolo, M.; Possanzini, L.; Campari, E.G.; Bonfiglioli, R.; Violante, F.S.; Bonfiglio, A.; Fraboni, B. Adaptable Pressure Textile Sensors Based on a Conductive Polymer. *Flex. Print. Electron.* **2018**, *3*, 034001. [CrossRef]
14. Kim, M.; Kim, H.; Park, J.; Jee, K.K.; Lim, J.A.; Park, M.C. Real-Time Sitting Posture Correction System Based on Highly Durable and Washable Electronic Textile Pressure Sensors. *Sens. Actuators A Phys.* **2018**, *269*, 394–400. [CrossRef]
15. Lee, J.; Kwon, H.; Seo, J.; Shin, S.; Koo, J.H.; Pang, C.; Son, S.; Kim, J.H.; Jang, Y.H.; Kim, D.E.; et al. Conductive Fiber-Based Ultrasensitive Textile Pressure Sensor for Wearable Electronics. *Adv. Mater.* **2015**, *27*, 2433–2439. [CrossRef]
16. Qi, K.; Wang, H.; You, X.; Tao, X.; Li, M.; Zhou, Y.; Zhang, Y.; He, J.; Shao, W.; Cui, S. Core-Sheath Nanofiber Yarn for Textile Pressure Sensor with High Pressure Sensitivity and Spatial Tactile Acuity. *J. Colloid Interface Sci.* **2020**, *561*, 93–103. [CrossRef]
17. Carvalho, H.; Tama, D.; Gomes, P.; Abreu, M.J.; Yao, Y.; Souto, A.P. Flexible Piezoresistive Pressure Sensors for Smart Textiles. *IOP Conf. Ser. Mater. Sci. Eng.* **2018**, *459*, 012035. [CrossRef]
18. Li, T.; Chen, L.; Yang, X.; Chen, X.; Zhang, Z.; Zhao, T.; Li, X.; Zhang, J. A Flexible Pressure Sensor Based on an MXene-Textile Network Structure. *J. Mater. Chem. C* **2019**, *7*, 1022–1027. [CrossRef]
19. Herbaut, A.; Simoneau-Buessinger, E.; Barbier, F.; Cannard, F.; Guéguen, N. A Reliable Measure of Footwear Upper Comfort Enabled by an Innovative Sock Equipped with Textile Pressure Sensors. *Ergonomics* **2016**, *59*, 1327–1334. [CrossRef] [PubMed]
20. Lim, S.J.; Bae, J.H.; Han, J.H.; Jang, S.J.; Oh, H.J.; Lee, W.; Kim, S.H.; Ko, J.H. Foldable and Washable Fully Textile-Based Pressure Sensor. *Smart Mater. Struct.* **2020**, *29*, 055010. [CrossRef]
21. Zhou, B.; Koerger, H.; Wirth, M.; Zwick, C.; Martindale, C.; Cruz, H.; Eskofier, B.; Lukowicz, P. Smart Soccer Shoe. In Proceedings of the 2016 ACM International Symposium on Wearable Computers—VISWC'16, New York, NY, USA, 12–16 September 2016; pp. 64–71. [CrossRef]
22. Meyer, J.; Arnrich, B.; Schumm, J.; Troster, G. Design and Modeling of a Textile Pressure Sensor for Sitting Posture Classification. *IEEE Sens. J.* **2010**, *10*, 1391–1398. [CrossRef]
23. Xu, W.; Huang, M.C.; Amini, N.; He, L.; Sarrafzadeh, M. ECushion: A Textile Pressure Sensor Array Design and Calibration for Sitting Posture Analysis. *IEEE Sens. J.* **2013**, *13*, 3926–3934. [CrossRef]

24. Zhou, B.; Cheng, J.; Sundholm, M.; Lukowicz, P. From Smart Clothing to Smart Table Cloth: Design and Implementation of a Large Scale, Textile Pressure Matrix Sensor. In *Lecture Notes in Computer Science (Including Subseries Lecture Notes in Artificial Intelligence and Lecture Notes in Bioinformatics)*; Springer Nature Switzerland: Cham, Switzerland, 2014; Volume 8350 LNCS, pp. 159–170. [CrossRef]
25. Cheng, J.; Sundholm, M.; Hirsch, M.; Zhou, B.; Palacio, S.; Lukowicz, P. Application Exploring of Ubiquitous Pressure Sensitive Matrix as Input Resource for Home-Service Robots. *Adv. Intell. Syst. Comput.* **2015**, *345*, 359–371. [CrossRef]
26. Cheng, J.; Sundholm, M.; Zhou, B.; Hirsch, M.; Lukowicz, P. Smart-Surface: Large Scale Textile Pressure Sensors Arrays for Activity Recognition. *Pervasive Mob. Comput.* **2016**, *30*, 97–112. [CrossRef]
27. Liu, M.; Pu, X.; Jiang, C.; Liu, T.; Huang, X.; Chen, L.; Du, C.; Sun, J.; Hu, W.; Wang, Z.L. Large-Area All-Textile Pressure Sensors for Monitoring Human Motion and Physiological Signals. *Adv. Mater.* **2017**, *29*, 1–9. [CrossRef]
28. Fraboni, B. Sensors Performance. *Sensors* **2019**, *19*, 4686. [CrossRef]
29. Kim, G.; Vu, C.C.; Kim, J. Single-Layer Pressure Textile Sensors Withwoven Conductive Yarn Circuit. *Appl. Sci.* **2020**, *10*, 2877. [CrossRef]
30. Kim, K.; Jung, M.; Jeon, S.; Bae, J. Robust and Scalable Three-Dimensional Spacer Textile Pressure Sensor for Human Motion Detection. *Smart Mater. Struct.* **2019**, *28*, 065019. [CrossRef]
31. Wang, Y.; Hua, T.; Zhu, B.; Li, Q.; Yi, W.; Tao, X. Novel Fabric Pressure Sensors: Design, Fabrication, and Characterization. *Smart Mater. Struct.* **2011**, *20*, 065015. [CrossRef]
32. Zhou, Z.; Li, Y.; Cheng, J.; Chen, S.; Hu, R.; Yan, X.; Liao, X.; Xu, C.; Yu, J.; Li, L. Supersensitive All-Fabric Pressure Sensors Using Printed Textile Electrode Arrays for Human Motion Monitoring and Human-Machine Interaction. *J. Mater. Chem. C* **2018**, *6*, 13120–13127. [CrossRef]
33. Luo, N.; Zhang, J.; Ding, X.; Zhou, Z.; Zhang, Q.; Zhang, Y.T.; Chen, S.C.; Hu, J.L.; Zhao, N. Textile-Enabled Highly Reproducible Flexible Pressure Sensors for Cardiovascular Monitoring. *Adv. Mater. Technol.* **2018**, *3*, 1–8. [CrossRef]
34. Lou, C.; Wang, S.; Liang, T.; Pang, C.; Huang, L.; Run, M.; Liu, X. A Graphene-Based Flexible Pressure Sensor with Applications to Plantar Pressure Measurement and Gait Analysis. *Materials* **2017**, *10*, 1068. [CrossRef]
35. Liu, W.; Liu, N.; Yue, Y.; Rao, J.; Luo, C.; Zhang, H.; Yang, C.; Su, J.; Liu, Z.; Gao, Y. A Flexible and Highly Sensitive Pressure Sensor Based on Elastic Carbon Foam. *J. Mater. Chem. C* **2018**, *6*, 1451–1458. [CrossRef]
36. Root, W.; Aguiló-Aguayo, N.; Pham, T.; Bechtold, T. Conductive Layers through Electroless Deposition of Copper on Woven Cellulose Lyocell Fabrics. *Surf. Coat. Technol.* **2018**, *348*, 13–21. [CrossRef]
37. Mayer, D. Optimierung Der Herstellung Funktionalisierter Viskosefasern Anhand Eines Verfahrenstechnischen Ansatzes. Master's Thesis, Technical University Munich, Chair of Biogene Polymers, Munich, Germany, 2017.

Publisher's Note: MDPI stays neutral with regard to jurisdictional claims in published maps and institutional affiliations.

© 2020 by the authors. Licensee MDPI, Basel, Switzerland. This article is an open access article distributed under the terms and conditions of the Creative Commons Attribution (CC BY) license (http://creativecommons.org/licenses/by/4.0/).

Article

Dielectric Characterization of Non-Conductive Fabrics for Temperature Sensing through Resonating Antenna Structures

Isidoro Ibanez-Labiano * and Akram Alomainy

School of Electronic Engineering and Computer Science, Queen Mary University of London, London E1 4NS, UK; a.alomainy@qmul.ac.uk
* Correspondence: i.ibanezlabiano@qmul.ac.uk; Tel.: +44-747-980-4581

Received: 13 February 2020; Accepted: 9 March 2020; Published: 11 March 2020

Abstract: Seamless integration of electronics within clothing is key for further development of efficient and convenient wearable technologies. Therefore, the characterization of textile and fabric materials under environmental changes and other parametric variations is an important requirement. To our knowledge, this paper presents for the first time the evaluation of dielectric characterization over temperature for non-conductive textiles using resonating structures. The paper describes the effects of temperature variations on the dielectric properties of non-conductive fabrics and how this can be derived from the performance effects of a simple microstrip patch antenna. Organic cotton was chosen as the main substrate for this research due to its broad presence in daily clothing. A dedicated measurement setup is developed to allow reliable and repeatable measurements, isolating the textile samples from external factors. This work shows an approximately linear relation between temperature and textile's dielectric constant, giving to fabric-based antennas temperature sensing properties with capability up to 1 degree Celsius at millimeter-wave frequencies.

Keywords: material characterization; smart clothing; temperature sensing; wearable technology

1. Introduction

Dielectric characterization of materials is crucial to understand the iteration between electromagnetic waves with matters [1,2]. In order to fully develop textile-based technology, the dielectric properties of fabrics need to be quantified. These properties are affected by external factors, such as moisture or relative humidity (RH) content and temperature [3]. Previous works [4] have studied the impact of temperature on the dielectric properties of textiles, but without considering the experimental results on several scenarios. There are several cases where textile-based devices must withstand changes in temperature without changing its overall performance, so a deep understanding of the response under different temperature conditions is needed [5]. In addition, an understanding of this phenomenon can be used for sensing applications. Temperature is a crucial parameter to be measured in several fields and applications such as infrastructures, system maintenance, the food industry or body sensing [4,6,7].

Recently developed technologies within the metamaterials discipline have looked into this topic as well. Where the field theory is used to relate electromagnetic fields with parameters such as the permittivity (ε) and how these change as a function of temperature [8,9]. According to microwave theory, material characterization can be performed using two complementary methods: resonant and non-resonant [10,11]. A general rule is to use non-resonant methods when the dielectric properties of a material over a frequency range are unknown and resonant ones for a specific set of discrete frequencies [12]. Among the resonant methods, resonators such as microstrip antennas are simple, low profile non-destructive solutions to evaluate the behavior of fabrics' dielectric properties through temperature. These antennas are mainly a sandwich structure composed of two kinds of materials: one dielectric for the substrate and another one conductive [13–15]. The principle of operation is using the shift of the antenna's resonant frequency (f_r) as a passive sensor, due to the effect of temperature changes on the dielectric properties of the fabric substrate [4]. In electromagnetism, the absolute permittivity, often known simply as permittivity explains how a material interacts when an electric field is applied to it

$$\varepsilon = \varepsilon_r \varepsilon_0 = \varepsilon' - j\, \varepsilon'' \qquad (1)$$

where ε_0 is the vacuum permittivity ($\sim 8.85419 \times 10^{-12}$ F·m^{-1}) and ε_r is the relative permittivity, also known as the dielectric constant, where ε' is the relative permittivity and ε'' is the loss index. The dielectric constant accounts for the molecules' polarization in the material, when an electric field is applied. The dielectric constant increases with temperature, due to the increased mobility of polar molecules, which allows them to align more easily with the electric field [16]. The higher the frequency of operation, the more sensitive the antenna is to temperature changes, due to the shorter wavelength. In this study, we did not consider the size variations due to thermal expansion of the antenna, since the porous nature of fabrics, makes this effect negligible [3]. In addition, we minimized the effect of moisture content, by using encapsulation inside an insulator box and limiting the time of each measurement to 15 min. Other precautions such as thermal sleeves were used during the test campaigns.

In addition to dielectric characterization through temperature, microstrip patches fabricated and tested within this paper can be used for passive temperature sensing within wearable applications. They provide advantages such as low cost, low profile, lightweight, integration into clothing and shielding effect of/from the body due to the full ground plane [17,18]. Within wearable systems, flexibility and conformability are fundamental characteristics in order to include the functionality of computers into individuals' daily lives. Allowing the user to benefit from the performance of the system without restricting the user activity and causing any behavior modification [19–21]. Different options for dielectric materials arise for building flexible wearable antennas, such as several types of papers, Kapton, polyethylene terephthalate (PET) [22], polydimethylsiloxane (PDMS) [23], liquid-crystal polymer (LCP) [24] and textiles [25]. Among them, fabrics withstand bending, twisting and stretching. Furthermore, they are thin and have a low dielectric constant (ε_r) and therefore they are good candidates for flexible wearable antennas as dielectric substrates [26]. In addition to temperature sensing textile-based resonator can plays a key role in the development of a wide range of applications, such as sports analytics [27], healthcare [28], gaming [29], and emergency services [30].

As far as the authors are aware, there is no experimental work carried out on measuring the relationship between dielectric constant and temperature on textile materials. For the first time, a thorough test campaign was carried out and extensive results are presented. Showing a linear relationship between ε_r and temperature for the three frequencies analyzed and independently of the fabric substrate used. A thermal threshold has been found at 50 °C, where the system gets into a saturation status increasing the frequency deviation of the measurements. Adding a constraint to the use of this technique.

2. Materials and Methods

In this section, two resonant methods for measuring the dielectric properties of the fabrics are addressed. Dielectric properties and physical structure of the textiles are provided for modelling steps in further sections; also, antenna design and fabrication for the three cases of study are explained.

2.1. Resonant Methods

One key characteristic of resonant methods is that they are more accurate than non-resonant ones at a single frequency or several discrete frequencies. Resonant methods could be classified into resonant perturbation and resonator techniques. In the former the material perturbates, passively, a resonant cavity, while in the later the material acts as a resonator forming part of the resonant structure [10–12].

2.1.1. Resonant Perturbation Method

In this method, the material to be measured was placed inside the cavity in an aperture and its dielectric properties were derived from the changes inside the cavity, by using conversion equations [31,32]. The changes in the cavity's resonant frequency and quality factor were caused by the insertion of the sample. This technique has high accuracy due to the control of the cavity's specifications and its initial conditions. As an initial step, in order to design the microstrip patch antenna, the dielectric properties of the four different fabrics considered in this study were calculated at ambient RH and temperature. For this purpose, a material characterization split cylinder (85072A, Keysight Technologies, Reading, UK), working at 10 GHz, and a material characterization SW (N1500A-003 MMS 2015, Keysight Technologies, Reading, UK) for data conversion were used.

2.1.2. Resonator Method

A resonator method consists of using a resonating structure, such as a ring or an antenna, to derive the dielectric properties from the S-parameters with a conversion technique [33]. For that purpose, a resonant microstrip patch antenna was designed with the permittivity value measured in the previous stage (Section 2.1.1.).

This method was selected due to its simplicity, it is low-cost and low profile, and it could be easily reproduced in any research laboratory. In addition, microstrip patch antennas are intrinsically a narrow bandwidth system, which is a beneficial characteristic for sensing [11,13], because the bandwidth acts as a probe within this method. Other resonator structures, such as rings have been widely used for material characterization [34]. On the other hand, microstrip antennas are a popular solution for long-range communications, allowing them to integrate remotely the feature of passive sensing.

Furthermore, as only the antenna is under specific conditions, it would avoid damaging any expensive piece of equipment making it an ideal option for any environmental test campaign.

After prototyping the antenna and measuring its insertion losses (S_{11}) the actual value of ε_r was derived based on the shift of the resonant frequency [10]. This method has good accuracy due to the narrow bandwidth nature of microstrip patch antennas, where a small variation on f_r is easier to recognise and to measure than for other antenna structures.

Dimensions of a microstrip patch antenna are calculated using following Equations (2)–(4) [13,35,36]

$$W_{ant} = (c/(2f_r)) \times (\sqrt{2/(\varepsilon_r + 1)}) \text{ and } L_{ant} = (c/(2f_r \times (\sqrt{(\varepsilon_{r,eff})}))) - 2\Delta L \qquad (2)$$

where W_{ant} is the width of the radiation patch, c is the speed of light in vacuum. L_{ant} is the physical length, $\varepsilon_{r,eff}$ is the effective permittivity of the substrate and ΔL is the additional line length because of fringing fields, which could be calculated from

$$\varepsilon_{r,eff} = ((\varepsilon_r + 1)/2) + ((\varepsilon_r - 1)/2) + (1 + (12 S_{ubsh})/W_{ant})^{-1/2} \qquad (3)$$

$$\Delta L = 0.412 S_{ubsh} \times ((\varepsilon_{r,eff} + 0.3)/(\varepsilon_{r,eff} - 0.258)) \times ((W_{ant}/S_{ubsh} + 0.264)/(W_{ant}/S_{ubsh} + 0.8)) \qquad (4)$$

where S_{ubsh} is the thickness of the fabric substrate.

As illustrated above, the resonant frequency of a microstrip antenna is sensitive to dielectric constant variations and according to Equations (2)–(4), if ε_r increases the resonant frequency of the antenna decreases. The theory of this research relays on previous studies [4,37], showing that the dielectric constant of materials increases with temperature. It was shown that the dielectric constant of a piece of Terylene film increased by 0.04 for a 20 °C temperature increase, from 20 °C to 40 °C.

2.2. Materials Characterization

Four of the most used fabrics (organic cotton, jeans, viscose and lycra) were tested in order to take into consideration a broad selection of daily use textiles (Figure 1a–d) [38,39]. In the pictures, the different porosity of the fabrics can be seen; these air voids have an impact on the variation of the effective permittivity. Some studies have shown that there is a linear variation of the relative permittivity depending on the infill percentage [40]. An increase in the density of porous within the fabric substrate implies more air voids trapped. The dielectric constant of the air ($\varepsilon_r = 1$) is lower than the fabrics, lowering the total dielectric constant and increasing the dissipation factor value (tan δ). The dissipation factor (DF) (often known as loss tangent, tan δ) is a ratio of the loss index (ε'') and the relative permittivity (ε')

$$DF = \tan \delta = \varepsilon''/\varepsilon' = 1/Q \tag{5}$$

where Q is the quality factor and it describes how underdamped a resonator is.

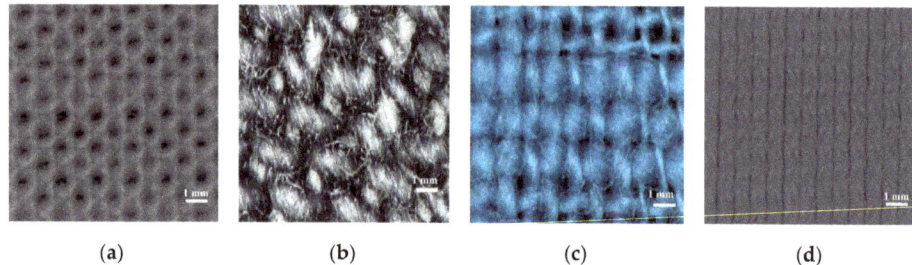

Figure 1. Microscope images for textiles substrates: (**a**) Cotton; (**b**) Jeans; (**c**) Viscose and (**d**) Lycra.

From the microscope images (Figure 1a–d), we extracted the morphology of the four textile substrates. The first three fabrics (organic cotton, jeans and viscose) were woven fabrics with multiple fibres crossing each other at different angles to form the grain, while the last one (lycra) was knitted, it was made up of a single yarn, looped continuously to produce a braided look. The two major fabric types were considered in this investigation.

For the second part of the research, we will focus on using organic cotton due to its presence in daily clothing. The values of dielectric constant and dissipation factor for the four textile substrates measured using the resonant method of cavity perturbation are given in Table 1 below.

Table 1. Dielectric properties of fabrics.

Fabric	ε_r	tan δ
Cotton	1.58	0.02
Jeans	1.62	0.018
Viscose	1.64	0.016
Lycra	1.68	0.008

2.3. Antenna Design and Fabrication

The antenna design falls into a modelled low-profile microstrip patch antenna that was slightly adapted for each of the three frequencies of operation considered in this case of study: 2.45 GHz (A), 9.45 GHz (B) and 38 GHz (C).

2.3.1. Case A at 2.45 GHz

The first case was an inset feed microstrip patch antenna (Figure 2a), designed to operate around 2.45 GHz (industrial, scientific and medical band, ISM) using the Equations (2)–(4). Figure 2a represents the layout of the proposed antenna for the four textile substrates used during the numerical analysis. The corresponding parameters dimensions in mm are illustrated in Figure 2b.

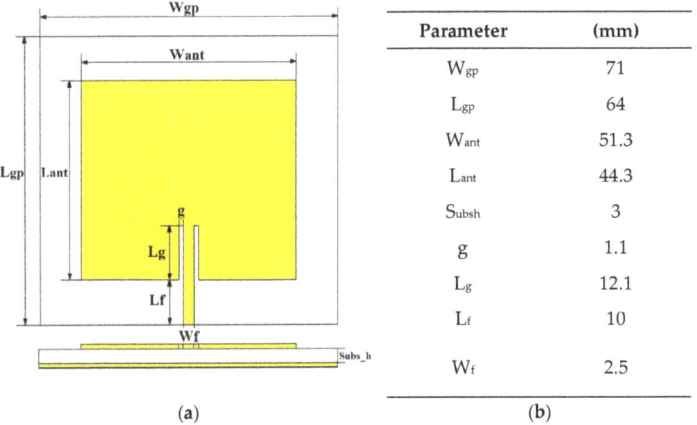

Figure 2. (a) Modelled textile antenna (b) Antenna parameter dimensions.

Regarding materials, adhesive copper tape (with conductivity $\sigma = 5.8 \times 10^7$ S·m^{-1}) was used to construct both the radiation patch and the full ground plane. In order to minimize errors both pieces were cut using a Graphtec Craft-Robo CC300 from Materials Engineering Laboratory at Queen Mary University of London (QMUL). These two pieces were manually attached with a thin layer of acrylic pressure-sensitive conductive adhesive with good heat resistance to the low-cost textile substrates. The impact of the thin layer of adhesive can be ignored for the purpose of this study. Finally, a standard SMA 50 Ohm (Ω) connector was soldered to both, the edge-fed of the antenna and the background plane. Final prototypes are depicted below (Figure 3a–e).

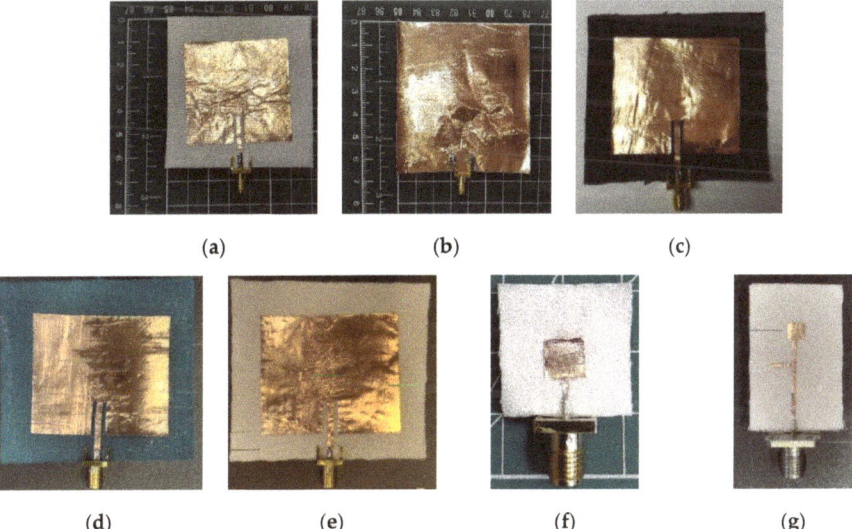

Figure 3. (a) Cotton top-view; (b) Cotton bottom-view; (c) Jeans; (d) Viscose; (e) Lycra; (f) Cotton Case B—9.45 GHz and (g) Cotton Case C—38 GHz.

2.3.2. Case B at 9.5 GHz

The second case of study was carried out in order to increase the sensitivity of the system. We kept the same simple approach of using a microstrip patch antenna. In this case, a microstrip feed was designed and fabricated to work at higher frequencies, around 9.5 GHz (Figure 3f) [13,41]. We moved the operational frequency band from the S-band (ISM, 2.45 GHz) towards the X-band (8–12 GHz). Same Equations (2)–(4) were used for designing the antenna, in which the antenna patch dimensions are 13.8 mm × 9.8 mm (width × length). These reduced dimensions imply resonating at higher in frequencies, due to the fact that the antenna size is related with the wavelength (λ) of operation, which in turn is inversely proportional to the resonant frequency

$$\lambda = c/f_r \tag{6}$$

For fabrication, the same process and techniques were followed and as substrate material, the same organic cotton was used. To interface with the vector network analyser (VNA) during the test campaigns the same connector was soldered.

2.3.3. Case C at 38 GHz

For the final case, a new antenna was designed and fabricated, following the same rationale of previous Section 2.3.2. The design proposed in [42] has been used for this frequency, optimizing the dimensions of the radiating square patch 2.7 mm × 2.8 mm (width × length) and the position of the stub through the feeding line (Figure 3g). An SSMA 2.92 mm edge-launch connector working at 38 GHz was soldered. This connector mates with the popular SMA connections from most of the laboratory pieces of equipment. In addition, this fabric (organic cotton) antenna was designed and fabricated to work at millimeter-waves (mmW), with the potential to be used within the range of the emerging 5G technology.

3. Results and Discussions

3.1. Numerical Analysis

For numerical analysis of the antenna, CST Studio Suite [43] was used to evaluate the time-domain characteristics of the antenna structure at the three different frequencies. The time-solver calculates the development of the electromagnetic fields through time at certain spatial spots and at discrete-time samples, using Maxwell's equations [44].

The antenna's performance was analyzed both in off body and on a body phantom. The phantom model consisted of a 44 mm thick four-layer block. The phantom was modeled as 1 mm of skin, 3 mm of fat, and 40 mm of muscle. The antenna under test (AUT) was placed on top of the four-layer block, leaving a 1 mm air gap in between (see Figure 4a,b). The dielectric properties and conductivity of the three different tissues have been obtained from [45] and are listed in Table 2.

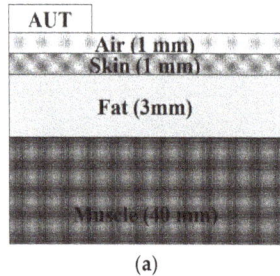

(a)   (b)

Figure 4. (a) 2D Body phantom model of four layers and (b) 3D body phantom model of four layers.

Table 2. Dielectric properties of human tissues at 2.45 GHz [45].

Tissue	ε_r	tan δ	σ (S/m)
Dry skin	38.007	0.28262	1.464
Fat	5.2801	0.14524	0.1045
Muscle	52.729	0.24194	1.7388

3.1.1. Case A at 2.45 GHz

The reflection coefficient describes how much of an electromagnetic wave is reflected due to a discontinuity in the transmission medium, it is often known as return loss or simply S_{11}. A comparison of the simulated S_{11} in off-body versus on-body is shown in (Figure 5a). As for insertion losses, the presence of a human body barely perturbed the antenna's performance. This is an expected result due to the use of a full ground plane, which isolates the antenna from the body. In fact, this is one of the reasons of choosing a microstrip patch model with a full ground plane, for wearable applications. In addition, the ground plane helped to focus the antenna's radiation on the broadside. The directivity (D) is a parameter that quantifies this ability to focus the radiation from the antenna. For this case, the microstrip patch had a computed directivity of 7.44 dBi towards the expected direction of propagation, and its realized gain (G) was 3.81 dB with an efficiency (eff) 43.4%

$$G = D \times eff \qquad (7)$$

Simulation results are shown, in 3D, in Figure 5b.

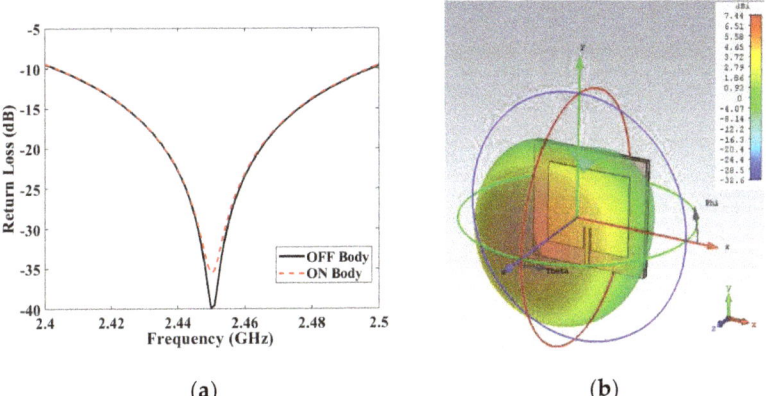

Figure 5. (a) S_{11} simulated in both off and on body and (b) antenna directivity (dBi) in spherical 3D coordinates.

3.1.2. Case B (9.5 GHz) and Case C (38 GHz)

For the next two frequencies cases: the case B at 9.5 GHz and the case C at 38 GHz, the analysis was focused on verifying the resonance frequency and the antenna behavior under general conditions. Since the impact of human body can be neglected, as shown before, we have omitted it for these two cases. The return loss for both frequencies and results are shown in (Figure 6a,b), respectively.

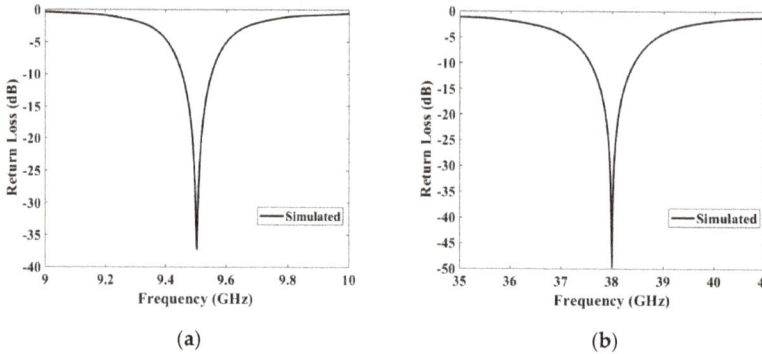

Figure 6. Return losses simulated for (**a**) S_{11} at 9.5 GHz and (**b**) S_{11} at 38 GHz.

3.2. Experimental Setup and Results

A test campaign was carried out at the Antennas Measurement Laboratory facilities of QMUL. To examine radiation patterns at far-field distances, the AUT was placed inside the anechoic chamber (Figure 7a). A diagram of the actual experimental setup used for characterizing the dielectric properties through temperature is depicted in Figure 7b.

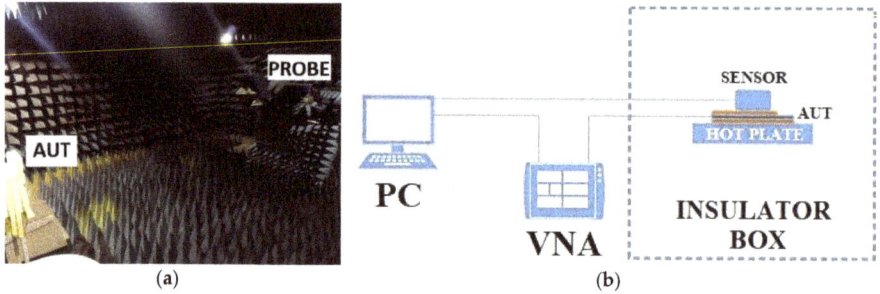

Figure 7. (**a**) Electromagnetic compatibility (EMC) anechoic chamber at Queen Mary University of London (QMUL) and (**b**) Measurement setup diagram.

The prototypes were placed inside a mobile antenna electromagnetic compatibility (EMC) screened anechoic chamber to examine the radiation patterns of the antenna under test (Figure 7a). The EMC chamber was equipped with two open boundary quad-ridge horn antennas (probe) operating from 400 MHz to 6 GHz (3164-06, ETS-Lindgren, TX, USA) and from 0.8 to 12 GHz (QH800, Satimo, Haydock-MSY, UK), allowing vertical and horizontal linear polarization measurements.

The AUT was located on top of a hot plate (RCT Basic, IKA, Oxford, UK) and a thermocouple as close as possible without interfering to measure hot plate's temperature (Figure 7b). Data from the thermocouple were correlated with the hot plate's internal thermometer to verify the antenna's temperature. To ensure the temperature stability and repeatability in the measurements, a due time of 15 min was allowed and 10 measurements (every ten seconds) were taken for each one of the temperatures. With these waiting periods, we guaranteed that the antenna was under the desired temperature in each step. In order to reduce the possible effects of external factors, and in particular of relative humidity (RH) variations, the setup was placed inside an insulator box made of foam. Coaxial cables close to the hot plate were protected with thermal insulator sleeves to avoid any damage to the equipment used and to minimize the impact on the measured magnitudes.

Return losses were measured with vector network analyzers (VNA) (Keysight Technologies, Reading, UK). A PNA-L N5230C for cases A and B, and PNA-X 5244A for case C. The VNA was calibrated at the end of the coaxial cable with an E-cal kit to suppress the effects of cables and connectors, and to have the same initial reference for all our measurements.

3.2.1. Case A at 2.45 GHz

First, a general test campaign for the organic cotton model at ISM frequency was carried out in order to evaluate the overall performance of the antenna's design. Comparison analysis between numerical and experimental performance was carried for the off body scenario.

Figure 8a shows the computed reflection coefficient of the textile antenna versus the fabricated prototype. The numerical estimation and experimental values of the prototype show a good agreement. The resonant frequency was slightly shifted (45 MHz) towards lower frequencies, due to fabrication tolerances.

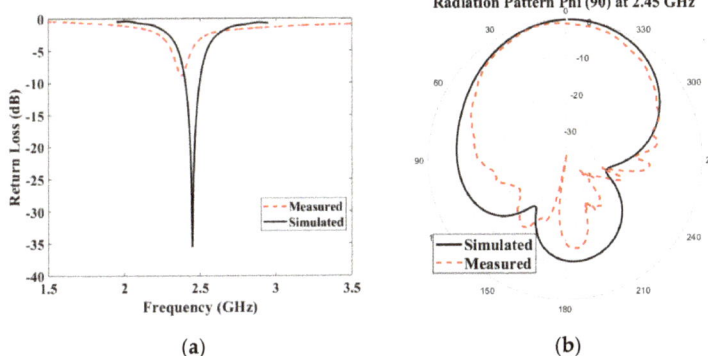

Figure 8. Antenna on cotton fabric at 2.45 GHz; (**a**) reflection coefficient S_{11} computed vs measured and (**b**) radiation pattern Phi (φ) 90 computed vs measured.

The radiation pattern properties in an off-body environment were measured in an anechoic chamber at the QMUL antennas laboratory, showing an expected behavior of a standard high Q-factor microstrip patch antenna. The E-plane cut shows that measurements match simulations fairly well, in terms of radiation pattern and directivity (Figure 8b), and behave as expected from a directional microstrip patch antenna.

Thermal Characterization at 2.45 GHz

The initial thermal characterization using the test setup exemplified in (Figure 7b) was performed for the woven and knitted textiles (four initial fabric substrates: cotton, jeans, viscose and lycra).

The thermal test campaign consists of taking ten measurements of the resonant frequency for each temperature step (20 °C, 30 °C, 40 °C, 50 °C to 60 °C). The average frequency shifts (in MHz) of each fabric are listed in Table 3. Results of all measurements are depicted in the graphs below, for organic cotton (Figure 9a), jeans cotton (Figure 9b), viscose (Figure 9c) and lycra (Figure 9d).

Table 3. Frequency shift over temperature sweep (20–60 °C per 10 °C) at 2.45 GHz (measured results).

Temperature Increment	Cotton	Jeans	Viscose	Lycra
10 Deg	9/11 MHz	9/11 MHz	9/11 MHz	9/11 MHz

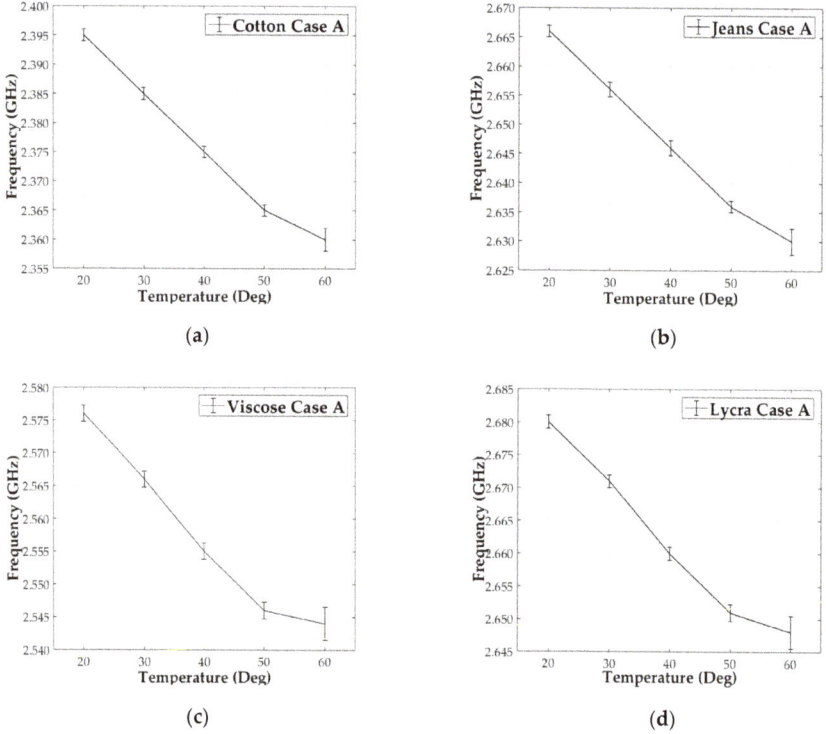

Figure 9. Measured frequency vs temperature representation. Case A at 2.45 GHz: 20–60 °C/10 °C steps for: (**a**) Cotton; (**b**) Jeans; (**c**) Viscose and (**d**) Lycra.

From the test campaign, a linear behavior between ε_r and temperature was observed. An average shift of 10 MHz per 10 °C increment was measured for all four textiles substrates up to a temperature of 50 °C, which according to Equations (2)–(4) is equivalent to approximately a 1.67×10^{-2} change in the dielectric constant for each step. All results are shown in the first row of the first column of Tables 4 and 5. For all four textiles, there was a 10% of frequency deviation (1 MHz) and thermal threshold at 60 °C where the resonant frequency tended to saturate. At the thermal threshold, the standard deviation of the resonant frequency showed a larger standard deviation, as well.

Table 4. Frequency shift over temperature sweep (measured results in MHz).

Temperature Increment	2.45 GHz	9.5 GHz	38 GHz
10 Deg	10 MHz	40 MHz	150 MHz
5 Deg	N/A	20 MHz	75 MHz
1 Deg	N/A	N/A	15 MHz

Table 5. Dielectric constant change over temperature sweep (measured results $\Delta\varepsilon_r$).

Temperature Increment	2.45 GHz	9.5 GHz	38 GHz
10 Deg	1.67×10^{-2}	1.67×10^{-2}	1.67×10^{-2}
5 Deg	N/A	8.35×10^{-3}	8.35×10^{-3}
1 Deg	N/A	N/A	1.67×10^{-3}

3.2.2. Case B at 9.5 GHz

The same thermal measurements were done for the second antenna, case B (Figure 10a). In this case, a 40 MHz decrement per 10 °C increase was measured. Following the same mathematical approach as in 3.2.1., the equivalent change in the dielectric constant is equal to 1.67×10^{-2}. The result matches the one from the previous case A independently on the final resonant frequency. In this case the frequency deviation is 2.5 MHz (6.25%). Final results are summarized in the first row of the second column of both tables, Tables 4 and 5.

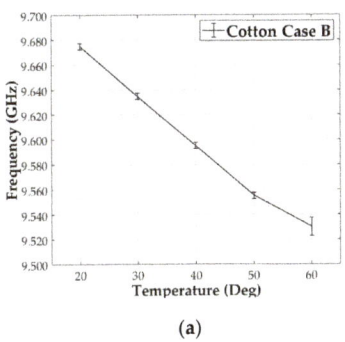 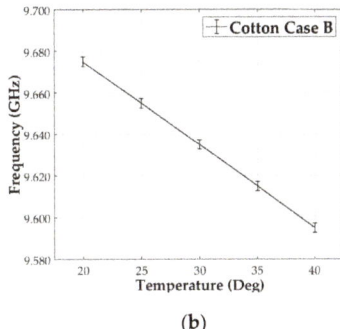

(a) (b)

Figure 10. Measured frequency vs temperature representation. Cotton Case B at 9.45 GHz for: (a) 20–60 °C/10 °C steps and (b) 20–40 °C/5 °C steps.

We performed a finer temperature sweep from 20 °C to 40 °C using 5 °C steps. A 20 MHz shift for each step were measured (Figure 10b), half from the 10 °C case, with a variation of 8.35×10^{-3} (ε_r). These results show that the relative change of the resonant frequency with temperature had a clear linear behavior. The set of results are listed in the second row of the second column of Tables 4 and 5.

3.2.3. Case C at 38 GHz

Finally, for the third case, the same procedure as in the previous ones was used. First, measurements from 20 to 60 °C in steps of 10 °C and second from 20 to 40 °C with increments of 5 °C (Figure 11a,b) were taken. Shifts of 150 MHz and 75 MHz respectively were measured, corresponding to $\Delta\varepsilon_r$ of 1.67×10^{-2} and 8.35×10^{-3}, with a frequency uncertainty of 5 MHz (3.33%) for this scenario. The increase in frequency shifts allowed a finer temperature sweep. In this case, an extra measurement was added to cover the ambient temperature range from 20 °C to 24 °C, with a 1 °C steps (Figure 11b). The resonance change measured for each step was of 15 MHz, $\Delta\varepsilon_r$ of 1.67×10^{-3}.

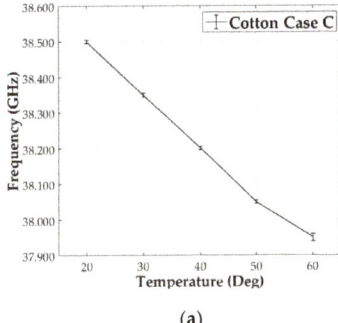 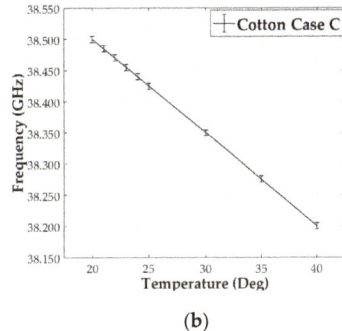

(a) (b)

Figure 11. Measured frequency vs temperature representation. Cotton Case C at 38 GHz for: (a) 20–60 °C/10 °C steps and (b) 20–40 °C/5 °C steps and 20–24 °C/1 °C steps.

Results for case C are in good agreement with previous cases, A and B, showing a linear behavior as expected, independently of the resonant f_r. The mmW prototype improves the sensitivity in an order of magnitude, up to 1 degree Celsius. It can be seen that increasing the sensing frequency, increases the frequency deviation in the measurement.

All the quantitative results for both frequency and dielectric constant are shown in the third column of Tables 4 and 5.

4. Conclusions

This paper demonstrated the efficiency and simplicity of the resonator method to accurately characterize flexible substrates, such as textiles, under environmental conditions. It also shows the cost-efficiency of the technique proposed. This enables a remote sensing scheme for services within harsh environments where equipment can be damaged or it cannot be placed.

As far as the author's knowledge goes, this paper presents for the first time measured results of dielectric properties variation of fabrics over temperature. First, different fabric substrates (cotton, jeans, viscose and lycra) were measured at 2.45 GHz over a temperature range from 20 to 60 °C, at 10 °C steps. As no essential difference was observed among the four textiles, a finer temperature characterization was carried out to focus on organic cotton. Temperature steps were reduced from 10 °C to 5 °C at 9.5 GHz and to 5 °C/1 °C at 38 GHz respectively.

From the test campaigns, it was observed that a linear relationship between the change in temperature and the change in dielectric constant exists and it is frequency independent. It was quantified to be $\Delta\varepsilon_r$ 1.67×10^{-3} per degree Celsius. This relation can be linearly extrapolated to any temperature value. For all the cases, within the four substrates and at the three different frequencies a saturation status behavior can be seen when heating up the substrate above 50 °C. Limiting the use of this technique up to that temperature range.

The textile antenna working at mmW (38 GHz) presents a substantial potential as a passive temperature sensor. Several applications such as food logistic or on-body sensing could benefit from its sensitivity up to one degree Celsius and its characteristics of a fabric-based device.

For future work would be to improve the test setup to remove some uncertainties in the measurements caused by environmental factors. See the impact of going even higher in frequency in terms of impact on physical properties. Looking into the thermal threshold, actual value and plausible cause, like rarefaction of the air trapped within the resonator structure.

Author Contributions: Conceptualization, I.I.-L. and A.A.; methodology, I.I.-L.; software, I.I.-L.; validation, I.I.-L. and A.A.; formal analysis, I.I.-L.; investigation, I.I.-L.; resources, I.I.-L. and A.A.; data curation, I.I.-L.; writing—original draft preparation, I.I.-L.; writing—review and editing, I.I.-L. and A.A.; visualization, I.I.-L. and A.A.; supervision, A.A.; project administration, I.I.-L. and A.A.; funding acquisition, I.I.-L. and A.A. All authors have read and agreed to the published version of the manuscript.

Funding: This research received no external funding.

Acknowledgments: This work was possible due to support from the School of Electronic Engineering and Computer Science, Queen Mary University of London.

Conflicts of Interest: The authors declare no conflict of interest.

References

1. Balanis, C.A. *Advanced Engineering Electromagnetics*; Wiley: New York, NY, USA, 1989; pp. 72–84.
2. Lunkenheimer, P.; Krohns, S.; Gemander, F.; Schamhl, W.W.; Loidl, A. Dielectric Characterization of a Nonlinear Optical Material. *Sci. Rep.* **2015**, *4*, 6020. [CrossRef] [PubMed]
3. Morton, W.E.; Hearle, J.W. *Physical Properties of Textile Fibres*, 4th ed.; The Textile Institute CRC Press Woodhead Publishing Ltd.: Shaston, UK, 2008.
4. Lin, X.; Seet, B.-C.; Joseph, F. Fabric antenna with body temperature sensing for BAN applications over 5G wireless systems. In Proceedings of the 9th International Conference on Sensing Technology (ICST), Auckland, New Zealand, 8–10 December 2015.

5. Sarita Maurya, S.; Yadava, L.R.; Yadav, K.R. Effect of temperature variation on microstrip patch antenna and temperature compensation technique. *Int. J. Wirel. Commun. Mob. Comput.* **2013**, *1*, 35–40. [CrossRef]
6. Moyo, P.; Brownjohn, J.M.W.; Suresh, R.; Tijn, S.C. Development of fiber Bragg grating sensors for monitoring civil infrastructure. *Eng. Struct.* **2005**, *27*, 1828–1834. [CrossRef]
7. Li, Q.; Zhang, L.-N.; Tao, X.-M.; Ding, X. Review of Flexible Temperature Sensing Networks for Wearable Physiological Monitoring. *Adv. Healthc. Mater.* **2017**, *6*, 1601371. [CrossRef] [PubMed]
8. Guler, U.; Chaudhuri, K.; Azzam, S.I.; Reddy, H.; Shalaev, V.; Boltasseva, A.; Kildishev, A. High Temperature Sensing with Refractory Plasmonic Metasurfaces. In Proceedings of the 12th International Congress on Artificial Materials for Novel Wave Phenomena—Metamaterials, Espoo, Finland, 27–30 August 2018.
9. La Spada, L.; Spooner, C.; Haq, S.; Yang, H. Curvilinear MetaSurfaces for Surface Wave Manipulation. *Nat. Sci. Rep.* **2019**, *9*, 3107. [CrossRef]
10. Sankaralingam, S.; Gupta, B. Determination of Dielectric Constant of Fabric Materials and Their Use as Substrates for Design and Development of Antennas for Wearable Applications. *IEEE Trans. Instrum. Meas.* **2010**, *59*, 3122–3130. [CrossRef]
11. Lesnikowski, J. Dielectric permittivity measurement methods of textile substrate of textile transmission lines. *Prz. Elektrotechniczny* **2012**, *88*, 148–151.
12. Khan, M.T.; Jilani, T.M.; Rehman, M.Z.; Khan, M.A.; Ali, M.S. A Brief Review of Measuring Techniques for Characterization of Dielectric Materials. *ITEE J.* **2012**, *1*, 1–5.
13. Ramesh, G. *Microstrip Antennas Design Handbook*; Artech House Antennas and Propagation Library: Boston, MA, USA, 2000.
14. Santas, J.G.; Alomainy, A.; Hao, Y. Textile antennas for on-body communications techniques and properties. In Proceedings of the IEEE 2nd EuCAP, New York, NY, USA, 11–16 November 2007.
15. Chen, S.J.; Kaufmann, T.; Fumeaux, C. Wearable textile microstrip patch antennas for multiple ISM band communications. In Proceedings of the IEEE APSURSI, Orlando, FL, USA, 7–9 July 2013.
16. La Spada, L.; Vegni, L. Electromagnetic Nanoparticles for Sensing and Medical Diagnostic Applications. *Materials* **2018**, *11*, 603. [CrossRef]
17. Vallozzi, L.; Vandendriessche, W.; Rogier, H.; Hertleer, C.; Scarpello, M.L. Wearable textile GPS antenna for integration in protective garments. In Proceedings of the IEEE 4th EuCAP, Barcelona, Spain, 12 April 2010.
18. Engku Embong, E.N.F.S.; Abdul Rani, K.N.; Rahim, H.M. The wearable textile-based microstrip patch antenna preliminary design and development. In Proceedings of the IEEE 3rd ICETSS, Bangkok, Thailand, 7–8 August 2017.
19. Simorangkir, R.-B.V.B.; Yang, Y.; Esselle, K.P. Robust implementation of flexible wearable antennas with PDMS-embedded conductive fabric. In Proceedings of the 12th EUCAP, London, UK, 9–13 April 2018.
20. Virili, M.; Rogier, H.; Alimenti, F.; Mezzanotte, P.; Roselli, L. Wearable Textile Antenna Magnetically Coupled to Flexible Active Electronic Circuits. *IEEE AWPL* **2014**, *13*, 209–212. [CrossRef]
21. Tronquo, A.; Rogier, H.; Hertleer, C. Robus planar textile antenna for wireless body LANs operating in 2.45 GHz ISM band. *IET Electron. Lett.* **2006**, *42*, 142–143. [CrossRef]
22. Guo, X.; Hang, Y.; Xie, Z.; Wu, C.; Gao, L.; Liu, C. Flexible and wearable 2.45 GHz CPW-fed antenna using inkjet-printing of silver nanoparticles on pet substrate. *Microw. Opt. Technol. Lett.* **2017**, *59*, 204–208. [CrossRef]
23. Simorangkir, B.V.B.R.; Kiourti, A.; Esselle, K. UWB Wearable Antenna with a Full Ground Plane Based on PDMS-Embedded Conductive Fabric. *IEEE AWPL* **2018**, *17*, 493–496. [CrossRef]
24. Sagor, H.M.; Abbasi, H.Q.; Alomainy, A.; Hao, Y. Compact and conformal ultra wideband antenna for wearable applications. In Proceedings of the 5th EuCAP, Rome, Italy, 11–15 April 2011.
25. Paracha, K.N.; Rahim, A.K.S.; Soh, J.P.; Mohsen, K. Wearable Antennas: A Review of Materials. *IEEE Access* **2019**, *7*, 56694–56712. [CrossRef]
26. Rais, N.H.M.; Soh, J.P.; Malek, F.; Ahmad, S.; Hasim, N.B.M.; Hall, P.S. A review of wearable antenna. In Proceedings of the IEEE LAPC, Loughborough, UK, 16–17 November 2009.
27. Minyoung, S. 12 Wearable sensors for athletes. In *Electronic Textiles: Smart Fabrics and Wearable Technology*; Tilak, D., Ed.; Woodhead Publishing: Cambridge, UK, 2015; pp. 257–273.
28. Corchia, L.; Monti, G.; De Benedetto, E.; Tarricone, L. Wearable Antennas for Remote Health Care Monitoring System. *IJAP* **2017**. [CrossRef]

29. Orefice, M.; Pirinoli, P.; Dassano, G. Electrically-small wearable antennas for emergency services applications. In Proceedings of the iWAT, Orlando, FL, USA, 29 February–2 March 2016.
30. Ching, C.C.; Stewart, M.K.; Hagood, D.E.; Rashedi, R.N. Representing and Reconciling Personal Data and Experience in a Wearable Technology Gaming Project. *IEEE TLT* **2016**, *9*, 342–353. [CrossRef]
31. Saeed, K.; Shafique, F.M.; Byrne, B.M.; Hunter, C.I. *Planar Microwave Sensors for Complex Permittivity Characterization of Materials and Their Application*; InTech Book: New York, NY, USA, 2012.
32. Chung, B.K. Dielectric constant measurement for thin material at microwave frequencies. *PIER* **2007**, *75*, 239–252. [CrossRef]
33. Bernard, P.A.; Gautray, J.M. Measurement of dielectric constant using a microstrip ring resonator. *IEEE Trans. Microw. Theory Tech.* **1991**, *39*, 592–595. [CrossRef]
34. Heinola, J.-M.; Tolsa, K. Dielectric characterization of printed wiring board materials using ring resonator techniques: A comparison of calculation models. *IEEE Trans. Dielectr. Electr. Insul.* **2006**, *13*, 717–726. [CrossRef]
35. Balanis, C.A. *Antenna Theory: Analysis and Design*; John Wiley & Sons: Hoboken, NJ, USA, 2012.
36. Pozar, D.M. *Microwave Engineering*; John Wiley & Sons: Hoboken, NJ, USA, 2009.
37. Kabacik, P.; Bialkowski, M. The temperature dependence of substrate parameters and their effect on microstrip antenna performance. *IEEE Trans. Antennas Propag.* **1999**, *47*, 1042–1049. [CrossRef]
38. Potey, P.-M.; Tuckley, K. Design of wearable textile antenna with various substrate and investigation on fabric selection. In Proceedings of the 3rd ICMAP, Dhanbad, India, 9–11 February 2018.
39. Vallozzi, L.; Hertleer, C.; Rogier, H. Latest developments in the field of textile antennas. In *Smart Textiles and Their Applications*; Woodhead Publishing: Cambridge, UK, 2016; pp. 599–626.
40. Bartlomiej Biernacki, B.; Zhang, S.; Whittow, W. 3D Printed Substrates with Graded Dielectric Properties and Their Applications to Patch Antennas. In Proceedings of the IEEE LAPC, Loughborough, UK, 14–15 November 2016.
41. Panwar, H.S.; Khan, F.; Khanna, P. Design & Analysis of Square Microstrip Patch Antenna. *IJRTE* **2013**, *2*, 227–3878.
42. Lin, X.; Seet, B.-C.; Joseph, F. Wearable humidity sensing antenna for BAN applications over 5G networks. In Proceedings of the IEEE 19th WAMICON, Clearwater, FL, USA, 9–10 April 2018.
43. Software, CST Studio Suite. 2017. Available online: https://www.3ds.com/products-services/simulia/products/cst-studio-suite/?utm_source=cst.com&utm_medium=301&utm_campaign=cst (accessed on 10 March 2020).
44. Lehner, G. *Electromagnetic Field Theory for Engineers and Physicists*; Springer: Berlin/Heidelberg, Germany, 2010.
45. Institute for Applied Physics. Dielectric Properties of Body Tissues in the Frequency Range 10–100 GHz Italian National Research Council. Available online: http://niremf.ifac.cnr.it/tissprop/ (accessed on 1 February 2020).

© 2020 by the authors. Licensee MDPI, Basel, Switzerland. This article is an open access article distributed under the terms and conditions of the Creative Commons Attribution (CC BY) license (http://creativecommons.org/licenses/by/4.0/).

Article

Textile Display with AMOLED Using a Stacked-Pixel Structure on a Polyethylene Terephthalate Fabric Substrate

Jae Seon Kim [1] and Chung Kun Song [2],*

[1] National Disaster Management Research Institute, Ulsan 44548, Korea; js9996@nate.com
[2] Department of Electronics Engineering, Dong-A University, Busan 49315, Korea
* Correspondence: cksong@dau.ac.kr

Received: 15 May 2019; Accepted: 20 June 2019; Published: 22 June 2019

Abstract: An active-mode organic light-emitting diode (AMOLED) display on a fabric substrate is expected to be a prominent textile display for e-textile applications. However, the large surface roughness of the fabric substrate limits the aperture ratio—the area ratio of the organic light-emitting diode (OLED) to the total pixel area. In this study, the aperture ratio of the AMOLED panel fabricated on the polyethylene terephthalate fabric substrate was enhanced by applying a stacked-pixel structure, in which the OLED was deposited above the organic thin-film transistor (OTFT) pixel circuit layer. The stacked pixels were achieved using the following three key technologies. First, the planarization process of the fabric substrate was performed by sequentially depositing a polyurethane and photo-acryl layer, improving the surface roughness from 10 µm to 0.3 µm. Second, a protection layer consisting of three polymer layers, a water-soluble poly-vinyl alcohol, dichromated-polyvinylalcohol (PVA), and photo acryl, formed by a spin-coating processes was inserted between the OTFT circuit and the OLED layer. Third, a high mobility of 0.98 $cm^2/V·s$ was achieved at the panel scale by using hybrid carbon nano-tube (CNT)/Au (5 nm) electrodes for the S/D contacts and the photo-acryl (PA) for the gate dielectric, enabling the supply of a sufficiently large current (40 µA @ V_{GS} = −10 V) to the OLED. The aperture ratio of the AMOLED panel using the stacked-pixel structure was improved to 48%, which was about two times larger than the 19% of the side-by-side pixel, placing the OLED just beside the OTFTs on the same plane.

Keywords: electronic textiles; AMOLED; OTFTs; OLEDs; textile displays; organic thin film

1. Introduction

Electronic textiles (e-textiles) are attracting much attention because they are expected to provide new functionality by integrating electronic devices into textiles [1,2]. Especially, as wearable electronics and the Internet of Things are emerging as prominent killer applications of IT technology [3], e-textiles are becoming an increasingly important technology because of their greater convenience and functionality than conventional hand-carried devices. It is expected that e-textiles will be applied in very diverse areas such as healthcare, sports, fashion, and the military [1,2].

Technologies related to e-textiles have evolved from a gadget style, where electronic circuit boards were simply attached onto textile surfaces [4–8], to on-cloth applications, where electronic devices are directly fabricated onto the textiles [1]. In the future, in-cloth textiles will be developed, where active devices such as organic thin-film transistors (OTFTs), organic light-emitting diodes (OLEDs), and organic photovoltaics will be implemented on a single fiber and the active fibers will be woven together to realize electronic functions in textiles [9–14].

A display is a key device needed in order to realize the ubiquitous features of e-textiles; information should be able to be obtained at any place and any time. Among the currently available displays, an

OLED is a unique device which can be implemented on textiles because of the attractive properties such as self-emission, the capability of low-cost solution processes, and the possible applications to flexible electronics, enabling to use a variety of substrates [15–19]. However, although articles about OLEDs on fabric substrates [20] and on a single fiber [21] have been published, reports regarding electronic display panels integrating OLEDs and OTFTs on a fabric substrate are rare.

Previously, the authors of this paper published an article regarding an active-mode organic light-emitting diode (AMOLED) panel fabricated on a fabric substrate [22]. In the paper, the AMOLED panel adopted a side-by-side pixel structure having an OLED beside the OTFTs' pixel circuit on the same plane. However, the aperture ratio—the area ratio of the OLED to the whole pixel—could not be larger than 20% due to the large area of the OTFTs required to drive the OLED and the rough surface of the fabric substrate.

In this paper, to improve the aperture ratio of the AMOLED display, a stacked-pixel structure placing the OLED above the OTFTs' circuit layer was developed for the fabric substrate. With the stacked-pixel structure, the aperture ratio was increased by approximately 2.5 times that of the side-by-side pixel structure.

2. Design of the AMOLED Panel with a Stacked-Pixel Circuit

In this study, the AMOLED panel adopted a standard pixel circuit, consisting of two OTFTs, one OLED, and one capacitor, as shown in Figure 1a. The switching OTFT (SW OTFT) was activated by applying a scan voltage (V_{scan}) to the gate of the SW OTFT. Then, information on the data line (V_{data}) was transferred to the storage capacitor (C_{st}) through the SW OTFT. The key function of the SW OTFT was to supply a sufficiently large current in the on-state to the C_{st}, and thus, to cause the voltage across the C_{st} to increase quickly during the scanning period. Having an extremely low off-state current in the off-state was also an important feature to sustain the voltage on the C_{st} during the time frame. The driving OTFT (DR OTFT) was activated by the voltage stored in the C_{st}. The DR OTFT should supply a large current to the OLED to light it up brightly.

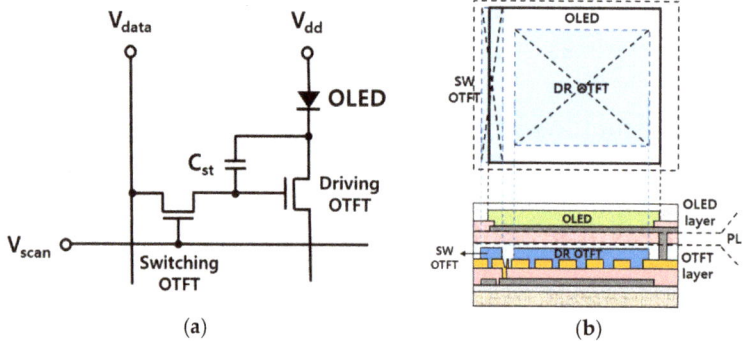

Figure 1. (a) The pixel circuit of the active-mode organic light-emitting diode (AMOLED) panel consisting of 2 organic thin-film transistors (OTFTs), 1 organic light-emitting diode (OLED), and 1 capacitor; and (b) the layout and cross-section of the stacked pixel.

Since the mobility of the OTFTs was less than 1 cm^2/V·s, a large ratio of the channel-width-to-length (W/L) was required to supply a large current, as described above. Therefore, the aperture ratio was generally less than 20% when the OTFTs and OLED were placed on the same plane; this is called a side-by-side structure [22].

In this study, the aperture ratio was enhanced by employing a stacked-pixel structure, where the OLED was placed above the OTFTs' pixel circuit, as shown in Figure 1b. In order to realize the stacked structure in the panel, several technological issues needed to be resolved. First, to supply a sufficiently large on-state current (I_{on}) to the enlarged OLED in the stacked pixel, the OTFTs' performance needed to be improved, because the OLED area relative to the OTFTs' circuit was larger compared to a side-by-side structure. Next, a protection layer (PL) should be inserted between the OTFTs' circuit layer and the OLED layer in order to protect the OTFTs' circuit from being damaged by the OLED processes performed above it. The PL must not affect the OTFTs below and should also have a self-patterning ability. Otherwise, a patterning process, such as photo-lithography, may seriously damage the OTFTs. In the next section, the fabrication processes will be described by focusing on these issues.

Based on a channel length of L = 20 μm as a minimum feature size, the AMOLED panel was designed as having 64 × 64 pixels. The channel width of the DR OTFT and the SW OTFT were designed to be W = 180 × L and 34 × L, respectively. Therefore, the pixel pitch was 1 mm × 0.77 mm, with an aperture ratio of 48%, where the OLED area was 0.76 mm × 0.49 mm, and a panel diagonal length of 3.2 inches. In the stacked pixel, the pixel pitch was reduced by about 70%, meanwhile, the aperture ratio was increased by 2.5 times compared with a side-by-side pixel with the same minimum feature size.

3. Fabrication

The fabrication processes are depicted in Figure 2a. Polyethylene terephthalate (PET) fabric was used as a substrate. The PET fabric was woven with PET fibers with a diameter of 200 μm. Before starting the processes, the PET fabric was pre-shrunken by heating at 150 °C for 3 h to avoid deformation as well as to minimize misalignment due to the shrinkage caused by the subsequent thermal processes. Then, the PET fabric was attached to a carrier glass using UV-detachable glue, and the edge sides of the substrate were strongly fixed on the glass with adhesive tape. This prevented the various chemicals from swelling the fabric during the subsequent processes and also the substrate from being deformed under high-temperature processes.

The rough surface of the PET fabric was smoothed by a special planarization process using a double layer of polyurethane (PU) and photo-acryl (PA) [22,23]. The surface roughness was reduced from 10 μm to 0.3 μm, which was suitable for fabricating devices on, because the high surface roughness was likely to disconnect thin devices, meanwhile, a small surface roughness can be continuously covered with thin OTFT and OLED devices. Subsequently, aluminum was evaporated on the smoothed fabric and patterned by photo-lithography for the gate electrodes of the OTFTs and scan bus lines. In addition, PA was spin coated for the gate dielectric layer of the OTFTs. The performance of the OTFTs strongly depended on the compositional ratio of the solvent in PA solution, and thus, a proper ratio was determined in order to produce a high performance. The results are discussed in the next section. Since the PA had a self-patterning ability, the layer was patterned for the gate area by exposing it to UV through a mask without a photoresist process.

As described in the previous section, the OTFT performance needed to be improved in order to supply a large on-current to the enlarged OLED in the stacked AMOLED. Therefore, the source and drain (S/D) contacts of the OTFTs used hybrid electrodes consisting of carbon nanotubes (CNTs) and Au because they produced the lowest contact resistance (2.9 K·cm) due to the work function modulation of CNTs with the deposition of Au on the CNTs. This resulted in a large on-state current. The detailed structure and the characteristics of the hybrid electrodes can be seen in Reference [24]. A CNT solution was spray-coated on the whole PA layer and then Au was evaporated with a thickness of 5 nm. The S/D electrodes were patterned using a photo-lithography process. By evaporating pentacene through a shadow mask for SW and DR OTFTs, the processes for the OTFTs' pixel circuit layer was completed.

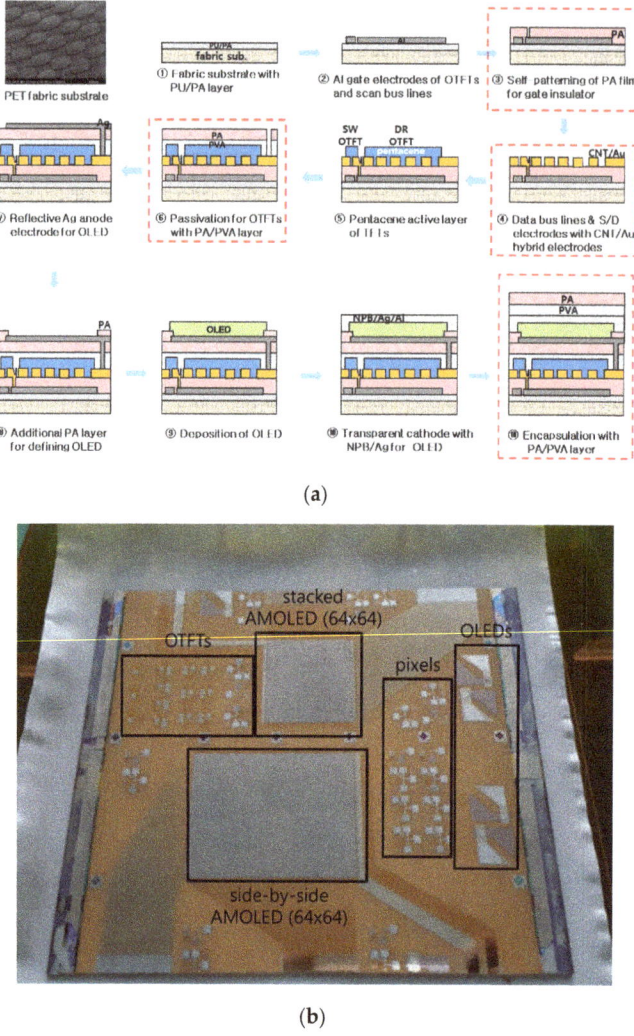

Figure 2. (**a**) The fabrication processes for the AMOLED panel using the stacked pixel and (**b**) a picture of two AMOLED panels, using the stacked and the side-by-side pixels, fabricated on a polyethylene terephthalate (PET) fabric substrate with the various test elements included.

As described in the previous section, a PL was deposited on the OTFTs' circuit layer to stack the OLED above it. To achieve the requirements as described in the previous section, the PL consisted of three polymer layers, including a water-soluble poly-vinyl alcohol (w-PVA), dichromated-PVA (d-PVA), and a PA. The w-PVA was applied to protect the pentacene OTFTs from being damaged by the organic solvents from the PA. The water solution did not affect the pentacene due to their different hydrophobicity. The d-PVA was used to pattern the w-PVA. The PA protected the double PVA layers and the OTFTs from the effects of the OLED process.

The w-PVA solution was prepared by mixing well 3 wt % PVA molecules with deionized (DI) water. The PVA solution was spin-coated on the panel containing the OTFTs' circuits at 1000 rpm for 20 s and dried for 30 min in air. Subsequently, the d-PVA solution, which was formulated by mixing ammonium dichromate of 0.03 wt % with the w-PVA solution, was spin-coated on the w-PVA film at 1000 rpm for 20 s and dried for 30 min in air. The total thickness of the PVA double layer was approximately 5 µm. The double PVA layers were exposed to UV for 1 min through a chrome mask and developed using DI water. The developed PVA film was baked at 60 °C for 10 min. Additionally, the PA solution with the same mixing ratio as the gate dielectric was spin-coated on the developed PVA film at 1000 rpm for 20 s and softly baked at 90 °C for 10 min. The thickness of the PA was about 1 µm. The PA film was patterned via-holes to interconnect the OTFTs to the OLED by exposure to UV for 30 s, developed for 40 s, and then hard-baked at 130 °C for 60 min.

Silver was evaporated on the patterned PA layer for the anode electrodes of the OLED, and another PA was spin-coated on the Ag electrodes and patterned to define the OLED area. Subsequently, the OLED layers were sequentially evaporated through a shadow mask, and the transparent cathode electrodes were evaporated above the OLED with 4,4′-bis(N-phenyl-1-naphthylamino) NPB) (40 nm)/Ag (20 nm)/Al (1 nm) layers. Finally, an encapsulation layer consisting of w-PVA and PA was spin-coated on the panel. The final AMOLED panel was detached by exposing it to UV through the carrier glass. Figure 2b shows the detached AMOLED panel picture, including the stacked AMOLED and the side-by-side AMOLED panel for comparison and the various test elements such as the discrete OTFTs, the OLED, and pixels.

4. Results and Discussion

The surface roughness of the fabric substrate was measured by AFM (Park System, Seoul, Korea) at each step of the planarization process. The surface roughness was plotted and also compared with the conventional plastic and glass substrate in Figure 3. The surface roughness of the PET substrate was reduced from 10 µm to 0.3 µm (as shown in Figure 3) after the planarization process, as described in the previous section. The PU layer decreased the roughness at the macro scale from 10 µm to 2 µm, and the PA decreased the roughness at the micro scale from 2 µm of the PET/PU to 0.3 m and also enhanced the process compatibility to the subsequent films by changing the hydrophobicity of the PET/PU as well. Although the surface roughness of the PET/PU/PA substrate (0.3 µm) was larger than that of poly carbonate (PC) plastic (0.025 µm) and glass (0.003 µm) substrates, it was comparable to the thickness of OTFTs (0.43 µm) and OLEDs (0.3 µm). Therefore, the devices could be fabricated on the planarized PET/PU/PA substrate.

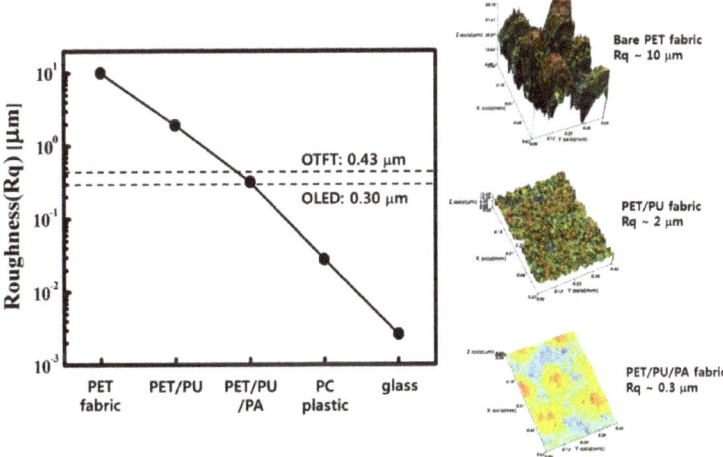

Figure 3. The surface roughness variation of the polyethylene terephthalate fabric substrate according to deposition of the polyurethane (PU) and photo-acryl (PA) layers, including the AFM images of bare PET, PET/PU, and PET/PU/PA fabric substrates.

The hybrid electrode of CNT/Au for the S/D contacts together with PA for the gate dielectric in the OTFTs was employed for the first time for an AMOLED panel to improve performance as well as to reduce the process steps in this work. To reduce the interface states density and to obtain a smoother gate surface, the PA solution was diluted from the as-purchased state by adding the solvent of propylene glycol monomethyl ether acetate (PGMEA). The electrical characteristics were varied with the mixing ratio of PA to PGMEA, as shown in Figure 4. Representative transfer curves are depicted according to the various mixing ratios of the PA solutions. The electrical parameters are summarized in Table 1, where the values were averaged out of sixteen OTFTs for each ratio.

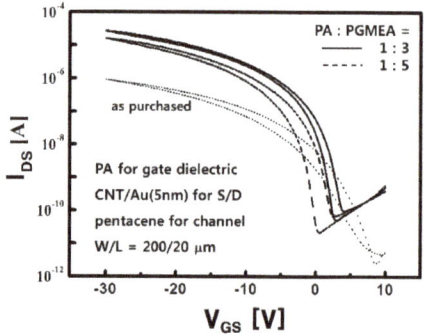

Figure 4. The transfer curves of the pentacene-OTFTs using photo-acryl (PA) for the gate dielectric with the various mixing ratios of PA to PGMEA, in which the CNT/Au (5 nm) electrodes were used for the S/D contacts.

Table 1. The average values of the electrical parameters of the pentacene OTFT using photo-acryl (PA) as the gate dielectric and the CNT/Au (5 nm) electrodes for the S/D contacts according to various mixing ratios of PA to the solvent PGMEA.

Mixing Ratio	μ_{FET} (cm^2/V·s)	I_{on} @ −30 V (μA)	I_{off} (pA/μm)	$I_{on/off}$	V_{on} (V)	SS (V/dec)	ε
1:1	0.02	0.91	0.02	1.9×10^5	8.6	2.70	2.70
1:3	0.98	26.5	0.26	5.1×10^5	4.2	1.65	2.57
1:5	0.72	15.9	0.11	7.6×10^5	2.6	1.30	2.36

μ_{FET}: field effect mobility; I_{on}: the on-state current; I_{off}: the off-state current: SS: sub-threshold slop: ε: dielectric constant.

The OTFTs with a ratio of 1:3 produced the largest mobility of 0.98 cm^2/V·s, which was 49 times larger than the 0.02 cm^2/V·s of the as-purchased PA, as shown in Table 1, and comparable to the 1.0 cm^2/V·s of the OTFTs using the polyvinylphenol (PVP) gate dielectric and the CNT/Au (5 nm) for the S/D contacts, as reported in Reference [24]. The required performance enhancement for the OTFTs, as well as the reduction in process steps in the stacked AMOLED panel, were successfully achieved using PA for the gate dielectric and the CNT/Au (5 nm) electrodes for the S/D contacts.

The high performance of the OTFTs was degraded after the PL was deposited on them. In Figure 5, the transfer curves of the DR and SW OTFTs without the PL are compared to those with the PL; the transfer curves were measured from separated test pixels. As shown in Table 2, the mobility was reduced by about 40%, from 0.87 cm^2/V·s and 0.75 cm^2/V·s to 0.54 cm^2/V·s and 0.49 cm^2/V·s for DR and SW OTFT, respectively. Although the water in the w-PVA of the PL was expected to protect the hydrophobic pentacene of the OTFTs from the subsequent processes, the developing process of the d-PVA layer and the depositing process of the PA layer seemed to damage the pentacene. However, the on-state currents at 25 V were sufficiently large with 228 μA and 7.97 μA for the DR and SW OTFT, respectively. As a result, the DR OTFT could drive the large OLED and the SW OTFT could quickly charge up the storage capacitor. In addition, the off-state current of the SW OTFT, 1.69 pA/μm, was small enough to keep the charge on the storage capacitor during the time frame. Even though the performance of the DR and SW OTFTs deteriorated after depositing the PL, the performance was still able to operate the AMOLED panel.

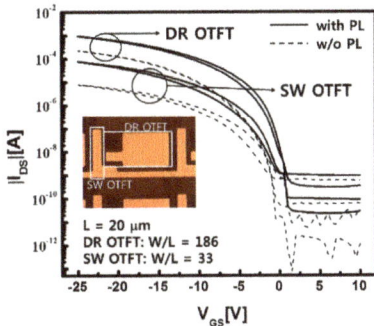

Figure 5. The transfer curves of the DR and SW OTFTs with and without the protection layer in the test pixel; the performance was degraded by the protection layer; however, the on-state current was still large enough to drive the enlarged OLED of the stacked pixel.

Table 2. The average values of the electrical parameters of the DR and SW OTFTs in the stacked pixel with and without the protection layer.

Device (W/L)	Protection Layer	μ_{FET} (cm^2/V·s)	I_{on} @ −25 V (μA)	I_{off} (pA/μm)	$I_{on/off}$	V_{on} (V)	SS (V/dec)
DR OTFT (186)	without	0.87	961	0.15	3.39×10^7	2.0	1.21
	with	0.54	228	0.09	1.31×10^7	0.0	2.30
SW OTFT (33)	without	0.75	76.8	3.27	7.11×10^5	1.0	2.76
	with	0.49	7.97	1.69	1.43×10^5	1.5	5.91

In Figure 6a, the structure of the phosphorescent OLED used in this paper is presented. It consists of multiple organic layers of 1,4,5,8,9,11-hexaazatriphenylene-hexacarbonitrile (HAT-CN) (10 nm)/NPB (30 nm)/4,4′-Cyclohexylidenebis[N,N-bis(4-methylphenyl)benzenamine] (TAPC)(10 nm)/4,4′-Bis(N-carbazolyl)-1,1′-biphenyl (CBP):Ir (ppy) (20 nm)/2,2′,″-(1,3,5-Benzinetriyl)-tris(1-phenyl-1-H-benzimidazole) (TPBi)(40 nm)/LiF (0.2 nm) between the cathode and anode electrodes. It also used a reflective anode of Ag (80 nm) and a transparent cathode of Al (1 nm)/Ag (20 nm)/NPB (40 nm) to implement the top emission on the opaque fabric substrate. The various types of OLED lights were also fabricated on the planarized PET/PU fabric substrate with an area of 20 cm × 20 cm, and a representative OLED light with an area of 7 cm × 7 cm is shown in Figure 6b. They successfully operated without an electrical short, reflecting that the surface roughness of the PET/PU substrate (0.3 μm) was smooth enough for OLEDs, even when the area was large. The OLEDs produced a phosphorescent green light with a wavelength of 534 nm and a luminance of 23,673 cd/m^2 at 7 V, as shown in Figure 6c.

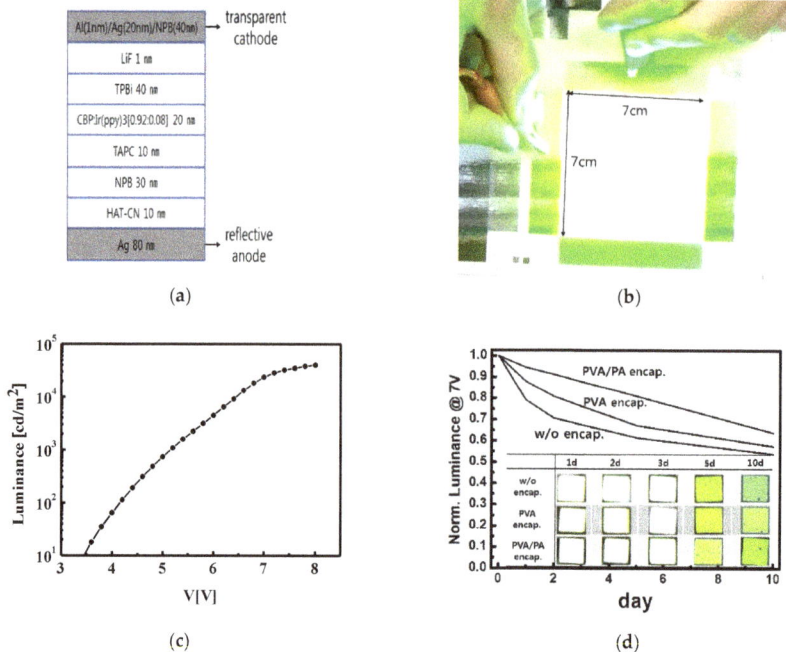

Figure 6. (a) The structure of the phosphorescent OLED, (b) an OLED light fabricated on a PET fabric substrate with an area of 7 cm × 7 cm, (c) the luminance characteristics of the top emitting phosphorescent OLED fabricated on the PET fabric substrate, and (d) the degradation of the OLED luminance with and without the PVA/PA encapsulation.

In order to protect the AMOLED panel from damage due to the air exposure during measurements, a temporary encapsulation consisting of PVA and PA double layers was spin-coated with a thickness of 1 μm for each layer on the final AMOLED panel. It was expected that the water in the PVA solution would not affect the hydrophobic organic layers of the OLEDs, and the thick polymer layers would protect the OLEDs from air exposure. As shown in Figure 6d, as a layer was added, the luminance decreased from the initial luminance of 19,895 cd/m^2 at 7 V with a PVA single layer encapsulation to 16,636 cd/m^2 with the PVA/PA double-layer encapsulation. However, the encapsulation retarded the degradation by keeping air from permeating into the OLEDs. The luminance of the OLEDs gradually decreased to 64% of the initial luminance with the PVA/PA encapsulation and to 58% with the PVA encapsulation after 10 days in air. Meanwhile, the bare OLED decreased to 54%. The AMOLED panel with the encapsulation maintained visible brightness for 10 days, although the brightness became dim with time, as shown in the inset of Figure 6d.

In Figure 7, two types of AMOLED panels are compared. The side-by-side pixels produced an aperture ratio of 19%, and the space between the OLEDs (1130 μm), where the DR and SW OTFTs were contained, was discernable even with bare eyes.

Figure 7. Comparison of the stacked pixels with the side-by-side pixels in the AMOLED fabricated on the PET fabric substrate; the aperture ratio of 48% with the stacked pixels (pixel pitch: 1.0 mm × 0.77 mm, OLED area: 0.76 mm × 0.49 mm) is clearly identified with 19% of the side-by-side case (pixel pitch: 1.6 mm × 1.6 mm, OLED area: 1.04 mm × 0.47 mm).

Meanwhile, in the stacked pixels, the pixel space could not be distinguished because the light coming out of a pixel overlapped those of the neighboring pixels. The overlapping light occurred due to the smaller pixel space (280 µm) and the brighter luminance of the OLEDs caused by the larger on-state current (40 µA @ V_{GS} = −10 V) of the DR OTFT. The aperture ratio was 48%, approximately 2.5 times larger than that of the side-by-side pixels.

The high-aperture ratio in the AMOLED panel was successfully achieved using the stacked pixel structure, which was possible due to the PL between the OTFTs' pixel circuit and the OLED, as well as the improved performance of the OTFTs by using CNT/Au (5 nm) hybrid electrodes for the S/D contacts and the self-patterning PA gate dielectric.

The washing problem is a large obstacle to overcome for practical applications of electronic textiles including AMOLED textile displays. Recently, articles concerning water resistant encapsulation layers for e-textiles have been reported [25,26]. They have successfully protected the underlaid OLEDs on a fabric without performance deterioration, even after being washed 10 times. Therefore, the washing issue can be resolved in the near future.

5. Conclusions

In this paper, a textile display of AMOLED was successfully fabricated on a PET fabric substrate. The aperture ratio was significantly enhanced using a stacked-pixel structure having an OLED on the OTFTs' pixel circuit. Three key technologies were employed: a planarization process on the rough PET substrate; the insertion of a protection layer (PL) between the OLED and the OTFTs' pixel circuit; and the use of hybrid CNT/Au (5 nm) electrodes for the S/D contacts together with PA for the gate dielectric of the OTFTs. The planarization process consisting of polyurethane and photo-acryl layer reduced the roughness from 10 µm to 0.3 µm, which was smooth enough for the devices. The PL consisted of three polymer layers: a water-soluble poly-vinyl alcohol (w-PVA), dichromated-PVA (d-PVA), and photo acryl (PA). This protected the bottom OTFTs from damage by the subsequent OLED process, and also enabled patterning for the interconnection between the bottom OTFTs and the top OLED without an additional lithography process. The hybrid CNT/Au (5 nm) electrodes used for the S/D contacts together with the PA for the gate dielectric noticeably increased the on-state current of the OTFTs, which could then provide a sufficiently large current to the enlarged OLED for high luminance. With those technologies, an aperture ratio of 48% was successfully achieved, which was 2.5 times larger than the 19% of the side-by-side pixel structure. In addition, it was also possible to eliminate several photo-lithography processes in the fabrication of the AMOLED panel, which might otherwise cause damage to the organic layers, by using PA for various layers, such as the gate dielectric of the OTFTs and for the PL and the encapsulation layer.

A sample of the AMOLED textile display panel was successfully demonstrated in this paper. I In the future, textile display technology will be advanced for low power consumption, and thus, will be able to integrate with flexible batteries. Furthermore, it is expected that the washing problem will be resolved in the near future. Therefore, the future use of practical AMOLED textile displays for commercial production is not hard to see.

Author Contributions: J.S.K. performed the experimental work, analyzed data, and started the writ-up. C.K.S. initiated and supervised the research work and improved the manuscript for submission and publication.

Funding: This research was funded by Dong-A University in Korea.

Conflicts of Interest: The authors declare no conflict of interest.

References

1. Park, S.; Jayaraman, S. Enhancing the quality of life through wearable technology. *IEEE Eng. Med. Biol. Mag.* **2003**, *22*, 41–48. [CrossRef] [PubMed]
2. Carpi, F.; De Rossi, D. Electroactive polymer-based devices for e-textiles in biomedicine. *IEEE Trans. Inf. Technol. Biomed.* **2005**, *9*, 295–318. [CrossRef] [PubMed]
3. Stoppa, M.; Chiolerio, A. Wearable electronics and smart textiles: a critical review. *Sensors* **2014**, *14*, 11957–11992. [CrossRef] [PubMed]
4. Post, E.R.; Orth, M. Smart Fabric, or "Wearable Clothing". In Proceedings of the 1st IEEE International Symposium on Wearable Computers, Cambridge, UK, 13–14 October 1997; p. 167.
5. Post, E.R.; Orth, M.; Russo, P.R.; Gershenfeld, N. E-broidery: Design and fabrication of textile-based computing. *IBM Syst. J.* **2000**, *39*, 840–860. [CrossRef]
6. Parker, R.; Riley, R.; Jones, M.; Leo, D.; Beex, L.; Milson, T. Stretch-an e-textile for large-scale sensor systems. In Proceedings of the International Interactive Textiles for the Warrior Conference, Cambridge, UK, 9–11 July 2002.
7. Katragadda, R.B.; Xu, Y. A novel intelligent textile technology based on silicon flexible skins. *Sens. Actuators A Phys* **2008**, *143*, 169–174. [CrossRef]
8. Cherenack, K.; Zysset, C.; Kinkeldei, T.; Münzenrieder, N.; Tröster, G. Woven electronic fibers with sensing and display functions for smart textiles. *Adv. Mater.* **2010**, *22*, 5178–5182. [CrossRef] [PubMed]
9. Han, J.W.; Meyyappan, M. Copper oxide transistor on copper wire for e-textile. *Appl. Phys. Lett.* **2011**, *98*, 192102. [CrossRef]
10. Maccioni, M.; Orgiu, E.; Cosseddu, P.; Locci, S.; Bonfiglio, A. Towards the textile transistor: Assembly and characterization of an organic field effect transistor with a cylindrical geometry. *Appl. Phys. Lett.* **2006**, *89*, 143515. [CrossRef]
11. Lee, J.B.; Heeney, M.; Tiemey, S.; Murphy, A.; Liu, J.; Frechet, J.M.J.; Subramanian, V. Polythiophene thin-film transistor array for gas sensing. In Proceedings of the 63rd Device Research Conference Digest, Santa Barbara, CA, USA, 20–22 June 2005; Volume 1, pp. 147–148.
12. González, R.; Pinto, N.J. Electrospun poly (3-hexylthiophene-2, 5-diyl) fiber field effect transistor. *Synth. Met.* **2005**, *151*, 275–278. [CrossRef]
13. Mattana, G.; Cosseddu, P.; Fraboni, B.; Malliaras, G.G.; Hinestroza, J.P.; Bonfiglio, A. Organic electronics on natural cotton fibres. *Org. Electron.* **2011**, *12*, 2033–2039. [CrossRef]
14. Hamedi, M.; Forchheimer, R.; Inganäs, O. Towards woven logic from organic electronic fibres. *Nat. Mater.* **2007**, *6*, 357–362. [CrossRef] [PubMed]
15. Saxena, K.; Singh, G.; Thallapaka, B.; Mehta, D.S. White organic light-emitting diodes based on a single-emissive layer using electrophosphorescent dopants in a fluorescent host. *J. Inf. Disp.* **2014**, *15*, 119–126. [CrossRef]
16. Kwon, J.H.; Kim, E.; Im, H.G.; Bae, B.S.; Chang, K.S.; Ko Park, S.H.; Choi, K.C. Metal-containing thin-film encapsulation with flexibility and heat transfer. *J. Inf. Disp.* **2015**, *16*, 123–128. [CrossRef]

17. Hung, L.S.; Chen, C.H. Recent progress of molecular organic electroluminescent materials and devices. *Mater. Sci. Eng. R* **2002**, *39*, 143–222. [CrossRef]
18. Kanno, H.; Hamada, Y.; Takahashi, H. Development of OLED with high stability and luminance efficiency by co-doping methods for full color displays. *IEEE J. Sel. Top. Quantum Electron.* **2004**, *10*, 30–36. [CrossRef]
19. Ikeda, H.; Sakata, J.; Hayakawa, M.; Aoyama, T.; Kawakami, T.; Kamata, K.; Iwaki, Y.; Seo, S.; Noda, Y.; Nomura, R.; et al. P-185: Low-Drive-Voltage OLEDs with a Buffer Layer Having Molybdenum Oxide. In *SID Symposium Digest of Technical Papers*; Wiley Publishing: Hoboken, NJ, USA, 5 July 2012; Volume 37, pp. 923–926.
20. Choi, S.; Kwon, S.; Kim, H.; Kim, W.; Kwon, J.H.; Lim, M.S.; Lee, H.S.; Choi, K.C. Highly flexible and efficient fabric-based organic light-emitting devices for clothing-shaped wearable displays. *Sci. Rep.* **2017**, *7*, 6424. [CrossRef] [PubMed]
21. O'Connor, B.; An, K.H.; Zhao, Y.; Pipe, K.P.; Shtein, M. Fiber shaped light emitting device. *Adv. Mater.* **2007**, *19*, 3897–3900. [CrossRef]
22. Kim, J.S.; Song, C.K. AMOLED panel driven by OTFTs on polyethylene fabric substrate. *Org. Electron.* **2016**, *30*, 45–51. [CrossRef]
23. Ryu, G.S.; Jeong, S.H.; Park, B.C.; Park, B.; Song, C.K. Fabrication of organic thin film transistors on Polyethylene Terephthalate (PET) fabric substrates. *Org. Electron.* **2014**, *15*, 1672–1677. [CrossRef]
24. Jeong, Y.; Jeong, J.; Kim, H.Y.; Jeong, H.J.; Song, C.K. Performance improvement of organic thin film transistors with carbon nanotube/metal hybrid electrodes for S/D contacts. *Org. Electron.* **2016**, *36*, 153–159. [CrossRef]
25. Jeong, E.G.; Jeon, Y.; Cho, S.H.; Choi, K.C. Textile-based washable polymer solar cells for optoelectronic modules: Toward self-powered smart clothing. *Energy Environ. Sci.* **2019**, *12*, 1878–1889. [CrossRef]
26. Kwon, J.H.; Jeon, Y.; Choi, K.C. Robust Transparent and Conductive Gas Diffusion Multibarrier Based on Mg-and Al-Doped ZnO as Indium Tin Oxide-Free Electrodes for Organic Electronics. *ACS Appl. Mater. Interfaces* **2018**, *10*, 32387–32396. [CrossRef] [PubMed]

© 2019 by the authors. Licensee MDPI, Basel, Switzerland. This article is an open access article distributed under the terms and conditions of the Creative Commons Attribution (CC BY) license (http://creativecommons.org/licenses/by/4.0/).

Article

Surface Modification of Polyester-Fabric with Hydrogels and Silver Nanoparticles: Photochemical Versus Gamma Irradiation Methods

Kathleen A. Montoya-Villegas [1], Alejandro Ramírez-Jiménez [1,2], Ángel Licea-Claverie [1,*], Sergio Pérez-Sicairos [1], Emilio Bucio [3], Johanna Bernáldez-Sarabia [4] and Alexei F. Licea-Navarro [4]

[1] Centro de Graduados e Investigación en Química, Tecnológico Nacional de México/Instituto Tecnológico de Tijuana, Tijuana 22000, Mexico; kathleen.montoya@tectijuana.edu.mx (K.A.M.-V.); alejandro.ramirezj@tectijuana.edu.mx (A.R.-J.); sperez@tectijuana.mx (S.P.-S.)

[2] CONACyT-Centro de Graduados e Investigación en Química, Tecnológico Nacional de México/Instituto Tecnológico de Tijuana, Tijuana 22000, Mexico

[3] Instituto de Ciencias Nucleares, Universidad Nacional Autónoma de México, Circuito Exterior, Ciudad Universitaria, Ciudad de Mexico 04510, Mexico; ebucio@nucleares.unam.mx

[4] Departamento de Innovación Biomédica, Centro de Investigación Científica y Educación Superior de Ensenada, Ensenada 22860, Mexico; jbernald@cicese.edu.mx (J.B.-S.); alicea@cicese.mx (A.F.L.-N.)

* Correspondence: aliceac@tectijuana.mx; Tel.: +52-664-623-4043

Received: 28 June 2019; Accepted: 26 July 2019; Published: 10 October 2019

Abstract: A Gamma irradiation and photochemical crosslinking/grafting of poly(2-hydroxyethyl methacrylate) (PHEMA) and poly(2-hydroxyethyl methacrylate-co-poly(ethylene glycol) methacrylate) (poly(HEMA-co-PEGMA)) hydrogels onto polyethyleneterephtalate fabric (PET) surfaces were evaluated, in order to obtain a hydrophilic homogeneous coating onto PET fabrics. The materials were characterized by FTIR-ATR, SEM, EDS, and thermal analysis. Furthermore, silver nanoparticles (AgNPs) were loaded by in situ reduction of $AgNO_3$, and its antibacterial activity against *Staphylococcus aureus* and *Escherichia coli* was determined. Results showed a ticker coating of hydrogel using gamma radiation and stronger in deep modification of the fibers; however, by the photochemical method, a thin coating with good coverage of PET surface was obtained. The differences in hydrophilicity, thermal properties, and antibacterial activity of the coated fabrics by using both methods were rather small.

Keywords: graft polymerization; surface modification; hydrogels; gamma irradiation; silver nanoparticles; antibacterial activity

---

1. Introduction

Materials for biomedical applications need to fulfill a series of mechanical, physical, chemical, and biological properties to have an adequate performance in contact with a living organism. It is not usual to find a material that meets all the requirements needed for a given application, so the development of new materials by molecular modification of basis materials remains a challenge. Currently there is a renewed interest in textile materials with antimicrobial properties for medical, healthcare, hygiene, and sports applications. Infections associated with medical devices, mainly medical surgical tools and supporting parts, cause at least 1.5–7.2% of post-operational complications. The attachment of microorganisms to the surface of the material is the main problem related to infection and this is the first step in the development of post-operational complications. In this context, chemical modification by grafting is a very good option for biocompatible surface preparation [1,2], however, in this way, the properties of the matrix, like mechanical or chemical resistance, may be slightly affected.

Combination of good mechanical properties of the film supports with the recognized biocompatibility of hydrogel may open novel possibilities. With respect to this, hydrogels obtained from the polymerization of 2-hydroxyethyl methacrylate (HEMA) are well-known in the biomedical and biotechnological fields because of their resemblance to biological tissues, high water content [3,4], low toxicity, high chemical stability to hydrolysis, hydrophilic character [5], and versatility to be copolymerized with functional monomers [6]. Generally, HEMA is thought to be a solvent-like monomer capable of homogenizing hydrophobic and hydrophilic phases [7]; these features mean it is biocompatible and possesses good properties for biomedical uses, for example for drug delivery systems and tissue engineering [8–10].

Grafting of polymers onto other polymers by gamma radiation is a suitable technique for surface modification of polymeric materials since this allows to introduce active functional groups on the polymer backbone [11–15]. Graft polymerization combines the features of the matrix, such as chemical and thermal resistance or specific mechanical properties, and introduces functional groups with advantageous new surface properties; this method is applicable for many substrates and combination of monomers and, unlike chemically initiated grafting, it does not require initiators, catalysts, or additives [16]. Moreover, the surface can be covered uniformly, and the polymer is sterilized by the radiation [17].

There are several methods of radiation grafting: (i) The direct (or mutual) grafting method in which the polymeric material is irradiated in contact with the monomer or monomers, although homopolymerization is a collateral effect; (ii) the pre-irradiation method, which involves the irradiation of the polymer matrix in the absence of air and then the grafting is initiated by macroradicals trapped in the irradiated polymer; the main disadvantage is a higher radiation dose than in the direct method, and polymer degradation may occur; and (iii) the pre-irradiation oxidative grafting method that consists of the pre-irradiation of the polymer in the presence of either air or oxygen, so that the macroradicals formed are converted to peroxides and/or hydroperoxides, then when the pre-irradiated polymer is heated, the peroxides decompose to give the macro-radicals, which react with the monomers obtaining a graft polymer [18–20]. The grafting yield depends on features of the polymers and monomers such as solubility and radiation stability (degradation and crosslinking) [21].

On the other hand, polyethyleneterephtalate (PET) is a polyester type polymer that has been used in biomedical devices mainly as sutures, cardiovascular grafts, sutures of artificial rings, and in heart valves. PET commercialized as Dacron^{TM} has been successfully used in vascular grafts in long diameter arteries [22,23]; however, the use of PET has been limited due to its low biocompatibility. Therefore, to obtain more biocompatible fabrics, surface modifications have been carried out; for example, Liu and coworkers modified PET surface by grafting of polyamides using UV treatment [24]; Ping and coworkers grafted poly(acrylic acid) by the gamma radiation method, and surfaces were loaded with AgNPs in order to obtain bacteriostatic surfaces [25]; Aubert-Viard and co-workers modified PET and polypropylene by chitosan immobilization, and they loaded iodide/iodate anions or silver cations in order to obtain antibacterial surfaces [26]; Lin and co-workers modified nonwoven PET fabrics by immobilization of antibacterial peptides ε-polylysine and natamicin obtaining excellent antibacterial efficiency [27]; Vesel and co-workers modified PET with heparine in order to obtain more hemocompatible surfaces [28]; Vesel and co-workers carried out another interesting modification introducing –SH functional groups [29] because it is well known that these may be used as anchoring sites for biomolecules. On the other hand; polyesters and polyamides have been treated to improve the binding efficiency of AgNPs due to the poor adhesion between AgNPs and the organic fabrics [30]. PET fabrics treated with silver antimicrobial agents have been extensively studied due to the fact that silver in different forms possesses activity against more than 650 pathogens (bacteria, fungi, and viruses) and has low toxicity towards mammalian cells [31]. Silver antimicrobial agents include silver coating [32], colloidal silver [33], and AgNPs using different methods: Plasma treatment [34,35], corona discharge [36], sonochemical methods [37], and photo reduction of Ag$^+$ ions on the surface of other nanoparticles [38]. Despite the different methods used, the challenges remain to modify the surface properties of PET to favor the interactions with the silver and the control of the attachment of AgNPs to avoid their release and the durability of the antimicrobial properties of the silver-modified PET fabrics [39].

The aim of this work was to evaluate the grafting of a hydrophilic gel of poly(2-hydroxyethyl methacrylate) (PHEMA) and of poly (2-hydroxyethyl methacrylate-co-polyethylene glycol methacrylate), poly(HEMA-co-PEGMA), onto PET fabrics using gamma-rays or UV treatment methods in order to obtain hydrophilic/antibacterial surfaces. PET fabrics grafted with poly (HEMA) and poly (HEMA-co-PEGMA) hydrogels were loaded with AgNPs taking advantage from its crosslinked structure, and the antibacterial activity of the modified fabrics against *E. coli* and *S. aureus* was evaluated.

2. Materials and Methods

2.1. Materials

Polyethyleneterephthalate non-textured fabric (ROGA-0715 Texlon Corp. Torrance, CA, USA), was cut into 9 cm × 9 cm samples for photochemical grafting and into 1 cm × 5 cm for radiation grafting; this was due to the size of the molds used in each reactor. These were then washed with hot water for 24 h in a Soxhlet apparatus, and finally, each sample was dried under vacuum until constant weight. Ethylene glycol dimethacrylate (EGDMA) 98%, 2-hydroxyethyl methacrylate (HEMA), poly (ethylene glycol) methacrylate Mn = 526 (PEGMA), and poly (ethylene glycol) methyl ether methacrylate Mn = 300 (PEGMA300), (Sigma-Aldrich Chemical Co., Toluca, México) were purified by passing through an inhibitor remover column (Aldrich Chemical Co., St. Louis, MO, USA) before use. Sodium borohydride (NaBH$_4$) 98%, 1, 2-diphey l-2, 2-dimethoxyethanone 99%, Irgacure® 651, and silver nitrate P. A. ACS 99% (AgNO$_3$) (all from Sigma-Aldrich Chemical Co., Toluca, México) were used as received

2.2. Photochemical Crosslinking/Grafting

HEMA (4.8 g, 37 mmol), Irgacure 651 (0.024 g, 0.085 mmol), EGDMA, and PEGMA at different molar ratios (Table 1) were placed into Schlenk flasks; deionized water (80 wt% with respect to HEMA) was used as solvent, and oxygen was displaced by bubbling argon for 6 min. In order to obtain homogeneous solutions, the solutions were sonicated using an ultrasonic bath Branson 2800 (Branson Ultrasonics, Danbury, CT, USA) for 5 min; afterwards, samples of PET fabric were introduced in the Schlenk flasks. Oxygen was thoroughly displaced by using three freeze-thaw cycles using a dry ice/acetone bath and argon flow. Afterwards, the flasks, filled with argon, were sealed, and stored for 12 h inside a refrigerator in order to allow the swelling of the fabric.

Table 1. Surface modification of polyethyleneterephtalate (PET) fabric with hydrogel using the photochemical method (λ = 350 nm); water was used as solvent.

Sample	EGDMA (% mol) [1]	PEGMA (% mol) [1]	Hydrogel (wt/wt %) [2]	Q_{water} [3]
UV_PET$_{100\_0\_0.5}$	0.50	–	96	1.7
UV_PET$_{97\_3\_0.5}$	0.50	3	49	2.0
UV_PET$_{95\_5\_0.5}$	0.50	5	67	2.1
UV_PET$_{93\_7\_0.5}$	0.50	7	64	2.1
UV_PET$_{95\_5\_0.25}$	0.25	5	96	2.3
UV_PET$_{95\_5\_1}$	1.00	5	85	2.0

[1] With respect to HEMA monomer; [2] calculated using the Equation (1); [3] calculated using the Equation (2).

Then, the fabrics were removed from the soaking solution and were placed between two glass plates of 10 cm × 10 cm, sealed with a silicone spacer (1 mm thickness), and the samples were irradiated using lamps of wavelength of 350 nm for 30 min inside a RMR 200 Rayonet Photochemical Chamber Reactor (Palisades Park, NJ, USA) under argon atmosphere. Residual monomers and other compounds were extracted by washing with successive ethanol/water mixtures with increasing ratio of deionized water; finally, the samples were dried under vacuum until constant weight.

2.3. Gamma Radiation Grafting

PET fabrics were exposed to ^{60}Co γ-source (Gammabeam 651 PT, MDS Nordion, Kanata, ON, Canada) at dose rates of around 10 to 12 kGy h^{-1} and doses between 50 and 70 kGy.

A mixture of 10 mL of HEMA:water or HEMA:PEGMA:water at different volume ratios (Table 2) were placed into a Schlenk flask, argon was then bubbled for 5 min; after that, PET fabrics were soaked with the mixtures and the mixtures were degassed by three freezing cycles with argon flow; afterwards, the flask was sealed and stored inside a refrigerator for 12 h. After that, the fabrics were quickly removed and placed between two glass plates of 2.5 cm × 7.5 cm, sealed with a silicone spacer (0.5 mm thickness), and finally, the samples were irradiated at doses of 50, 60, or 70 kGy. In order to extract the residual monomer and homopolymer formed during the grafting reaction, the samples were soaked four times for 2 h in ethanol-water 25:75 (V/V, %), followed by ethanol-water 50:50 (V/V, %) ethanol-water 75:25 (V/V, %), and ethanol for 12 h each. Afterwards, the samples were dried until constant weight.

Table 2. Surface modification of PET fabric with hydrogel using the direct gamma irradiation method; water was used as solvent.

Sample	Dose (kGy)	PEGMA (% Vol.)	Solvent (% Vol.)	Hydrogel (wt/wt%) [1]	Q_{water} [2]
γ60_PET$_{100\_0\_50}$	60	–	50	257	2.5
γ60_PET$_{90\_10\_50}$	60	10	50	400	1.8
γ60_PET$_{90\_10\_40}$	60	10	40	400	1.8
γ60_PET$_{90\_10\_30}$	60	10	30	400	1.8
γ60_PET$_{80\_20\_50}$	60	20	50	213	1.9
γ60_PET$_{70\_30\_50}$	60	30	50	285	2.9
γ50_PET$_{90\_10\_50}$	50	10	50	285	1.8
γ70_PET$_{90\_10\_50}$	70	10	50	285	1.8

[1] Calculated using the Equation (1); [2] calculated using the Equation (2).

The grafting yield (GY) in both methods was calculated using Equation (1):

$$GY = 100\% \, (W_g - W_0)/W_0 \quad (1)$$

where W_0 and W_g represent the weights of the initial and the grafted fabric, respectively.

2.4. Characterization of Grafted Hydrogels on PET Fabrics

FTIR-ATR spectra were recorded using a Perkin-Elmer Spectrum 400, FTIR-/FT-NIR spectrometer (Perkin Elmer Cetus Instruments, Norwalk, CT, USA) with 8 or 16 scans between 650 and 4000 cm^{-1}. Thermal decomposition of samples was evaluated undo nitrogen flow of 50 mL min^{-1} between 20 and 600 °C at a heating rate of 20 °C min^{-1} using a TGA SDT 2960 Simultaneous DSC-TGA equipment (TA Instruments, New Castle, DE, USA). Glass transition temperature (Tg) and melting point (Tm) were determined by differential scanning calorimetry (DSC) using a TA-Instrument modulated DSC equipment (DSC 2929) (TA Instruments, New Castle, DE, USA); two heating cycles were recorded in modulation mode under a nitrogen flow of 60 mL min^{-1} at a heating rate of 5 °C min^{-1} with an amplitude of ±0.5 °C over a modulation period of 60 s; in both cycles, the temperature was equilibrated at −10 °C during 5 min before heating to 200 °C for the first cycle, to erase the thermal history, and up to 280 °C for the second one, for measurement. The thickness of the hydrogel thin-film on fabric surface was measured using SEM images of cross-sections. For the cross-section analysis, samples were fractured in liquid nitrogen and fixed on sample holder with a double-sided graphite tape. Analyses were carried-out with a TESCAN VEGA 3 SEM-microscope (Brno, Czech Republic), at an acceleration voltage of 25 kV with a secondary electron detector. Previous to SEM analysis, all samples were sputter-coated with gold for 180 s at 18 mA using a SPI-MODULE sputter coater. The theoretical thickness of coating was 200 Å, according to the manufacturer's technical information.

Water absorption equilibrium was monitored by immersion of pristine and modified PET fabrics of 2.5 cm × 2.5 cm into distilled water for 96 h. The excess of water on the materials was removed with filter paper and the swollen samples were weighed. The mass-swelling degree in water (Q_{water}) of the hydrogel coating was determined using Equation (2).

$$Q_{water} = 1 + ((W_S \times \varrho_h)/(W_h * \varrho_s)) \qquad (2)$$

W_s and W_h represent the weights of the adsorbed water and the mass of the dry hydrogel, respectively, where the weight of the fabric without hydrogel was subtracted; and ϱ is the density in g/mL. For this work the value of the polymer density PHEMA ($\varrho = 1.15$ g/mL) was taken considering a linear polymer with molecular weight of 20,000 g/mol and $\varrho = 1.15$ g/mL at T = 25 °C, for the polymer PEGMA ($\varrho = 1.105$ g/mL) was taken from a polymer with molecular weight of 300 g/mol at T = 25 °C. For the density of the hydrogel (ϱ_h) the feed composition before crosslinking was taken as a first approximation. The density of water ($\varrho_s = 1.0$ g/mL) was used for all calculations. Measurements were performed in triplicate and the average value is reported.

2.5. In Situ Synthesis of Silver Nanoparticles

Silver nanoparticles (AgNPs) were synthesized by chemical reduction in situ of $AgNO_3$ using $NaBH_4$ as reducing agent. Equation (2) shows the chemical reaction:

$$2AgNO_3 + 2NaBH_4 \rightarrow 2Ag° + H_2 + B_2H_6 + 2NaNO_3. \qquad (3)$$

Pristine and hydrogel grafted PET fabrics of 1.5 cm × 1.5 cm were swollen in distilled water for five days; after that, samples were soaked in 5 mL of a $AgNO_3$ solution (0.005 M); one day later, the pieces of fabrics were transferred into 5 mL of $NaBH_4$ solution (0.01 M) at 0 °C. The fabrics were stirred gently for 2 h while cooling with an ice bath; finally, the solution was decanted, and the fabrics were washed with deionized water.

The presence of AgNPs on modified PET fabrics was determined semi-quantitatively by energy dispersive X-ray spectroscopy (EDS), by mapping measurements on surface and cross-sections using a Bruker XFlash Detector 4010 (Berlin, Germany). For surface analysis, samples of 5 mm × 5 mm were cut and fixed on sample holder with a double-sided graphite tape and for cross-section analysis, samples were prepared as described previously for SEM analysis. EDS information was processed using Quantax 200 ESPRIT 1.9 software (Bruker Nano GmbH, Berlin, Germany).

2.6. Antibacterial Activity

Bacterial strains were obtained from bacterial culture collection of the Biomedical Innovation Department, at Scientific Research and High Education Center from Ensenada. The Gram (−) strain was *E. coli*, and the Gram (+) strain was *S. aureus*. The antimicrobial activity of hydrogels was performed by modification of the agar disk diffusion method of Collins et al. [40]. Approximately 106 colony-forming units of each bacterium were inoculated on Luria–Bertani (LB) plates. To verify the exponential growth phase of each strain, we measured OD600 values of the bacterial strains over time. The inhibition of the bacterial growth was assessed by triplicate using samples grafted by the photochemical (UV) or gamma irradiation treatments, with or without AgNPs. Silver nanoparticles and silver-free samples were used as a control, and carbencillin (500 µg/mL), a broad-spectrum antibiotic, was used as a positive control. The inhibition zone (mm) was defined as the area in which no bacterial growth was detected. The plates were incubated for 24 h at 37 °C.

3. Results and Discussion

3.1. Grafting of Hydrogels onto PET-Fabrics

Grafts of HEMA or HEMA/PEGMA were introduced by the photochemical or the direct gamma irradiation methods; grafting percentage was calculated using the Equation (1). Tables 1 and 2 show the grafting conditions and the hydrogel percentage obtained by the photochemical and the direct gamma irradiation methods, respectively. Results showed that when using the photochemical method, the grafting yield was between 49% and 96%, whereas using the gamma radiation method, it was between 210% and 400%, both related to the weight of the unmodified PET-fabric. The high grafting percentages obtained by gamma irradiation are not good for the application of PET as fabric owing to their increased thickness; even at 50 kGy of irradiation dose, the percentage of grafted hydrogel was too high (285%). These high values may have resulted due to the higher monomer concentrations used, as well as owing to the high radiochemical yield of the monomers and the solvent used. One option to tailor down the grafting percentage could be the use of a mold with a spacer of adequate thickness or the use of a modified grafting method: The so-called oxidative pre-irradiation method previously described [41].

An important step in the synthesis of hydrogels on the PET fabric is the initiation of the polymerization. In this stage, with the photochemical method, free radicals are generated by UV radiation evolving from the photoinitiator (Irgacure ® 651), since this dissociates after the absorption of photons, generating benzoyl and dimethoxybenzoyl radicals, the latter tending to form a methyl radical; both benzoyl radicals and methyl radicals are involved in the polymerization reaction, however the benzoyl radical is more stable than the methyl radical [42].

Scheme 1 shows the reaction mechanism for the hydrogels cross-linked with EGDMA, initiated by the photochemical dissociation of the photoinitiator. In the direct gamma irradiation method, the PET backbone can produce two types of free radicals under γ-ray radiation, but $-[CH_2\_\dot{C}HOOCC_6H_4COO]-$ is the predominant free radical [25,43]. The main procedure possibly involved in the graft copolymerization induced by γ-rays can be described as shown in Scheme 2.

The reactions were carried out in aqueous medium to favor a rapid formation of free radicals on the reagents used, where the reaction is initiated by the formation of free radicals by the radiolysis of deionized water, after the formation of radicals, the coupling of the different species of radicals is favored, forming grafts of polymer networks attached to the PET fabric. Please note that in this case, EGDMA crosslinker was not added.

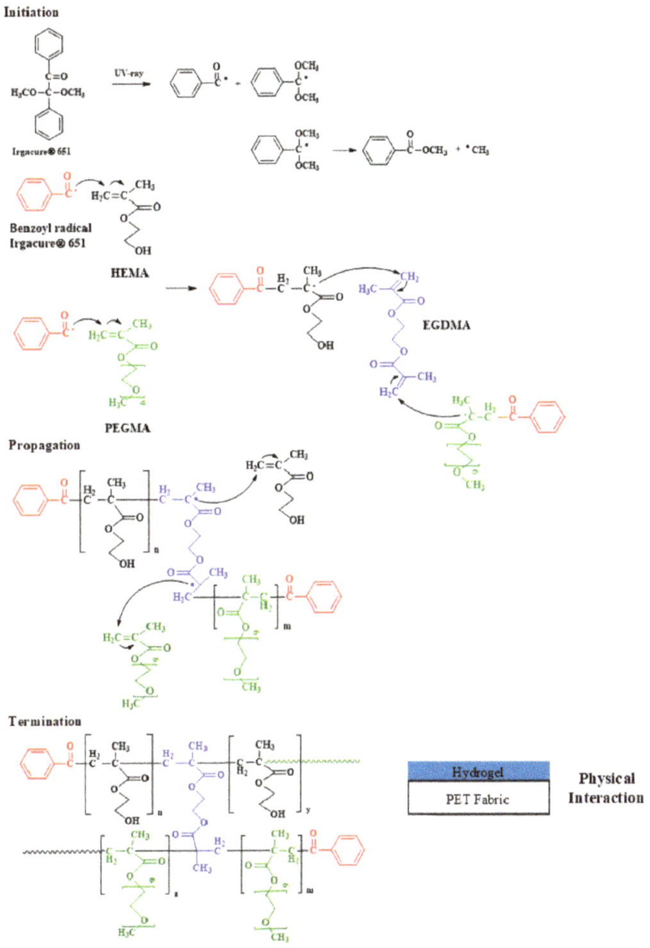

Scheme 1. Mechanism of photochemical cross-linking of 2-hydroxyethyl methacrylate (HEMA) and poly (ethylene glycol) methacrylate (PEGMA) onto PET fabric.

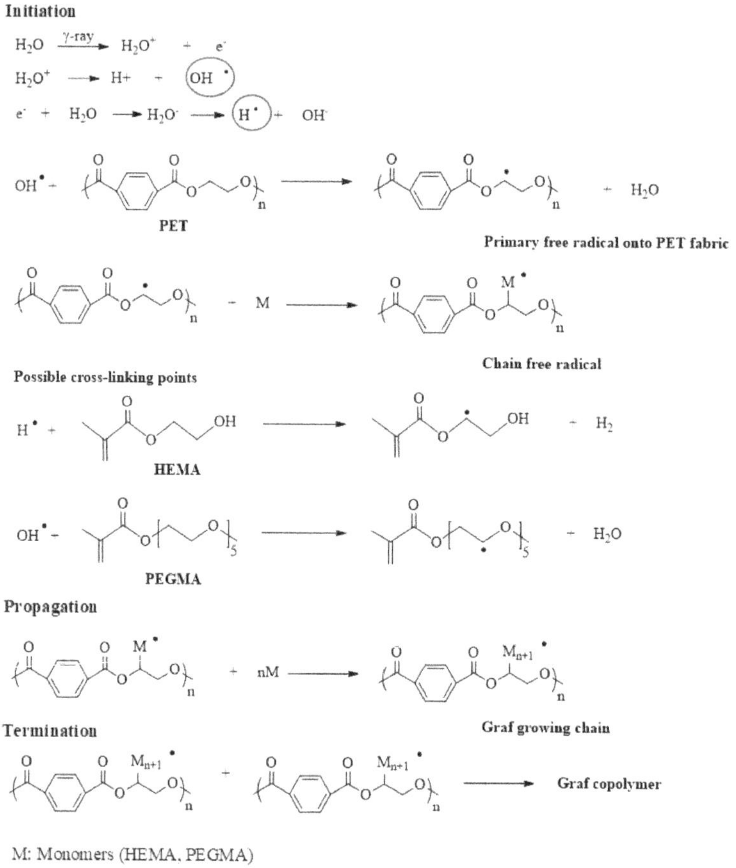

Scheme 2. Possible mechanism of grafting HEMA and PEGMA onto PET fabric by γ-ray induced graft copolymerization.

3.1.1. Characterization of Hydrogel-Grafted PET-Fabrics

Water Absorption

Tables 1 and 2 (last row) show the results of water absorption at the equilibrium for PET fabrics coated with hydrogel by the photochemical method and by gamma irradiation. All samples absorbed water in relatively high values (roughly 2 times their weight). It can be recognized that increased crosslinker content resulted in a lower water absorption, while increasing the PEGMA content resulted in higher water absorption for the hydrogel coatings prepared by the photochemical method (Table 1). In the case of hydrogel coatings prepared by gamma irradiation, the effect of increased water absorption due to the increased PEGMA content was also observed, however the sample prepared without PEGMA also showed a relatively high water absorption (Table 2). Since the crosslinking in the latter case is induced by the gamma rays (no chemical crosslinker added), it is possible that coatings without PEGMA resulted in lower crosslinking degree (higher water absorption). The effect of the coating thickness on water absorption was eliminated by subtracting the PET fabric weight in each case, see experimental section Equation (2).

FTIR-ATR

FTIR-ATR of modified fabrics shows bands centered at 3400 cm^{-1} corresponding to the hydroxyl groups from HEMA; this band was not observed in the pristine PET-fabric spectrum. A very strong band among 1713–1722 cm^{-1} corresponding to asymmetric vibrations of carbonyl groups and others around 1240 and 1095 cm^{-1} corresponding to C-O vibrations from ether groups were observed (Figure 1).

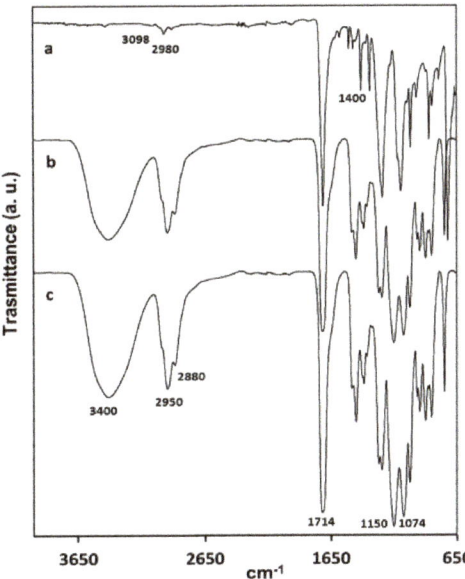

Figure 1. FTIR-ATR spectra of (**a**) unmodified PET, (**b**) UV PET$_{95:5:0.5}$ 67% hydrogel, and (**c**) γ60 PET$_{90:10:50}$ 400% hydrogel.

The bands due to the double bond from methacrylate groups were not observed, confirming the polymerization of the monomers included in the synthesis process.

Thermal Stability

The thermal stability was measured by TGA; unmodified PET showed a one-step decomposition at an average temperature of 424 °C (Figure 2a), whereas pure hydrogel of PHEMA obtained as a by-product from UV modification showed a loss of volatile compounds at 120 °C (<5%) and thermal decompositions at 285, 374, and 434 °C average temperatures (Figure 2e). Pure PHEMA hydrogel obtained by gamma irradiation showed similar steps of weight loss at average temperatures of 118, 286, 402, and 436 °C (Figure 3c); however, the percentages of weight loss were different. These differences in thermal decomposition may be due to crosslinking induced by gamma radiation forming C-C bonds increased the thermal stability of the hydrogel [21]. PHEMA decomposition occurred via a complex depolymerization mechanism [44]; therefore fabrics grafted with PHEMA by the photochemical method showed three decomposition steps similar to its hydrogel at 278, 360, and 445 °C (Figure 2b), whereas PET modified by gamma irradiation also showed three decomposition steps at 249, 311, and 444 °C (Figure 3b).

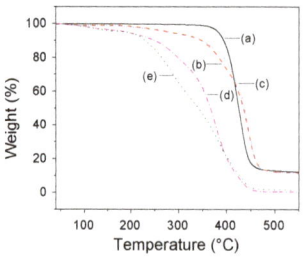
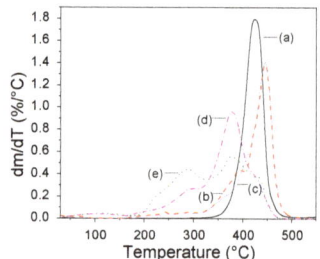

Figure 2. Weight loss and 1st derivate curves of (**a**) unmodified PET, (**b**) UV PET$_{95:5:0.5}$ (67% hydrogel), (**c**) UV PET$_{100:0:0.5}$ (96% hydrogel), (**d**) UV poly(HEMA-co-PEGMA) hydrogel, and (**e**) UV PHEMA hydrogel.

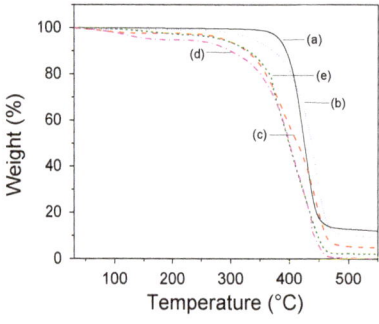
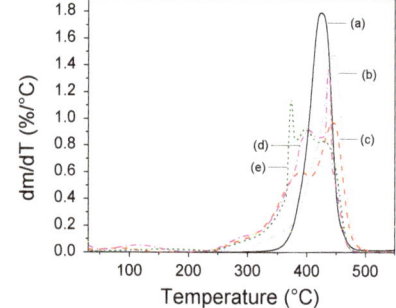

Figure 3. TGA curves of (**a**) unmodified PET, (**b**) γ60 PET$_{100:0:50}$ (257% hydrogel), (**c**) γ60 PHEMA hydrogel, (**d**) γ60 PET$_{90:10:50}$ (400% hydrogel), and (**e**) γ60 poly (HEMA-co-PEGMA) hydrogel.

These results allowed us to confirm the grafting of PHEMA onto PET fabrics. On the other hand, when poly (HEMA-co-PEGMA) was grafted, small differences in thermal stability were observed; pure hydrogel from the photochemical method showed the same values of thermal decompositions at 287, 378, and 440 °C but with different percentages of weight loss (Figure 2d), whereas pure poly (HEMA-co-PEGMA) hydrogel, obtained by gamma irradiation (Figure 3e), showed decomposition steps at 286, 376, 398, and 432 °C. PET modified with poly (HEMA-co-PEGMA) by the photochemical method showed three steps of decomposition at 242, 389, and 446 °C and PET modified by gamma irradiation also showed three steps of decomposition at 247, 389, and 445 °C. Results confirmed that the hydrogel were attached to PET fabrics.

DSC Analysis

PET is a semi-crystalline polymer with a Tg at 80 °C and a Tm at 253 °C (Figure 4a); fabrics modified with PHEMA or poly(HEMA-co-PEGMA) by photochemical (UV) method and gamma irradiation showed small variations, which are highlighted in Tables 3 and 4, respectively. Pure hydrogels are amorphous polymers; the one obtained by UV treatment with 5% of PEGMA showed a Tg at 83 °C (Figure 4c), whereas the one obtained by gamma irradiation with 10% of PEGMA showed a Tg at 96 °C (Figure 5c). The modified fabrics showed different Tg values; in Tables 3 and 4, we can observe that the higher the PEGMA percentage, the lower the Tg and by UV treatment, the higher the EGDMA (crosslinker) content, the higher the Tg value.

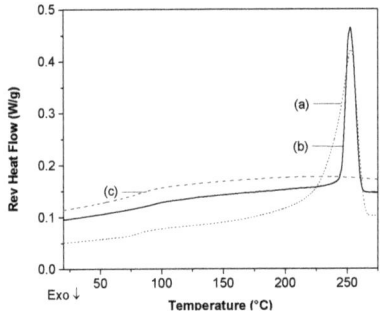

Figure 4. Differential scanning calorimetry (DSC) curves of (**a**) PET, (**b**) UV PET$_{95:5:0.25}$ 96% hydrogel, and (**c**) UV PHEMA hydrogel.

Table 3. Results by DSC for PET fabrics modified with hydrogel, obtained by the photochemical method.

Sample	HEMA:PEGMA (% mol)	EGDMA (% mol)	Hydrogel (wt/wt %)	T_g (°C)	T_m (°C)
Unmodified PET fabric	-	-	-	80	253
UV_PET$_{100\_0\_0.5}$	100:0	0.5	96	79	254
UV_PET$_{97\_3\_0.5}$	97:3	0.5	49	108	255
UV_PET$_{95\_5\_0.5}$	95:5	0.5	67	98	253
UV_PET$_{93\_7\_0.5}$	93:7	0.5	64	58	255
UV_PET$_{95\_5\_0.25}$	95:5	0.25	96	91	253
UV_PET$_{95\_5\_1}$	95:5	1	85	110	254

Table 4. Results by DSC for PET fabrics modified obtained by the direct gamma irradiation method.

Sample	Dose (kGy)	H:P (v/v %)	Hydrogel (wt/wt%)	T_g (°C)	T_m (°C)
Unmodified PET fabric	-	-	-	80	253
γ60_PET$_{100\_0\_50}$	60	100:0	257	95	255
γ60_PET$_{70\_30\_50}$	60	70:30	400	54	255
γ60_PET$_{80\_20\_50}$	60	80:20	400	76	255
γ60_PET$_{90\_10\_50}$	60	90:10	400	91	255
γ60_PET$_{90\_10\_40}$	60	90:10	213	96	254
γ60_PET$_{90\_10\_30}$	60	90:10	285	98	254
γ50_PET$_{90\_10\_50}$	50	90:10	285	88	255
γ70_PET$_{90\_10\_50}$	70	90:10	285	93	255

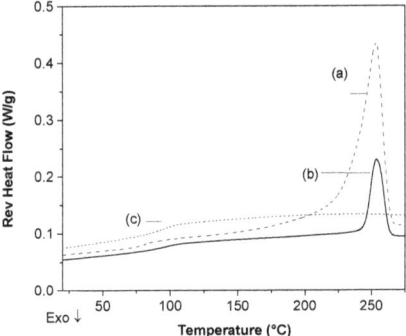

Figure 5. DSC curves of (**a**) PET, (**b**) γ60 PET$_{90:10:30}$ 400% hydrogel, and (**c**) γ60 poly(HEMA-co-PEGMA) hydrogel.

On the other hand, the higher the gamma radiation dose at a given HEMA:PEGMA ratio, the higher the grafting percentage and the higher the T_g value; this as a result of a higher crosslinking effect. In the case of the melting temperature, it was observed that the T_m was maintained between 253 and 255 °C, which indicated that the crystalline regions were not affected by the grafting methods; therefore, the grafting was mainly performed in the amorphous regions of PET, independently of the method used.

SEM

Cross-section of PET samples grafted with PHEMA and poly(HEMA-co-PEGMA) using UV method and gamma irradiation were analyzed by SEM and their thicknesses were measured. Samples modified by using the UV method with PHEMA showed a hydrogel cover of roughly 50 μm and the sample modified with poly(HEMA-co-PEGMA) (95:5%) using the same method had a cover of roughly 60 μm (Figure 6a,b). Samples grafted using gamma irradiation with PHEMA showed a thick cover of around 450 μm and of 600 μm for a sample with poly(HEMA-co-PEGMA) (90:10%) (Figure 6c,d).

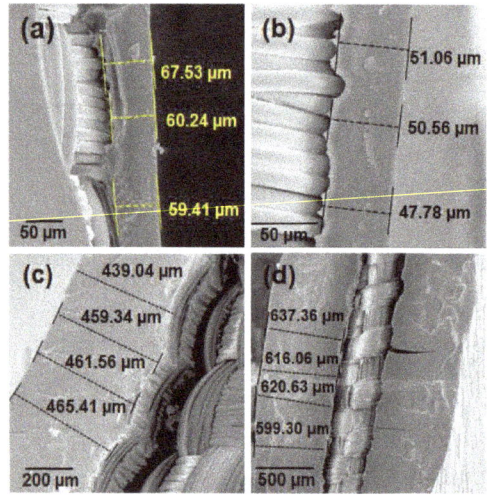

Figure 6. SEM images of (**a**) UV PET$_{100:0:0.5}$, (**b**) UV PET$_{95:5:0.5}$, (**c**) γ60 PET$_{100:0:50}$, and (**d**) γ60 PET$_{90:10:30}$.

Diameters of individual fabric fibers were also measured obtaining mean thickness of roughly 15.0 μm for pristine fibers and of 15.2 μm and 15.0 μm for samples modified by UV method with PHEMA and poly(HEMA-co-PEGMA) (95:5%) respectively (Figure 7), these constant diameters indicated that there was only a surface grafting of the hydrogel. On the other side, by gamma irradiation, an increase in thickness of fibers was observed; for example, the single fibers of the PET sample grafted with PHEMA had diameters around 21 μm; these results indicated that grafting was also performed inside of the fibers [41]. Also, a clear change on fabric fibers surface was observed (Figure 7e) when gamma irradiation was used.

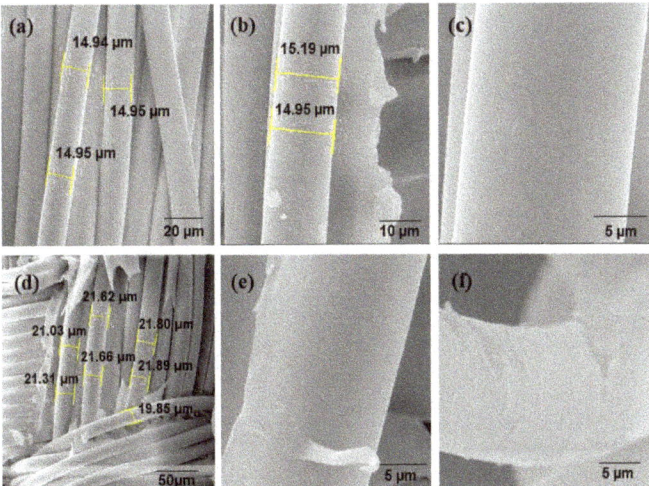

Figure 7. SEM images of fiber PET-fabric: (**a**) Unmodified PET, (**b**) and (**c**) UV PET$_{95:5:0.5}$, (**d**) and (**e**) γ60 PET$_{100:0:50}$, (**f**) γ60 PET$_{90:10:30}$.

3.2. Imparting PET-Fabrics Antibacterial Properties

3.2.1. Incorporation of Silver Nanoparticles

The in situ synthesis of AgNPs favors the interaction between the NPs and the hydrogel obtained on PET fabrics surface. As described above, immobilization of Ag$^+$ was carried out within the hydrogel and the PET fibers because silver ions from AgNO$_3$ solution have affinity toward R-COOH and R-OH groups [45]. The silver ions reduction reaction was carried out using NaBH$_4$, this reduction was optically evident since this was accompanied by a color change in the modified fabrics. In Figure 8a,b, the PET fabric with coatings obtained by the photochemical and the gamma irradiation methods, respectively, are observed; white transparent coatings were observed for both samples, however the sample obtained by gamma irradiation presented a thicker coating than the one obtained by the photochemical method. On the other hand, the fabrics modified with AgNPs changed their color. Figure 8c,d shows the back and front of the fabric without hydrogel and with hydrogel obtained by the photochemical method. A yellow color on the front of the fabric is clearly observed, indicating the presence of dispersed AgNPs. Figure 8e,f shows the back and front of the fabric without hydrogel and with hydrogel obtained by gamma radiation. The front cover is brown due to the fact that this side of the fabric has a thick layer of hydrogel, which provides the ability to trap more AgNPs allowing the agglomeration of these within the hydrogel cover. The back of the fabric shows also brownish translucent color from the AgNPs accumulated on the front side.

Figure 9 shows the images obtained by SEM along with mapping of silver and elemental analysis (spectrum) from EDS of PET fabrics loaded with AgNPs. From the SEM images, it is not clear where the AgNPs are located; some clustering that can be observed do not correspond with the mapping of silver, which is distributed along the whole sample, both in the surface and also in the cross-section. In Figure 9a,b, some examples of hydrogel-coated PET samples by UV-method (surface and cross-section) are shown. The apparent size of the silver mapping spots do not represent a true size. The same can be concluded from Figure 9c,d, which shows some examples of hydrogel-coated PET samples by the γ60-method. The distribution of AgNPs on the surface of coatings obtained by both the photochemical and the gamma radiation methods were similar; meanwhile, it was evident that the cross-sections demonstrate a high concentration of AgNPs in the PET fabrics prepared using both grafting methods, independently of the thickness of the grafted hydrogel on the surface. The size of the obtained AgNPs was not directly measured on the fabrics, however as an approximation, the same synthetic protocol was repeated but without the presence of the PET fabrics. The obtained AgNPs showed a surface plasmon resonance at a wavelength of 400 nm, which, according to the literature, corresponds to sizes below 20 nm [46].

Figure 8. Images of PET fabrics grafted with hydrogels without silver nanoparticles (AgNPs) by (**a**) the photochemical method, (**b**) the gamma irradiation method; and photos of both sides of PET fabrics grafted with hydrogel and AgNPs by method: (**c**,**d**) Photochemical, back and front view, respectively, and (**e**,**f**) gamma irradiation, back and front view, respectively.

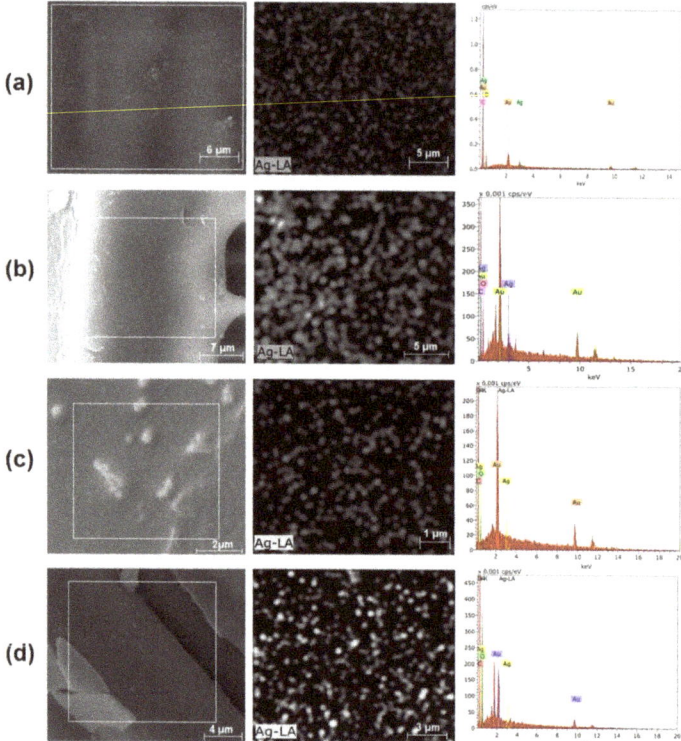

Figure 9. SEM images, EDS mapping of silver, and EDS spectrum from PET fabric modified with hydrogel and AgNPs: (**a**) surface of UV $PET_{95:5:0.5}$, (**b**) cross-section of UV $PET_{100:0:0.5}$, (**c**) surface of $\gamma 60\ PET_{90:10:30}$, and (**d**) cross-section of $\gamma 60\ PET_{100:0:50}$.

3.2.2. Antibacterial Test

The antibacterial activity of each sample was determined by measurements of inhibition halo against using two representative bacteria, *S. aureus* and *E. coli* (Table 5). The samples grafted by both methods without AgNPs did not show bacterial inhibition, whereas the samples with AgNPs showed inhibition against the two types of bacteria. By the photochemical method, it was evident that the inhibition was slightly stronger against *E. coli*, which is Gram-negative bacteria, whereas by the gamma radiation method, the samples showed inhibition against *E. coli* in all cases and against *S. aureus* in most of them; the inhibition halo was clearly larger against *E. coli*. Previous studies have suggested that antimicrobial effects of AgNPs may be associated with characteristics of certain bacterial species.

Table 5. Antibacterial results of modified PET fabrics.

Sample	Inhibition Distance (mm)		Inhibition Distance with Ag NP's (mm)		Hydrogel (%)
	S. aureus	*E. coli*	*S. aureus*	*E. coli*	
UV_PET$_{100:0:0.5}$	0	0	0.83 ± 0.3	1.00 ± 0.0	96
UV_PET$_{97:3:0.5}$	0	0	0.67 ± 0.6	1.33 ± 0.6	49
UV_PET$_{95:5:0.5}$	0	0	0.67 ± 0.6	0.67 ± 0.6	67
UV_PET$_{93:7:0.5}$	0	0	0.83 ± 0.3	1.83 ± 0.3	64
UV_PET$_{95:5:0.25}$	0	0	1.00 ± 0.5	1.33 ± 0.6	96
γ60_PET$_{100:0:50}$	0	0	0	2.0 ± 0.0	257
γ60_PET$_{90:10:50}$	0	0	0.83 ± 0.6	2.33 ± 0.6	400
γ60_PET$_{80:20:50}$	0	0	0	3.67 ± 0.8	213
γ60_PET$_{70:30:50}$	0	0	1.33 ± 1.3	3.33 ± 0.6	285
γ70_PET$_{90:10:50}$	0	0	0.67 ± 0.5	2.17 ± 0.8	285

Due to the structural difference in the composition of the membrane of Gram-positive and Gram-negative bacteria, AgNPs have significantly less effect on the growth of Gram-positive bacteria [47]. In our case, the comparison between both types of grafted fabrics (obtained by UV method or by direct gamma irradiation) demonstrated that there was a clear difference in the effectiveness against both types of bacteria when the gamma irradiation method was used for grafting. This may result from the fact that the thicker coating of hydrogel contains a larger amount of AgNPs and, therefore, a higher content on Ag$^+$ ions are presumably slowly released from the AgNPs surface. In this investigation, no tests at different contact times were carried out to corroborate the mentioned above; only tests at 24 h were carried out because the biofilm reaches its maturation at this time [48]. The AgNPs size within the samples is another important parameter to be considered in a future work. Besides, according to Taglietti et al., the antimicrobial effect could also be due to a direct contact between bacteria and the AgNPs [39,49].

4. Conclusions

Grafting of PHEMA and poly(HEMA-*co*-PEGMA) hydrogels onto the surface of PET fabrics was achieved by two methods: Photochemical crosslinking and gamma irradiation. Photochemical crosslinking used less energy to achieve a smooth and thin (~50 microns) coverage of the PET fabric not affecting the fiber diameter. The hydrogel cover changed the thermal behavior of the PET fabric, increasing its *Tg* with increased crosslinking of the hydrogel and decreasing the *Tg* with the increase in the PEGMA content. The grafted hydrogel was able to contain silver nanoparticles distributed randomly and able to inhibit bacterial growth against *S. aureus* and *E. coli*. Gamma irradiation was used to graft more profoundly PHEMA and poly(HEMA-*co*-PEGMA) into PET fabric, resulting in a thicker coverage of the surface of the fabric (600 microns), and therefore increasing the diameter of the single fibers, with similar impact on thermal properties of PET than the photochemical method; the gamma irradiation modified PET was also able to contain randomly distributed silver nanoparticles. The hydrogel coating containing silver nanoparticles was able to inhibit the growth of *S. aureus* to a similar extent than the photochemical method and inhibit better the growth of E. coli than the other method. Hydrogel-coated fabrics by both methods were highly hydrophilic.

Author Contributions: Conceptualization, Á.L.-C. and E.B.; methodology, K.A.M.-V., E.B., S.P.-S., and J.B.-S.; validation, K.A.M.-V., A.R.-J., and A.F.L.-N.; formal analysis, Á.L.-C.; investigation, K.A.M.-V. and J.B.-S.; resources, Á.L.-C., E.B., and A.F.L.-N.; writing—original draft preparation, K.A.M.-V. and A.R.-J.; writing—review and editing, S.P.-S., Á.L.-C., and A.F.L.-N.; supervision, Á.L.-C., E.B., and A.F.L.-N.; project administration, Á.L.-C., E.B., and A.F.L.-N.; funding acquisition, Á.L.-C., E.B., and A.F.L.-N.

Funding: This investigation was supported by grants from the National Council of Science and Technology of México [CONACYT-CNPq 2011-174378] and [CONACYT-CB-2012-178709], by a grant from the Tecnológico Nacional de México [TNM-5635.15-P] and from CICESE [685-101].

Acknowledgments: Technical support by Pedro Navarro-Vega and Felipe López-Saucedo is gratefully acknowledged.

Conflicts of Interest: The authors declare no conflict of interest. The funders had no role in the design of the study; in the collection, analyses, or interpretation of data; in the writing of the manuscript, or in the decision to publish the results.

References

1. Bhattacharya, A.; Misra, B.N. Grafting: A versatile means to modify polymers: Techniques, factors and applications. *Prog. Polym. Sci.* **2004**, *29*, 768–814. [CrossRef]
2. Neděla, O.; Slepička, P.; Švorčík, V. Surface Modification of Polymer Substrates for Biomedical Applications. *Materials* **2017**, *10*, 1115. [CrossRef]
3. Wichterle, O.; Lim, D. Hydrophilic Gels for Biological Use. *Nature* **1960**, *185*, 117–118. [CrossRef]
4. Hoffman, A.S. Hydrogels for Biomedical Applications. *Ann. N.Y. Acad. Sci.* **2001**, *944*, 62–73. [CrossRef] [PubMed]
5. Gregonis, D.E.; Russell, G.A.; Andrade, J.D.; Visser, A.C. Preparation and properties of stereoregular poly (hydroxyethyl methacrylate polymers and hydrogels. *Polymer* **1978**, *19*, 1279–1284. [CrossRef]
6. Tomic, S.L.; Mićic, M.M.; Dobić, S.N.; Filipović, J.M.; Suljovrujić, E.H. Smart poly (2-hydroxyethyl methacrylate/itaconic acid) hydrogels for biomedical application. *Radiat. Phys. Chem.* **2010**, *79*, 643–664. [CrossRef]
7. Sharrock, P.; Grégoire, G.J. HEMA reactivity with demineralized dentin. *J. Dent.* **2010**, *38*, 331–335. [CrossRef] [PubMed]
8. Mei, Y.; Wu, T.; Xu, Ch.; Langenbach, K.J.; Elliot, J.T.; Vogt, B.D.; Beers, K.L.; Amis, E.J.; Washburn, N.R. Tuning Cell Adhesion on Gradient Poly (2-hydroxyethyl methacrylate)-Grafted Surfaces. *Langmuir* **2005**, *21*, 12309–12314. [CrossRef]
9. Brahim, S.; Narinesingh, D.; Guiseppi-Eli, A. Synthesis and Hydration Properties of pH-Sensitive p(HEMA)-Based Hydrogels Containing 3-(Trimethoxysilyl)propyl Methacrylate. *Biomolecules* **2003**, *4*, 497–503. [CrossRef] [PubMed]
10. Lee, K.Y.; Mooney, D.J. Hydrogels for Tissue Engineering. *Chem. Rev.* **2001**, *101*, 1869–1879. [CrossRef]
11. Nurkeeva, Z.S.; Aal, A.S.; Kupchishin, A.I.; Khutoryanskiy, V.V.; Mun, G.A.; Beksyrgaeva, A.G. Radiation grafting from binary monomer mixtures. II. Vinyl ether of monoethanolamine and N-vinylpyrrolidone. *Radiat. Phys. Chem.* **2003**, *68*, 793–798. [CrossRef]
12. Meléndez-Ortiz, H.I.; Bucio, E.; Burillo, G. Radiation-grafting of 4-vinylpyridine and N-isopropylacrylamide onto polypropylene to give novel pH and thermo-sensitive films. *Radiat. Phys. Chem.* **2009**, *78*, 1–7. [CrossRef]
13. Contreras-Garcia, A.; Burillo, G.; Aliev, R.; Bucio, E. Radiation grafting of N, N'-dimethylacrylamide and N-isopropylacrylamide onto polypropylene films by two-step method. *Radiat. Phys. Chem.* **2008**, *77*, 936–940. [CrossRef]
14. Ramírez-Fuentes, Y.S.; Bucio, E.; Burillo, G. Radiation-induced grafting of N-isopropylacrylamide and acrylic acid onto polypropylene films by two step method. *Nucl. Instrum. Meth. B* **2007**, *265*, 183–186. [CrossRef]
15. Álvarez-Lorenzo, C.; Bucio, E.; Burillo, G.; Concheiro, A. Medical devices modified at the surface by γ-ray grafting for drug loading and delivery. *Expert Opin. Drug Del.* **2010**, *7*, 173–185. [CrossRef]

16. Vahdat, A.; Bahramia, H.; Ansaria, N.; Ziaie, F. Radiation grafting of styrene onto polypropylene fibres by a 10 MeV electron beam. *Radiat. Phys. Chem.* **2007**, *76*, 787–793. [CrossRef]
17. Hernández-Martínez, A.R.; Bucio, E. Novel pH and Temperature-Sensitive Behavior of Binary Graft DMAEMA/PEGMEMA onto LDPE Membranes. *Des. Monomer Polym.* **2009**, *12*, 543–552. [CrossRef]
18. Bucio, E.; Contreras-García, A.; Meléndez-Ortiz, H.I.; Muñoz-Muñoz, F.D.; Alvarez-Lorenzo, C.; Concheiro, A. Smart polymers for biomedical applications and graft synthesis by gamma-rays. In *Smart Polymeric Materials for Biomedical Applications*; NOVA Science Publishers: New York, NY, USA, 2010; pp. 277–306.
19. Ramírez-Jiménez, A.; Álvarez-Lorenzo, C.; Concheiro, A.; Bucio, E. Radiation-grafting of 2-hydroxyethylmethacrylate and oligo (ethylene glycol) methyl ether methacrylate onto polypropylene films by one step method. *Radiat. Phys. Chem.* **2012**, *81*, 27–32. [CrossRef]
20. Stannet, V.T. Radiation grafting state-of-the-art. *Radiat. Phys. Chem.* **1990**, *35*, 82–87. [CrossRef]
21. Charlesby, A. *Atomic Radiation and Polymers*; Pergamon Press: New York, NY, USA, 1960.
22. Li, J.; Lin, F.; Li, L.; Li, J.; Liu, S. Surface Engineering of Poly (ethylene terephthalate) for Durable Hemocompatibility via a Surface Interpenetrating Network Technique. *Macromol. Chem. Phys.* **2012**, *213*, 2120–2129. [CrossRef]
23. Liu, Y.X.; He, T.; Gao, C.Y. Surface modification of poly (ethylene terephthalate) via hydrolysis and layer-by-layer assembly of chitosan and chondroitin sulfate to construct cytocompatible layer for human endothelial cells. *Colloids Surf. B* **2005**, *46*, 117–126. [CrossRef] [PubMed]
24. Liu, S.; Zhao, N.; Rudenja, S. Surface Interpenetrating Networks of Poly (ethylene terephthalate) and Polyamides for Effective Biocidal Properties. *Macromol. Chem. Phys.* **2010**, *211*, 286–296. [CrossRef]
25. Ping, X.; Wang, M.; Ge, X. Surface modification of poly (ethylene terephthalate) (PET) film by gamma-ray induced grafting of poly (acrylic acid) and its application in antibacterial hybrid film. *Radiat. Phys. Chem.* **2011**, *80*, 567–572. [CrossRef]
26. Aubert-Viard, F.; Martin, A.; Chai, F.; Neut, C.; Tabary, N.; Martel, B.; Blanchemain, N. Chitosan finishing nonwoven textiles loaded with silver and iodide for antibacterial wound dressing applications. *Biomed. Mater.* **2015**, *10*, 015023. [CrossRef] [PubMed]
27. Lin, S.; Wang, Z.; Qi, J.C.; Wu, J.H.; Tiang, T.; Hou, L.L.; Hao, L.M.; Yang, J.Q. One-pot fabrication and antimicrobial properties of novel PET nonwoven fabrics. *Biomed. Mater.* **2011**, *6*, 045009. [CrossRef] [PubMed]
28. Kolar, M.; Mozetič, M.; Stana-Kleinschek, K.; Fröhlich, M.; Turk, B.; Vesel, A. Covalent Binding of Heparin to Functionalized PET Materials for Improved Haemocompatibility. *Materials* **2015**, *8*, 1526–1544. [CrossRef]
29. Vesel, A.; Kovac, J.; Primc, G.; Junkar, I.; Mozetic, M. Effect of H2S Plasma Treatment on the Surface Modification of a Polyethylene Terephthalate Surface. *Materials* **2016**, *9*, 95. [CrossRef]
30. Radetić, M.J. Functionalization of textile materials with silver nanoparticles. *Mater. Sci.* **2013**, *48*, 95–107. [CrossRef]
31. Singh, M.; Singh, S.; Prasad, S.; Gambhir, I.S. Nanotechnology in medicine and antibacterial effect of silver nanoparticles. *Dig. J. Nanomater. Biostruct.* **2008**, *3*, 115–122.
32. Chen, Y.H.; Su, C.H.; He, J.L. Antibacterial silver coating on poly (ethylene terephthalate) fabric by using high power impulse magnetron sputtering. *Surf. Coat. Tech.* **2013**, *232*, 868–875. [CrossRef]
33. Lee, H.J.; Jeong, S.H. Bacteriostasis of Nanosized Colloidal Silver on Polyester Nonwovens. *Text. Res. J.* **2004**, *74*, 442–447.
34. Deng, X.; Nikiforov, A.Y.; Coenye, T.; Cools, P.; Aziz, G.; Morent, R.; Geyter, N.; Leys, C. Antimicrobial nano-silver non-woven polyethylene terephthalate fabric via an atmospheric pressure plasma deposition process. *Sci. Rep.* **2015**, *5*, 10138. [CrossRef]
35. Slepicka, P.; Slepickova-Kasalkova, N.; Siegel, J.; Kolska, Z.; Bacakova, L.; Svorcik, V. Nano-structured and functionalized surfaces for cytocompatibility improvement and bactericidal action. *Biotech. Adv.* **2015**, *33*, 1120–1129. [CrossRef]
36. Ilić, V.; Šaponjić, Z.; Vodnik, V.; Molina, R.; Dimitrijević, S.; Jovančić, P.; Nedeljković, J.; Radetić, M. Antifungal efficiency of corona pretreated polyester and polyamide fabrics loaded with Ag nanoparticles. *J. Mat. Sci.* **2009**, *44*, 3983. [CrossRef]
37. Perelshtein, I.; Applerot, G.; Perkas, N.; Guibert, G.; Mikhailov, S.; Gedanken, A. Sonochemical coating of silver nanoparticles on textile fabrics (nylon, polyester and cotton) and their antibacterial activity. *Nanotechnology* **2008**, *19*, 245705. [CrossRef] [PubMed]

38. Milosević, M.; Radoicić, M.; Šaponjić, Z.; Nunney, T.; Marković, D.; Nedeljković, V.; Radetić, M. In situ generation of Ag nanoparticles on polyester fabrics by photoreduction using TiO$_2$ nanoparticles. *J. Mater. Sci.* **2013**, *48*, 5447–5455. [CrossRef]
39. D'Agostino, A.; Taglietti, A.; Grisoli, P.; Dacarro, G.; Cucca, L.; Patrini, M.; Pallavicini, P. Seed mediated growth of silver nanoplates on glass: Exploiting the bimodal antibacterial effect by near IR photo-thermal action and Ag+ release. *RSC Adv.* **2016**, *6*, 70414–70423. [CrossRef]
40. Collins, C.H.; Lynes, P.M.; Granje, J.M. Antimicrobial sensibility and assay test. In *Microbial Methods*; Collin, C.H., Lynes, P.M., Eds.; Butterworth: London, UK, 1989; pp. 155–168.
41. Contreras-Garcia, A.; Ramírez-Jiménez, A.; Bucio, E. Grafting polymerization induced by gamma-Rays. In *Gamma Rays Technology, Applications and Health Implications*; NOVA Science Publishers: New York, NY, USA, 2013; pp. 287–320.
42. Kaczmarek, H.; Galka, P. Effect of Irgacure 6551 initiator on poly (methyl methacrylate) photostability studied by UV-vis spectroscopy. *Open Proc. Chem. J.* **2008**, *1*, 8–11. [CrossRef]
43. Campbell, D.; Araki, K.; Turner, D.T. ESR study of free radicals formed by γ-irradiation of poly (ethylene terephthalate). *J. Polym. Sci. Part A Polym. Chem.* **1966**, *4*, 2597–2606. [CrossRef]
44. Demirelli, k.; Coşkun, M.; Kaya, E. A detailed study of thermal degradation of poly (2-hydroxyethyl methacrylate). *Polym. Degrad. Stabil.* **2001**, *72*, 75–80. [CrossRef]
45. Yuranova, T.; Rincon, A.G.; Bozzi, A.; Parra, S.; Pulgarin, C.; Albers, P.; Kiwi, J.J. Antibacterial textiles prepared by RF-plasma and vacuum-UV mediated deposition of silver. *J. Photochem. Photobiol. A* **2003**, *161*, 27–34. [CrossRef]
46. Solomon, S.; Bahadory, M.; Jeyarajasingam, A.V.; Rutkowsky, S.A.; Boritz, C. Synthesis and Study of Silver nanoparticles. *J. Chem. Educ.* **2007**, *84*, 322–325.
47. Franci, G.; Falanga, A.; Galdiero, S.; Palomba, L.; Rai, M.; Morelli, G.; Galdiero, M. Silver Nanoparticles as Potential Antibacterial Agents. *Molecules* **2015**, *20*, 8856–8874. [CrossRef] [PubMed]
48. Zhao, G.Q.; Ye, L.H.; Huang, Y.C.; Yang, D.K.; Li, L.; Xu, G. In vitro model of bacterial biofilm formation on polyvinyl chloride biomaterial. *Cell Biochem. Biophys.* **2011**, *61*, 371. [CrossRef] [PubMed]
49. D'Agostino, A.; Taglietti, A.; Desando, R.; Bini, M.; Patrini, M.; Dacarro, G.; Cucca, L.; Pallavicini, P.; Grisoli, P. Bulk Surfaces Coated with Triangular Silver Nanoplates: Antibacterial Action Based on Silver Release and Photo-Thermal Effect. *Nanomaterials* **2017**, *7*, 7. [CrossRef] [PubMed]

© 2019 by the authors. Licensee MDPI, Basel, Switzerland. This article is an open access article distributed under the terms and conditions of the Creative Commons Attribution (CC BY) license (http://creativecommons.org/licenses/by/4.0/).

Article

Polyethylene Terephthalate Textiles Enhance the Structural Maturation of Human Induced Pluripotent Stem Cell-Derived Cardiomyocytes

Mari Pekkanen-Mattila [1],*, Martta Häkli [1], Risto-Pekka Pölönen [1], Tuomas Mansikkala [1], Anni Junnila [1], Elina Talvitie [1], Janne T Koivisto [2], Minna Kellomäki [1] and Katriina Aalto-Setälä [3]

[1] BioMediTech, Faculty of Medicine and Health Technology, Tampere University, 33140 Tampere, Finland; martta.hakli@tuni.fi (M.H.); risto-pekka.polonen@tuni.fi (R.-P.P.); leo.mansikkala@oulu.fi (T.M.); junnila.anni@gmail.com (A.J.); elina.t.talvitie@gmail.com (E.T.); minna.kellomaki@tuni.fi (M.K.)
[2] Microelectronics Research Unit, University of Oulu, FI-90014 Oulu, Finland; janne.koivisto@oulu.fi
[3] Finland and Heart Hospital, Tampere University Hospital, 33100 Tampere, Finland; katriina.aalto-setala@tuni.fi
* Correspondence: mari.pekkanen-mattila@tuni.fi; Tel.: +358-50-4377153

Received: 15 May 2019; Accepted: 30 May 2019; Published: 3 June 2019

Abstract: Human-induced pluripotent stem cell-derived cardiomyocytes (hiPSC-CMs) have the potential to serve as a model for human cardiomyocytes. However, hiPSC-CMs are still considered immature. CMs differentiated from hiPSCs more resemble fetal than adult cardiomyocytes. Putative factors enhancing maturation include in vitro culture duration, culture surface topography, and mechanical, chemical, and electrical stimulation. Stem cell-derived cardiomyocytes are traditionally cultured on glass surfaces coated with extracellular matrix derivatives such as gelatin. hiPSC-CMs are flat and round and their sarcomeres are randomly distributed and unorganized. Morphology can be enhanced by culturing cells on surfaces providing topographical cues to the cells. In this study, a textile based-culturing method used to enhance the maturation status of hiPSC-CMs is presented. Gelatin-coated polyethylene terephthalate (PET)-based textiles were used as the culturing surface for hiPSC-CMs and the effects of the textiles on the maturation status of the hiPSC-CMs were assessed. The hiPSC-CMs were characterized by analyzing their morphology, sarcomere organization, expression of cardiac specific genes, and calcium handling. We show that the topographical cues improve the structure of the hiPSC-CMs in vitro. Human iPSC-CMs grown on PET textiles demonstrated improved structural properties such as rod-shape structure and increased sarcomere orientation.

Keywords: textile; PET; biomaterials; iPS-cells; cardiomyocytes; maturation; gene expression

---

1. Introduction

Cardiovascular diseases are the leading cause of death worldwide [1]. Cardiotoxicity is one of the main causes of withdrawal of drugs from the market [2]. Traditionally, new cardiac drugs and the cardiotoxicity of cardiac and non-cardiac drugs have been tested with rodent cardiomyocytes as well as with transfected non-cardiac cells [3–5]. However, the results of these experiments are not always applicable to humans. Therefore, more accurate human cardiomyocyte models are needed for preclinical analysis of drugs as well as for basic research and disease modeling of human cardiac diseases [6]. Human-induced pluripotent stem (hiPS) cells can be reprogrammed from any somatic cell by introducing the pluripotency factors [7] and these cells can be differentiated into functional cardiomyocytes with multiple methods, as recently reviewed [8]. However, these cells have been criticized as being immature and more resembling fetal than adult cardiomyocytes

(CMs) [9]. Compared to adult human CMs, hiPSC-CMs are small in size, round or multi-angular, and typically single-nucleated, whereas adult CMs are rod-like and large, with 25%–57% of the cells multi-nucleated [10,11]. The aspect ratio can be used as an indicator of the cell shape. Due to the round shape, hiPSC-CMs have an aspect ratio of (2–3): 1, whereas adult CMs are clearly longitudinally-oriented with an aspect ratio of (5–9):1. Sarcomeres of the hiPSC-CMs are disorganized and short (<2 μm), the sarcoplasmic reticulum is poorly developed, and the sarcolemma exhibits no transverse tubules [10,12]. The electrophysiological properties and the gene expression of the hiPSC-CMs differ from adult CMs [12].

The contractile ability of CMs is enabled by multiple sarcomere units that are integrated in the cytoskeleton of the cell [13]. The efficiency of the CM contraction relies on the proper orientation and length of the sarcomeres and on the rod-like shape of the cell [14]. Thus, the structural maturation of the cells has been the focus when studying maturation methods for hiPSC-CMs. Multiple methods and strategies have been suggested to improve the maturation of hiPSC-CMs in vitro, including topographical cues, substrate stiffness, medium additives, mechanical and electrical stimulation, genetic manipulation, and co-culture with other cell types [10,15]. However, a deeper understanding of the maturation process of hiPSC-CMs is still required to develop platforms to promote the maturation of the cells and producing hiPSC-CMs more resembling adult CMs.

Various scaffolds have been studied to orient hiPS-CMs, such as electrospun-aligned fiber textiles [16,17] and micro-grooved culture substrates [18]. Most of the cell cultures in vitro are coated on flat surfaces, which provide a two-dimensional (2D) environment for the cells. 2D culture forces cell polarization by providing cell-extra cellular matrix (ECM) connections on only one side of the cells [19,20]. Cells in 2D are usually flat because they try to spread out on the surface. By providing a three-dimensional (3D) environment with proper topographical cues and an extracellular matrix, cells can create more cell–ECM connections, which potentially affect cell proliferation and even differentiation and maturation [20]. Stiffness of a flat, 2D culture substrate affects cell properties, proliferation, and differentiation [21]. To increase the cardiac functionality and maturity, hiPS-CMs have been cultured with, for example, endothelial cells, and this has shown to enhance cardiomyocyte proliferation and functionality [22]. Co-culture of hiPS-CMs with cardiac fibroblasts has improved the structural and functional properties of the cells [23,24]. A similar maturation-enhancing effect was observed when cardiomyocytes were cultured on top of the vascular-like network produced from endothelial cells and fibroblasts [25,26]. Stem-cell-derived cardiomyocytes aligned according to the vascular structures of the network and their sarcomere structures were more oriented.

Textiles create a 3D culture environment and provide topographical support for different types of cells. Having a highly interconnective porous structure, textiles enable access of media and nutrients to the cells inside the material. Weaving is a conventional and basic textile technique that can also be used to fabricate tissue engineering scaffolds. Weaving enables the formation of textile structures with controllable properties, such as porosity, orientation, morphology, and mechanical properties. These parameters can be modified, for example, by changing the number of filaments, filament diameter, and weaving patterns. Biostable polyethylene terephthalate (PET) is one of the most used polyesters, and has many applications including in biomedical applications, for example as hernia meshes. It has also been used in many cell culture studies. The raw material of fibers can be changed too, and in the future, biodegradable textiles could be used as a vehicle for implantation of cardiomyocyte sheets for myocardial ischemia or scar repair applications [27–29].

In this study, PET textiles were used as culture substrates for hiPSC-CMs. The aligned textile fibers were hypothesized to provide sufficient topographical cues to improve the maturation state of hiPSC-CMs. The PET textiles had different weaving patterns, including a plain weave and a plain weave derivative, which altered their topography and other properties. They were coated with different biological compounds: Geltrex^{TM} (Thermo Fisher Scientific) and gelatin were used. The CMs were characterized by cell morphology, sarcomere organization, expression of cardiac specific genes, and calcium handling properties.

2. Materials and Methods

2.1. 1 Polyethylene Terephthalate Textile

Five different PET textiles were used (PET 1–5, Figure 1). They differed in color, texture of the fibers used as warp and weft, single filament (fiber) thickness, textile density, and pattern of the textile according to the details listed in Table 1. The average single filament thickness (measured from immunostaining images using autofluorescence of the fibers and ImageJ software used in the textiles varied between 20.4 and 24.4 μm without significant differences. All the textiles were of narrow fabric type, i.e., they were woven by a narrow-weaving loom having aligned fibers in their structure as warps. Perpendicular to the warps, the interweaving wefts formed the structure for the textile according to the pattern followed. The textile pattern of PET 1–4 was the same, plain weave, but other parameters varied. PET 5 was an in-house-designed plain weave derivative (pattern drawings in Figure 1). The textiles were woven narrow fabrics and the width of the textiles was 9 mm. The textiles were cut to pieces of 7–8 mm before final sterilization and cell seeding. Textiles were washed with ethanol (3–4 times washing with excess amounts of alcohol), followed by thorough drying before heat treatment. All the textiles were heat treated to stabilize the textile structure for the cell culture experiments.

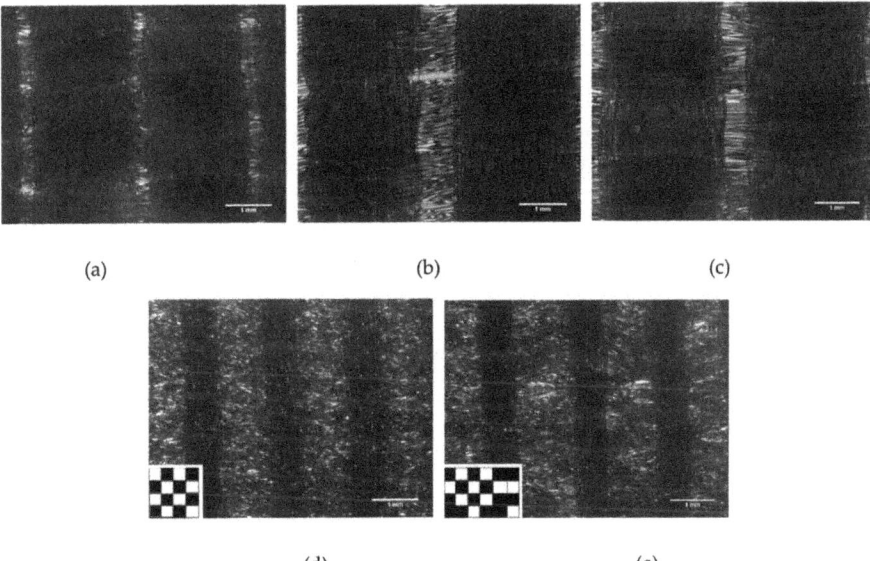

Figure 1. Structures of the polyethylene terephthalate (PET)-based textiles types 1–5 used in the present study, imaged with a Zeiss Axio Vert.A1 microscope (bright field) and AxioCam MRc5 camera using 5× objective. PET types 1–3 (**A–C**, respectively) were commercial textiles produced by Inka Oy, Killinkoski, Finland. PET types 4 and 5 (**D,E**, respectively) were produced at Tampere University, Tampere, Finland. Warp beams were provided by Finn-Nauha Oy, Haapamäki, Finland (yarn from Sinterama, Biella, Italy). The weaving type of the type 4 and 5 PET textiles were plain weave and plain weave derivative, respectively, as shown in lower left corner of the images.

Table 1. Details, manufacturers, and warp and weft type and diameter of a single filament (mean + SD) of the different PET textiles. Types 1, 2, and 3 were commercial textiles produced by Inka Oy, Killinkoski, Finland. Types 4 and 5 were produced at Tampere University, Tampere, Finland. Warp beams were provided by Finn-Nauha Oy, Haapamäki, Finland (yarn from Sinterama, Biella, Italy).

Textile Type and Details	Figure	Manufacturer	Warp/Weft	Single Filament ø (µm)
PET 1 Colorless, heat treated	1A	Inka Oy, Killinkoski, Finland	Textured/textured	24.4 ± 1.82
PET 2 Colorless, heat treated	1B	Inka Oy, Killinkoski, Finland	Straight/textured	23.2 ± 1.43
PET 3 Colorless, heat treated	1C	Inka Oy, Killinkoski, Finland	Textured/textured	22.9 ± 1.91
PET 4 Blue, heat treated	1D	Yarn: Finn-Nauha Oy, Haapamäki, Finland Textile: Tampere University of Technology	Straight/straight	20.4 ± 1.53
PET 5 Blue, heat treated	1E	Yarn: Finn-Nauha Oy, Haapamäki, Finland Textile: Tampere University of Technology	Straight/straight	22.0 ± 1.4

2.2. Textile Coating

The PET textiles were coated to create a thin layer on the textile to increase attachment without compromising textile topography. Five different surface coatings were tested in the optimization phase of the present study: Geltrex^{TM}, Gelatin, dopamine-bound gelatin coating, plasma-treatment of the textile without any coating, and plasma-treatment and gelatin-coating. Two replicates of each coating were used in each experiment. The glass coverslips were used as the control surface for hiPS-CMs. During the optimization phase of the experiment, all five PET types were used, and all different coating methods were tested for each PET type. In the experiment phase, only the gelatin coating was used for PET type 5.

To improve the attachment of the coating, as well as the attachment of the cells, PET textile was plasma-treated prior to gelatin coating. The plasma treatment was performed with plasma system Pico, Model 2, standard system controlled via PC and Windows CE operating system (control type C: PCCE control) and with reactive ion etching electrode. The electrode was type E (stainless steel), the generator was type D (13.56 MHz, 0-100W) (Diener electronic GmbH, Ebhausen, Germany) and the vacuum pump was Leybold 19 SC5D (Leybold Vacuum GmbH, Cologne, Germany). The gas used in the plasma treatment was O_2 and PET textiles were treated for 2 min in 0.4 mbar pressure with 50 W.

Prior to coating, the textiles and the coverslips were disinfected by washing with 70% ethanol (Altia, Rajamäki, Finland) and left to dry properly (1–2 h) before coating in the laminar hood.

Geltrex^{TM} and gelatin were used as coating materials. Geltrex^{TM} (Thermo Fisher Scientific, Waltham, Massachusetts, USA) was thawed and diluted 1:100 in KnockOut Dulbecco's Modified Eagle Medium (DMEM) (Thermo Fisher Scientific, Waltham, Massachusetts, USA). We pipetted 500 µL and 150 µL of diluted Geltrex^{TM} on PET textiles and coverslips, respectively, which were incubated at 37 °C for 1 hour. Excess coating was aspirated just prior to the cell plating in all cases.

Gelatin coating was performed in three different ways. In the first method, Type A porcine gelatin (Sigma-Aldrich, Saint Louis, Missouri, USA) was dissolved in phosphate buffered saline (PBS) to form 0.1% solution. We pipetted 500 µL and 150 µL of 0.1% gelatin solution on PET textiles and coverslips, respectively, which were incubated in room temperature for an hour. The second method involved using plasma treatment before gelatin coating. The third method was used to improve the attachment of gelatin to a polymer [27]. Dopamine hydrochloride (Sigma-Aldrich, Saint Louis, Missouri, USA) was used to crosslink gelatin with the PET fibers. The PET fibers were incubated in 2 g/L dopamine solution for 24 h on a shaking bed at room temperature. After washing with distilled water, the samples were incubated in 5% (w/v) Gelatin type A (Sigma-Aldrich, Saint Louis, Missouri, USA) solution for 24 h at 37 °C. After incubation in gelatin, the samples were washed overnight in distilled water at 37 °C to remove non-chemically bound gelatin.

2.3. Cell Culture and Differentiation of hiPSC-CMs

The hiPSC line UTA.04602, produced from dermal fibroblasts of a healthy individual and cultured as previously described [30], was used in the study. The ethical committee of Pirkanmaa Hospital

District (Tampere, Finland) approved collection of biopsies for generating patient-specific hiPSC lines and written informed consent was obtained from all the donors (Aalto-Setälä R08070). The hiPSCs were cultured in mTeSR1 medium (STEMCELL Technologies, Vancouver, Canada) on a Geltrex™ (Thermo Fisher Scientific, Waltham, Massachusetts, USA)-coated surface. The culture medium was changed three times a week for the cells and they were passaged for a one-week culture using Versene (Thermo Fisher Scientific, Cibco, Billings, Montana, USA).

Small molecule differentiation was achieved as previously described [31] with small exceptions. In short, the differentiation was initiated when the hiPSC-culture was 100% confluent (day 0) by changing the mTeSR1 medium to insulin-free RPMI/B27 (Thermo Fisher Scientific, Cibco, Billings, Montana, USA) medium containing 8 µM CHIR99021 (Tebubio, BPS Bioscience, San Diego, California, USA) and 0.5% penicillin/streptomycin. After 24 hours, the medium was changed to fresh insulin-free RPMI/B27 medium. On day three, half the medium was collected from the wells and mixed with fresh insulin-free RPMI/B27 medium. IWP-4 (Tocris, Bristol, England) was mixed with the medium so that the final concentration was 5 µM. The rest of the old medium was exchanged to IWP-4-containing medium. On days five and seven, the medium was changed to fresh insulin-free RPMI/B27 medium, and from day 10 forward, half the medium was changed three times a week to fresh RPMI/B27 medium with insulin (Thermo Fisher Scientific, Cibco, Billings, Montana, USA).

2.4. hiPS-CM Dissociation and Magnetic-Activated Cell Sorting

hiPSC-CMs were dissociated using two methods. For the PET coating optimization phase, the hiPS-CMs were dissociated using Collagenase A and suspended into a suspension medium containing KnockOut DMEM with 10% fetal bovine serum (Biosera, Nuaille, France), 1% non-essential amino acids (NEAA), 1% GlutaMAX-I (100×) (all from Thermo Fisher Scientific, Cibco, Billings, Montana, USA), and 0.5% penicillin/streptomycin (Lonza, Basel, Switzerland) [32].

To improve the purity of the hiPSC-CM population in the following experiments with PET 5, the cardiomyocytes were dissociated and separated from other cell types using magnetic-activated cell sorting (MACS) on day 21–27 of the differentiation. The cells were dissociated using a Multi Tissue Dissection Kit 3 (Miltenyi Biotec, Bergisch Gladbach, Germany) following the manufacturer's instructions. MACS sorting was performed using PSC-Derived Cardiomyocyte Isolation Kit, human (Miltenyi Biotec, Bergisch Gladbach, Germany). After cell sorting, the cells were resuspended in the suspension medium described above and the cells were plated on the gelatin-coated PET 5 textiles and gelatin-coated glass coverslips, which were used as controls.

2.5. Calcium Imaging

Calcium imaging was performed on day 12 after plating the cells to the PET 5 textiles. Ten independent PET 5 samples and two control samples were analyzed. Imaging was performed as previously described [33]. Shortly, the cells were loaded with 4 µM Fluo 4 AM (Thermo Fisher Scientific, Waltham, Massachusetts, USA) for 30 minutes at 37 °C. The sample was placed into an imaging chamber (RC-25, Warner Instruments, Hamden, Connecticut, USA) and the chamber was placed onto an Olympus XI71 microscope (Olympus, Tokyo, Japan) and connected to a perfusion system. Cells were perfused with 37 °C pre-heated perfusate solution consisting of 137 mM NaCl, 5 mM KCl, 1.2 mM $MgCl_2$, 0.44 mM KH_2PO_4, 4.2 mM $NaHCO_3$, 2 mM $CaCl_2$, 1 mM Na pyruvate, 5 mM D-glucose, and 20 mM HEPES dissolved in distilled water (pH adjusted to 7.4 with NaOH).

The adrenaline response of CMs on PET 5 was evaluated with 1 µM adrenaline (Sigma Aldrich, Saint Louis, Missouri, USA) from six independent samples. Baseline was recorded and hiPSC-CMs were treated with adrenaline for one minute and their response was recorded. Before a new baseline measurement, the adrenaline was washed off for at least two minutes. The recordings were performed with an Olympus XI71 microscope (Olympus, Tokyo, Japan) using ANDOR iXon+ camera and an Olympus UApo 20× 0.75 NA air objective and Live Acquisition software (TILL Photonics, Munich, Germany).

For calcium imaging analysis, single-beating hiPSC-CMs were selected as regions of interest and the analysis of fluorescence ($\Delta F/F_0$) videos were recorded using TILL Photonics Offline Analysis. The clampfit data analysis module of Axon pClamp 10 Electrophysiology Data Acquisition & Analysis software was used for peak detection (Molecular Devices, San Jose, California, USA). The studied peak parameters included peak duration, rise time from 10% to 90%, decay time from 90% to 10%, and peak frequency.

2.6. Immunocytochemistry

Immunocytochemistry was performed on day 10–11 after plating the cells on the PET 5 textile samples. The samples were fixed with 4% paraformaldehyde, blocked with 10% normal donkey serum (Biowest, Nuaille, France) solution, and stained with goat anti-cardiac troponin T (1:1,000, Abcam) and mouse anti-MyBPC3 (1:400, Santa Cruz Biotechnology, Dallas, Texas, USA) at 4 °C overnight. Donkey anti-goat Alexa Fluor 568 and donkey anti-mouse Alexa Fluor 488 (1:800, Thermo Fisher Scientific, Waltham, Massachusetts, USA) were used as secondary antibodies. The cell nuclei were stained using Vectashield mounting medium with DAPI (Vector Laboratories, Burlingame, California, USA). Fluorescence was visualized with a Nikon A1R+ Laser Scanning Confocal Microscope (Nikon, Tokyo, Japan) using a Nikon Apo 60× 1.40NA oil objective and with Zeiss Axio Imager.M2 with ApoTome.2 and AxioCamHRm3 camera using a Zeiss EC Plan-Neofluar 40× 1.30NA oil objective.

2.7. Analysis of Cell Alignment and Sarcomere Orientation

The orientation and sarcomere length of the hiPSC-CMs cultured on the PET textiles were analyzed from microscopy images using a spectral analysis tool, CytoSpectre [34]. CytoSpectre allows quantification of orientation and size distribution of cellular structures by using Fourier transform. In this study, the circular variance and wavelength of the detailed spectral component were used to determine the sarcomere orientation and modal sarcomere length of the hiPSC-CMs, respectively. Circular variance ranges from zero to one, with zero describing perfect anisotropy and one describing perfect isotropy. CytoSpectre determines the axes of the cell, which can be used to determine the aspect ratio. Prior to the analysis, the images were processed with ImageJ for masking.

2.8. Quantitative Reverse Transcription-Polymerase Chain Reaction (qRT-PCR)

hiPSC-CMs were prepared for qRT-PCR on day 1 and day 11 after plating the cells as previously described [35] to study the expression of several cardiac related genes. PET 5 and control samples (cells from glass coverslips) were collected from six independent experiments ($n = 6$). Two replicate samples from each independent experiment were collected. The cells were lysed with lysis solution of a CellsDirect One-Step qRT-PCR Kit (Invitrogen, Carlsbad, California, USA) following the manufacturer's protocol. The lysis was stored at −80 °C until genomic DNA degradation with DNase I and reverse transcription-specific target amplification (RT-STA) using the CellsDirect One-Step qRT-PCR Kit. Biomark HD (Fluidigm Corporation, San Francisco, California, USA) was used to perform the real-time qPCR according to the manufacturer's protocol. All samples were run as duplicates in Fluidigm Dynamic array-plates and the 2-$\Delta\Delta$CT [36] method was used to calculate relative expression. TATA-box binding protein (*TBP*), eukaryotic translation elongation factor 1 alpha 1 (*EEF1A1*), and glyceraldehyde-3-phosphate dehydrogenase (*GAPDH*) were used as endogenous control genes for data normalization. In assessment of the relative expression, day one samples were used as a calibrator for the data. These samples were similar to the controls samples, the cells were plated to glass coverslips, but cells were lysed one day after plating. Cells were collected from four coverslips ($n = 4$). The TaqMan assays used are listed in Table 2.

Table 2. TaqMan assays used in the Quantitative Reverse Transcription-Polymerase Chain Reaction (qRT-PCR).

Gene	Description	Function	TaqMan Assay ID
MYL2	Myosin regulatory light chain 2	Sarcomeric gene	Hs00166405_m1
MYL7	Myosin regulatory light chain 7	Sarcomeric gene	Hs01085598_g1
MYL9	Myosin regulatory light chain 9	Sarcomeric gene	Hs00697086_m1
MYH6	Myosin heavy chain 6	Sarcomeric gene	Hs01101425_m1
MYH7	Myosin heavy chain 7	Sarcomeric gene	Hs01110632_m1
TNNC1	Slow skeletal and cardiac type troponin C1	Sarcomeric gene	Hs00896999_g1
TNNT2	Cardiac type troponin T2	Sarcomeric gene	Hs00165960_m1
ACTN2	α-actinin 2	Sarcomeric gene	Hs00153809_m1
TTN	Titin	Sarcomeric gene	Hs00399225_m1
MYBPC3	Myosin binding protein C, cardiac	Sarcomeric gene	Hs00165232_m1
TPM1	α-tropomyosin	Sarcomeric gene	Hs00165966_m1
KCNH2	Potassium voltage-gated channel subfamily H member 2	Potassium channel	Hs04234270_g1
KCNH6	Potassium voltage-gated channel subfamily H member 6	Potassium channel	Hs00229215_m1
KCNA10	Potassium voltage-gated channel subfamily A member 10	Potassium channel	Hs1563550_s1
KCND3	Potassium voltage-gated channel subfamily D member 3	Potassium channel	Hs00542597_m1
KCNQ1	Potassium voltage-gated channel subfamily Q member 1	Potassium channel	Hs00923522_m1
HCN4	Hyperpolarization activated cyclic nucleotide-gated potassium channel 4	Potassium channel	Hs00975492_m1
SCN5A	Voltage-gated sodium channel, V type, alpha subunit	Sodium channel	Hs00165693_m1
CACNA1C	Voltage-dependent calcium channel, L type, alpha 1C subunit/CaCNA1.2	Calcium channel	Hs00167681_m1
SLC8A1	Solute carrier family 8, member 1/NCX1	Sodium-calcium exchanger	Hs01062258_m1
PLN	Phospholamban	Protein kinase substrate	Hs01848144_s1
ATP2A2	ATPase, calcium transporting, cardiac muscle, slow twitch 2/ SERCA2a	Calcium ATPase	Hs00544877_m1
EEF1A1; EE+	Eukaryotic translation elongation factor 1 alpha 1	Housekeeping gene	Hs00265885_g1
GAPDH	Glyceraldehyde-3-phosphate dehydrogenase	Housekeeping gene	Hs02758991_g1
TBP	TATA-box binding protein	Housekeeping gene	Hs00427620_m1

2.9. Statistical Analysis

The statistical significance of the differences in circular variance, sarcomere length, and height to width ratio of the hiPSC-CMs was assessed by Mann–Whitney U test where $p < 0.05$ was considered statistically significant. Assessing the statistical significance of differences in gene expression levels was performed using the Kruskal–Wallis test with Bonferroni correction. When comparing the calcium baseline measurements to adrenaline measurements, related samples' Wilcoxon Signed Rank Test was used. $p < 0.05$ was considered statistically significant. The data are presented as mean ± standard deviation.

3. Results

3.1. Attachment of the hiPSC-CMs to the PET Textiles

Five PET textiles (Figure 1) were tested as a scaffold for the hiPS-CMs. None of the fiber-related parameters (thickness range of the fibers and straight vs. textured quality of the fibers) or the weave pattern changed the behavior of the cells, but all the studied PETs (1–5) supported the growth of the hiPS-CMs in a similar manner (data not shown). PET 5, with a plain weave derivative pattern, was chosen for the following experiments. A combination of plain weave derivative pattern and the reed density used produced the most variating topography for the studied PET textile samples (Figure 1). PET 5 was also blue and had slight autofluorescence, which made the fibers visible with fluorescent imaging.

Gelatin has been used as a basic coating material for hiPS-CMs culturing in our laboratory; therefore, it was used also in the above-mentioned PET textile screening study. However, the number of the attached hiPS-CMs remained relatively low. To improve the cell attachment on the PET textile, coating with commercial basement membrane matrix Geltrex™ was also tested. In addition, plasma treatment prior to gelatin coating and dopamine-bound gelatin were tested. There were no clear differences in the cell attachment (Figure S1) or structural maturation state of the hiPS-CMs (Table S1) with the coating material or plasma treatment. Thus, after testing multiple PET textile types and coatings, PET 5 with normal gelatin coating was chosen for further experiments.

3.2. hiPSC-CM Morphology, Sarcomere Orientation, and Sarcomere Length

hiPSC-CMs cultured on PET 5 and glass coverslips were immunolabeled with Troponin T and Myocin binding protein C3 (MyBPC3) antibodies. Qualitative analysis revealed that the cells aligned according to the PET 5 textile fibers and exhibited clearly elongated structures and increased sarcomere orientation, as shown in Figure 2. The orientation of the CM sarcomeres was significantly higher on PET 5 ($n = 98$) compared to controls ($n = 174$), which was indicated by the lower circular variance (0.611 ± 0.162 and 0.882 ± 0.069, respectively; $p < 0.05$). Table 3 shows examples of the distribution of the sarcomeres in hiPSC-CM cultured on PET 5 and coverslip. Sarcomeres in CMs grown on PET 5 were more oriented than those in the controls. The difference in sarcomere length (Table 3) between PET 5 and control samples was not significant and was 1.736 ± 0.187 μm and 1.749 ± 0.122 μm on average, respectively. The shape of CMs was determined using the aspect ratio, and CMs grown on PET 5 had significantly higher aspect ratios than controls (4.915 ± 2.263 and 1.567 ± 0.455, respectively; $p < 0.05$), indicating that the cells exhibited a more rod-like structure essential for efficient contraction (Figure 2). However, confocal imaging revealed that the hiPS-CMs are still flat and wrap around single PET fibers (Figure 3).

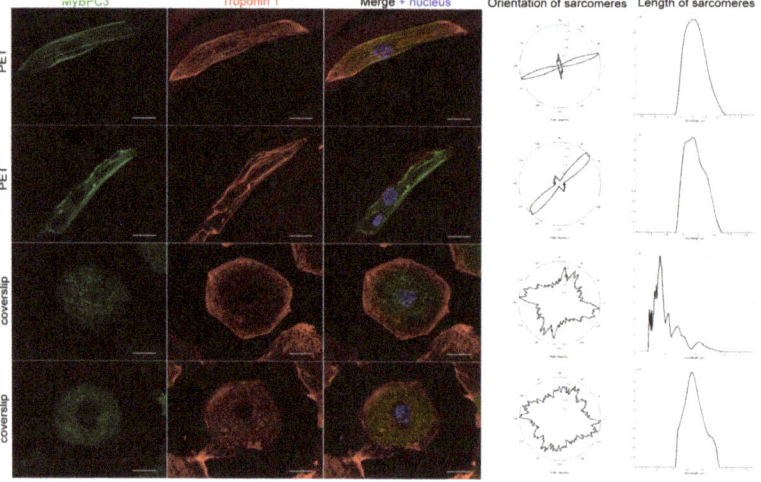

Figure 2. Two representative examples of the structure of human-induced pluripotent stem cell-derived cardiomyocytes (hiPSC-CMs) cultured on gelatin-coated polyethylene terephthalate (PET)-based textiles (PET) textiles and coverslips (controls). The hiPS-CMs were immunostained with myosin binding protein C (MyBPC3) (green) and Troponin T (red). The nuclei of the cells were stained with DAPI (blue). Scale bar is 25 μm. On PET 5, the cells and their sarcomeres clearly aligned according to the fibers of the textile, whereas the control cells exhibited no longitudinal axis or sarcomere orientation to one direction. Orientations of the sarcomeres were analyzed with CytoSpectre. The analysis results of sarcomere orientation and length of sarcomeres confirmed that the orientation of the sarcomeres improved when the cells were cultured on PET 5 textiles, but the sarcomere length distribution in the cells did not differ significantly.

Table 3. The data of the CytoSpectre analysis of cells grown on PET 5 and glass coverslips (control). The orientation of the CM sarcomeres was significantly higher on PET 5 compared to control ($p < 0.05$) as indicated by the average circular variance. The difference in sarcomere length between PET 5 and control samples was not significant. However, the shape of CMs was determined using the aspect ratio and CMs grown on PET 5 had a significantly higher ratio than the control ($p < 0.05$).

Sample	Average Circular Variance (0–1)	Average Modal Sarcomere Length (μm)	Average Aspect Ratio (Length to Width)	Number of Cells Analyzed
PET 5	0.611 ± 0.162	1.736 ± 0.187	4.915 ± 2.263	98
Control	0.882 ± 0.069	1.749 ± 0.122	1.567 ± 0.455	174

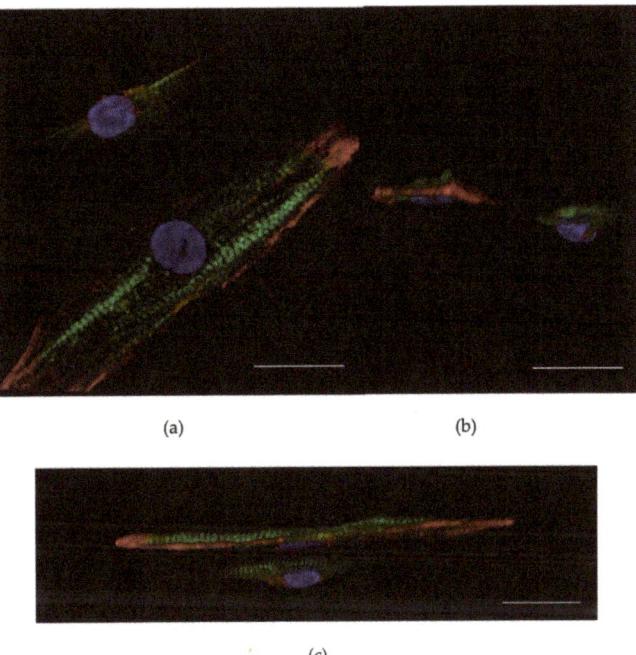

(a) (b)

(c)

Figure 3. (a–c) Confocal images from single human induced pluripotent stem cell-derived cardiomyocytes (hiPSC-CMs) on the polyethylene terephthalate PET type 5 textiles, with Troponin T (red), myosin binding protein C, MYBPC (green), nuclear stain DAPI (blue) from different projections. Images reveal that the hiPS-CMs were aligned with the PET fibers. However, the cells wrapped around the single PET 5 fibers and exhibited a relatively flat structure. The sarcomeres of the cells were clearly oriented along the fibers. Scale bar is 25 μm.

3.3. Calcium Handling

Differences in the calcium handling properties between hiPSC-CMs cultured on PET 5 textiles and coverslips were analyzed after 12 days of culture (the age of the cells was 33–39 days after initiation of differentiation). Cells exhibiting normal calcium transients were distinguished from those exhibiting arrhythmias and analyzed separately. The structure of the PET 5 did not hinder the Ca^{2+} imaging. There was no significant difference in the Ca^{2+} peak duration between CMs cultured on PET 5 ($n = 160$) or the control ($n = 40$) plates (582 ± 229 ms and 590 ± 202 ms, respectively; Table 4). However, there were statistically significant differences in rise and decay times, which were 112 ± 49 ms and 295 ± 131 for PET 5 samples and 90 ± 41 ms and 324 ± 96 ms for control samples ($p < 0.05$), respectively. This indicates that the release of calcium was slower while the uptake of calcium was faster for CMs grown on PET 5 compared to the control. Additionally, the amplitude of the peaks was significantly lower in CMs grown on PET 5 compared to controls (0.048 ± 0.037 ΔF/F$_0$ and 0.067 ± 0.030 ΔF/F$_0$ for PET and control samples, respectively; $p < 0.05$), indicating that control cells released more calcium during contraction cycles. The beating frequency was significantly higher in CMs grown on PET 5 compared to controls (0.93 ± 0.52 Hz and 0.75 ± 0.34 Hz, respectively; $p < 0.05$).

Table 4. Functionality of the hiPS-CMs cultured on PET 5 analyzed using Ca2+ imaging. The structure of the PET 5 textile did not hinder the Ca2+ imaging and the calcium handling properties were assessable from the hiPSC-CMs cultured on PET 5 textiles. Culturing on the PET 5 textiles slightly altered the calcium handling properties of the hiPSC-CMs; however, no significant changes were observed.

Sample	Peak Duration (ms)	Peak Amplitude (ΔF/F$_0$)	Rise Time from 10% to 90% (ms)	Decay Time from 90% to 10% (ms)	Peak Frequency (Hz)	Cell Number
PET 5	582 ± 229	0.048 ± 0.037	112 ± 49	295 ± 131	0.93 ± 0.52	160
Control	590 ± 202	0.067 ± 0.030	90 ± 41	324 ± 96	0.75 ± 0.34	40

The response of hiPS-CMs cultured on PET 5 to adrenaline and its effect on calcium handling properties was studied ($n = 43$) (Table 5 and Figure 4). There were statistically significant differences in the peak parameters between baseline and adrenaline measurements ($p < 0.05$). The peak duration decreased by 4.8% (648 ± 101 ms at baseline vs. 614 ± 87 after adrenaline). The peak amplitude decreased by 61.4% (0.0360 ± 0.0183 ΔF/F$_0$ at baseline vs. 0.0310 ± 0.0139 ΔF/F$_0$ after adrenaline). The rise time increased by 6.1% (115 ± 31 ms at baseline vs. 122 ± 34 ms after adrenaline). The decay time decreased by 7.6% (328 ± 73 ms at baseline vs. 303 ± 63 ms after adrenaline). The beating frequency increased by 21.2% (0.709 ± 0.254 Hz at baseline vs. 0.859 ± 0.242 Hz after adrenaline) as expected.

Table 5. Adrenaline significantly increased beating frequency and decreased peak duration in CMs grown on PET 5 textiles.

Sample	Peak Duration (ms)	Peak Amplitude (ΔF/F$_0$)	Rise Time from 10% to 90% (ms)	Decay Time from 90% to 10% (ms)	Peak Frequency (Hz)
Baseline	648 ± 101	0.0360 ± 0.0183	115 ± 31	328 ± 73	0.709 ± 0.254
Adrenaline	614 ± 87	0.0310 ± 0.0139	122 ± 34	303 ± 63	0.859 ± 0.242

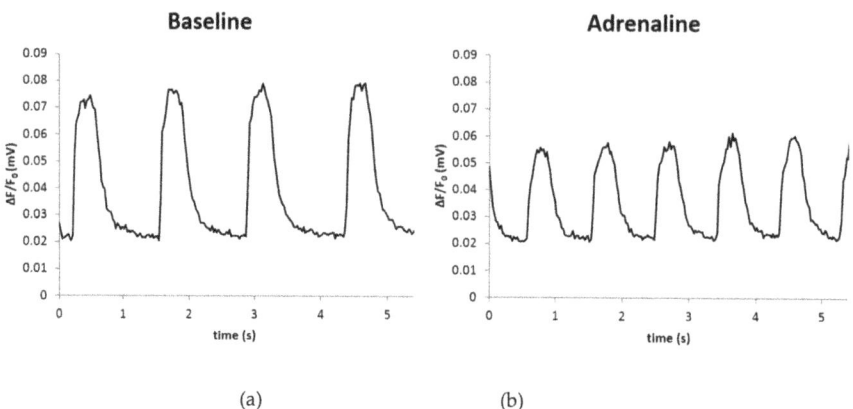

(a)　　　　　(b)

Figure 4. Adrenaline significantly increased beating frequency and decreased peak duration (a) when compared to the baseline (b) in hiPS-CMs grown on PET 5 textiles.

3.4. Expression of Cardiac-Specific Genes

The expression of cardiac specific genes was analyzed using qRT-PCR. Data from PET 5 ($n = 6$) and control coverslip ($n = 6$) samples collected at day 11 (the age of the cells was 32–38 days after differentiation initiation) were calibrated with the day 1 ($n = 4$) samples. Two biological replicates were analyzed from each sample, and all the samples were run as triplicates. TATA-box binding protein (*TBP*), eukaryotic translation elongation factor 1 alpha 1 (*EEF1A1*), and glyceraldehyde-3-phosphate dehydrogenase (*GAPDH*) were used as endogenous control genes for normalization. Overall, high variation was observed between the experiments. The expression levels of the genes coding for the contractile proteins, such as Troponin T (*TNNT2*), myosin binding protein C (*MYBPC*), and cardiac alpha actinin (*ACTN2*) had an increasing trend after the 11 days of culturing on both the PET 5 and control hiPS-CMs. However, only expression of TNNT2 was significantly higher in the hiPSC-CMs cultured on PET 5 compared to controls on glass coverslips ($p < 0.05$, Figure 5). The expression levels of the genes coding for cardiac ion channels were similar for the CMs cultured on glass coverslips and PET 5 (Figure S2).

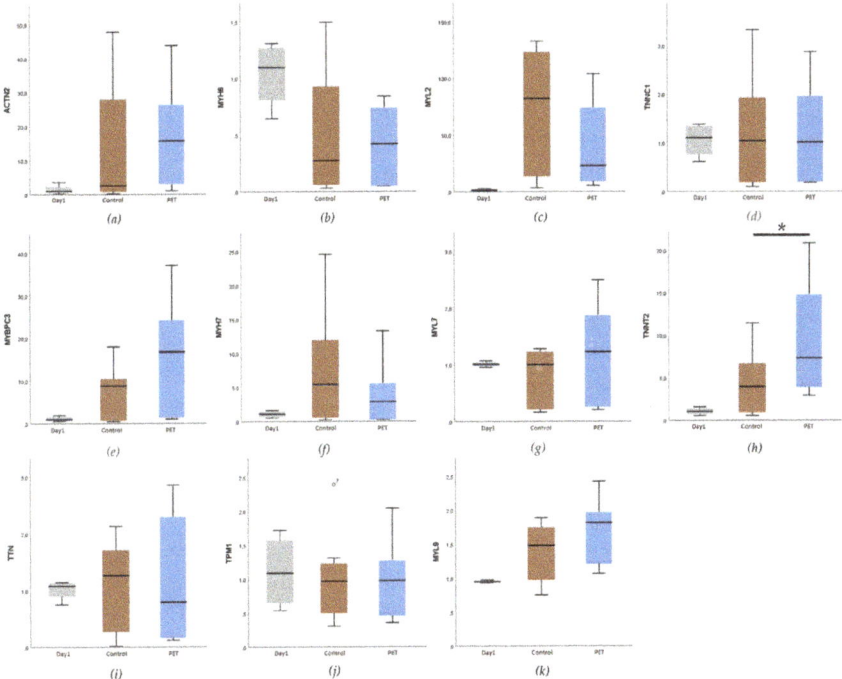

Figure 5. Expression levels of genes coding for the cardiac specific structural proteins: (**a**) α-actinin 2 (*ACTN2*), (**b**) myosin heavy chain 6 (*MYH6*), (**c**) myosin regulatory light chain 2 (*MYL2*), (**d**) slow skeletal and cardiac type troponin C1 (*TNNC1*), (**e**) myosin binding protein C, cardiac (*MYBPC3*), (**f**) myosin heavy chain 7 (*MYH7*), (**g**) myosin regulatory light chain 7 (*MYL7*), (**h**) cardiac type troponin T2 (*TNNT2*), (**i**) titin (*TTN*), (**j**) tropomyosin (*TPM1*), and (**k**) myosin regulatory light chain 9 (*MYL9*). Only the expression level of (**h**) *TNNT2* was significantly higher in hiPSC-CMs cultured on gelatin coated PET 5 textiles compared to the control sample ($p < 0.05$), marked with *.

4. Discussion

Several studies have reported the positive effects of a structured substrate on the maturation of hiPSC-CMs [16–18,37–39]. These findings have suggested that topographical cues can align the hiPSC-CMs and improve their morphology. In this study, we assessed the possibilities of textile structures being used as a culturing scaffold for hiPSC-CMs. The differentiated hiPSC-CMs were cultured on textile constructs, and their structure, sarcomere orientation, cardiac function, as well as expression of cardiac specific genes were analyzed. Based on our results, culturing hiPSC-CMs on PET textiles improved their structural properties, such as elongation and sarcomere orientation, as well as improved the expression of sarcomeric genes such as *TNNT2*.

Textiles as scaffolds have beneficial properties for cell culture experiments. First, they provide topographical cues for the cells and enable the transportation of nutrients through the porous material. Secondly, textiles provide support to form tissue-like three-dimensional structures by superimposing multiple textile layers with one or multiple types of cells. In the present study, PET-based textiles were chosen as the scaffold for hiPSC-derived CMs, since PET is one of the most used polyesters in biomedical applications [40].

The hiPSC-CMs cultured on the PET textile were clearly elongated along the fibers of the textile and the sarcomeres were more aligned than in CMs cultured on the standard gelatin coated glass coverslips. A similar alignment of stem cell-derived CMs and their sarcomeric structures was reported when cultured on micro-grooved substrates [18]. Compared to the micro-grooved cultures, PET culturing had a similar effect on the hiPS-CMs even though the cells were wrapped around the fibers of PET textiles, and therefore exhibited rather flat and rounded shapes compared to the more 3D structure of the CMs cultured in the microgrooves [18]. Rao et al. suggested that the alignment of the sarcomeric structures and the orientation of the cell along the fibers are enhanced because the focal adhesion complexes of the cells are formed parallel to the microgrooves. Therefore, the contraction of the cell is directed along the grooves and this orients the sarcomeres as well as the whole cell along the grooves. Similar phenomena might cause the orientation of hiPSC-CMs on PET fibers. The hiPS-CMs focal adhesion complexes might be formed parallel to the fibers, and due to this, the sarcomeres are oriented toward the same direction as the fibers. In addition to the microgrooves, nanogrooves have been reported to facilitate the alignment of cardiomyocytes [41]. Both of these studies speculated that the edges of the grooves are especially crucial in the formation of the focal adhesion complexes. Our data also showed that single 22.0 ± 1.4-μm-diameter fiber without any edges had a similar orienteering effect on the hiPS-CMs.

Culturing of hiPS-CMs on PET textile did not have an effect on the sarcomere length, which was approximately 1.7 μm and still shorter compared to the average of 2.2 μm in primary adult human CMs [41]. The sarcomere lengths reported for stem cell-derived cardiomyocytes have ranged from 1.4–1.7 μm [42] and some studies have reported that structured culture substrates that provide topographical cues have increased the sarcomere length. However, the sarcomere length has not reached the length of primary CMs [16,39,43]. An elongated structure and sarcomere orientation are important in terms of the efficiency of the CM contraction as the magnitude the CMs can contract in one direction increases with increased elongation and sarcomere orientation [14].

As mentioned above, topographical cues have been shown to align and improve the morphology of hiPSC-CMs toward a more mature phenotype. However, their effects on functionality and gene expression of cardiac specific genes have been controversial. In the present study, PET culturing enhanced the expression of the sarcomeric gene *TNNT2* compared to the hiPSC-CMs cultured on a flat surface. No statistically significant changes were observed in the expression of other cardiac-specific structural genes or in the genes encoding cardiac ion channels. Similar results have been reported earlier; topographical cues have oriented the cells but there has been no significant improvement at the gene expression level [18,43]. However, culturing of iPS-CMs on electrospun fibers had positive effects on the gene expression levels of genes coding for cardiac structural proteins as well as ion channels when compared to culturing on cell culture plastic. Interestingly, positive effects were seen observed regardless of the alignment of the electrospun fibers [16]. The time scale in these studies had been similar, cells were cultured on the scaffolds for two weeks. The longer culturing time of the hiPS-CMs on the surfaces providing topographical cues might enhance the expression level of cardiac

cardiac-specific genes in addition to the structural maturation. Furthermore, the controversy in gene expression studies might be due to the other cell types present in the cultures. Even though hiPS-CMs were sorted in the present study, there are other cell types left remained in the cultures with in varying quantities. Due to the lack of cardiac cardiac-specific markers for normalization, this phenomenon can cause the variation in qPCR studies.

According to our previous study in which hiPSC-CMs were cultured with a vascular-like network formed by human foreskin fibroblasts and human umbilical vein endothelial cells, the construct improved the hiPS-CM structure toward a more rod-like shape [25]. hiPSC-CMs had more elongated morphology and aligned with the tubular structures of the vascular-like network. Thus, our hypothesis in this present study was whether textile fibers would have similar orientating effect on the hiPSC-CMs as the tubular vascular structures. Our results support this hypothesis: Culturing of hiPSC-CMs on the PET textile had a positive effect on the morphology of the cells.

Previous studies analyzing the effect of CM orientation and anisotropy on calcium handling of hiPSC CMs have reported inconsistent results [16,18,21,38,43,44]. According to our data, CMs cultured on PET textiles had slightly altered calcium handling properties, but no significant changes were observed. We also studied the adrenaline response of hiPSC-CMs. Adrenaline significantly increased beating frequency and decreased Ca^{2+} peak duration in CMs grown both on PET and on control coverslips. Adrenaline decreased the calcium transient decay time, indicating improved calcium reuptake. Therefore, the beta-adrenergic pathway is functional in the hiPS-CMs grown on PET 5 and the cells respond to adrenaline as expected.

Throughout the study, the attachment of the dissociated hiPS-CMs to the textile structure was poor. To improve the attachment, multiple coating materials were tested. Regardless of the coating material, the level of attachment remained the same. A portion of the cells slid through the textile fibers without attaching to the substrate. After PET textile removal from the cell culture well, a high number of vital hiPS-CMs was observed (data not shown). However, regardless of the low number of cells attached, the cells that were attached to the material remained viable for an extended period of time.

Culturing on PET textiles supports the formation of an oriented sarcomere structure as well as alignment of the hiPSC-CMs, thus inducing the structural maturation of these cells. However, the textile culturing had only minor effects on the expression levels of the cardiac-specific genes. Additionally, the functionality was comparable to the culturing on gelatin-coated glass surfaces. Notably, compared to the micro- and nano-grooved culture surfaces [18,41], one cross-sectionally round PET fiber had similar effects on the hiPS-CMs' structure. It was earlier speculated that the edges of the grooves are especially important in the orientation of the cells. Here, we showed that the cross-sectionally round fibers of the textile structure support the cells and have a similar orientation effect on hiPS-CMs.

According to the results of the present study, for disease modeling studies as well as for drug screening and toxicology experiments, culturing of hiPS-CMs on textile structures would be beneficial. Compared to 2D cultures, culturing of hiPS-CMs would produce more standardized cultures, so the hiPS-CM population is more homogeneous in terms of cell structure and orientation. Therefore, for example, the effects of potential drug molecules on the cell structure and sarcomeres could be more reliably studied. However, more optimization is needed for the textile material, and the time scale for cell culturing on the scaffolds should be extended in future studies. Softer and more elastic textile material may be more suitable for hiPS-CMS. Material optimization is left for future studies.

5. Conclusions

In the present study, cardiomyocytes differentiated from the hiPSCs were cultured on PET textiles, and their structural properties, expression of the cardiac specific genes, and calcium handling properties were assessed. Based on the results, culturing hiPSC-CMs on the PET textiles improved their structural properties, such as elongation and sarcomere orientation, as well as improved the expression of the sarcomeric genes such as *TNNT2*. However, no statistically significant changes in the expression of the genes encoding cardiac ion channels or in the calcium handling properties of the hiPS-CMs were observed, and only minor changes were observed in their functionality as suggested by Ca^{2+} transients.

Supplementary Materials: The following are available online at http://www.mdpi.com/1996-1944/12/11/1805/s1, Figure S1: Immunostaining of hiPS-CMs on different coating materials, Figure S2: The expression levels of the genes coding cardiac ion channels, Table S1: Data describing the structural maturation state of the hiPS-CMs with the tested coating materials.

Author Contributions: Conceptualization, M.P.-M., J.K., M.K., and K.A.-S.; Data curation, M.P.-M.; Funding acquisition, M.K., and K.A.-S.; Investigation, M.P.-M., M.H., R.-P.P., T.M., A.J., E.T., and J.K.; Methodology, M.P.-M.; Project administration, M.P.-M., M.K., and K.A.-S.; Resources, E.T.; Supervision, M.P.-M., M.K., and K.A.-S.; Visualization, M.H.; Writing—original draft, M.P.-M., M.H., and R.-P.P.; Writing—review & editing, M.P.-M., J.K., M.K., and K.A.-S.

Funding: This research was funded by Business Finland, Päivikki and Sakari Sohlberg foundation, Pirkanmaa Regional Fund of the Finnish Cultural Foundation, the Finnish Foundation for Cardiovascular Research, Academy of Finland and Pirkanmaa Hospital District.

Acknowledgments: The authors acknowledge Henna Lappi and Markus Haponen for their technical expertise with the stem cell cultures and cardiac differentiation, as well as Kirsikka Stenlund for the design of the weave pattern. The authors acknowledge the Tampere facility of Electrophysiological Measurement, Tampere Imaging Facility (TIF), Tampere iPSC core facility and the Tampere CellTech Laboratories for their service.

Conflicts of Interest: The authors declare no conflict of interest.

References

1. GBD 2016 Causes of Death Collaborators. Global, regional, and national age-sex specific mortality for 264 causes of death, 1980–2016: A systematic analysis for the Global Burden of Disease Study 2016. *Lancet* **2017**, *390*, 1151–1210. [CrossRef]
2. Onakpoya, I.J.; Heneghan, C.J.; Aronson, J.K. Post-marketing withdrawal of 462 medicinal products because of adverse drug reactions: A systematic review of the world literature. *BMC Med.* **2016**, *14*, 10. [CrossRef] [PubMed]
3. Kannankeril, P.J.; Roden, D.M. Drug-induced long QT and torsade de pointes: Recent advances. *Curr. Opin. Cardiol.* **2007**, *22*, 39–43. [CrossRef]
4. Carlsson, L. In vitro and in vivo models for testing arrhythmogenesis in drugs. *J. Intern. Med.* **2006**, *259*, 70–80. [CrossRef] [PubMed]
5. Thomsen, M.B.; Matz, J.; Volders, P.G.A.; Vos, M.A. Assessing the proarrhythmic potential of drugs: Current status of models and surrogate parameters of torsades de pointes arrhythmias. *Pharmacol. Ther.* **2006**, *112*, 150–170. [CrossRef] [PubMed]
6. Sala, L.; Bellin, M.; Mummery, C.L. Integrating cardiomyocytes from human pluripotent stem cells in safety pharmacology: Has the time come? *Br. J. Pharmacol.* **2017**, *174*, 3749–3765. [CrossRef] [PubMed]
7. Takahashi, K.; Tanabe, K.; Ohnuki, M.; Narita, M.; Ichisaka, T.; Tomoda, K.; Yamanaka, S. Induction of Pluripotent Stem Cells from Adult Human Fibroblasts by Defined Factors. *Cell* **2007**, *131*, 861–872. [CrossRef]
8. Di Baldassarre, A.; Cimetta, E.; Bollini, S.; Gaggi, G.; Ghinassi, B.; Di Baldassarre, A.; Cimetta, E.; Bollini, S.; Gaggi, G.; Ghinassi, B. Human-Induced Pluripotent Stem Cell Technology and Cardiomyocyte Generation: Progress and Clinical Applications. *Cells* **2018**, *7*, 48. [CrossRef]
9. Smith, A.S.T.; Macadangdang, J.; Leung, W.; Laflamme, M.A.; Kim, D.H. Human iPSC-derived cardiomyocytes and tissue engineering strategies for disease modeling and drug screening. *Biotechnol. Adv.* **2017**, *35*, 77–94. [CrossRef]
10. Denning, C.; Borgdorff, V.; Crutchley, J.; Firth, K.S.A.; George, V.; Kalra, S.; Kondrashov, A.; Hoang, M.D.; Mosqueira, D.; Patel, A.; et al. Cardiomyocytes from human pluripotent stem cells: From laboratory curiosity to industrial biomedical platform. *Biochim. Biophys. Acta* **2016**, *1863*, 1728–1748. [CrossRef]
11. Lundy, S.D.; Zhu, W.-Z.; Regnier, M.; Laflamme, M. a Structural and functional maturation of cardiomyocytes derived from human pluripotent stem cells. *Stem Cells Dev.* **2013**, *22*, 1991–2002. [CrossRef] [PubMed]
12. Robertson, C.; Tran, D.D.; George, S.C. Concise Review: Maturation Phases of Human Pluripotent Stem Cell-Derived Cardiomyocytes. *Stem Cells* **2013**, *31*, 829–837. [CrossRef] [PubMed]

13. Dias, T.P.; Pinto, S.N.; Santos, J.I.; Fernandes, T.G.; Fernandes, F.; Diogo, M.M.; Prieto, M.; Cabral, J.M.S. Biophysical study of human induced Pluripotent Stem Cell-Derived cardiomyocyte structural maturation during long-term culture. *Biochem. Biophys. Res. Commun.* **2018**, *499*, 611–617. [CrossRef]
14. Kuo, P.L.; Lee, H.; Bray, M.A.; Geisse, N.A.; Huang, Y.-T.; Adams, W.J.; Sheehy, S.P.; Parker, K.K. Myocyte shape regulates lateral registry of sarcomeres and contractility. *Am. J. Pathol.* **2012**, *181*, 2030–2037. [CrossRef] [PubMed]
15. Ronaldson-Bouchard, K.; Ma, S.P.; Yeager, K.; Chen, T.; Song, L.; Sirabella, D.; Morikawa, K.; Teles, D.; Yazawa, M.; Vunjak-Novakovic, G. Advanced maturation of human cardiac tissue grown from pluripotent stem cells. *Nature* **2018**, *556*, 239–243. [CrossRef] [PubMed]
16. Han, J.; Wu, Q.; Xia, Y.; Wagner, M.B.; Xu, C. Cell alignment induced by anisotropic electrospun fibrous scaffolds alone has limited effect on cardiomyocyte maturation. *Stem Cell Res.* **2016**, *16*, 740–750. [CrossRef] [PubMed]
17. Parrag, I.C.; Zandstra, P.W.; Woodhouse, K.A. Fiber alignment and coculture with fibroblasts improves the differentiated phenotype of murine embryonic stem cell-derived cardiomyocytes for cardiac tissue engineering. *Biotechnol. Bioeng.* **2012**, *109*, 813–822. [CrossRef] [PubMed]
18. Rao, C.; Prodromakis, T.; Kolker, L.; Chaudhry, U.A.R.; Trantidou, T.; Sridhar, A.; Weekes, C.; Camelliti, P.; Harding, S.E.; Darzi, A.; et al. The effect of microgrooved culture substrates on calcium cycling of cardiac myocytes derived from human induced pluripotent stem cells. *Biomaterials* **2013**, *34*, 2399–2411. [CrossRef] [PubMed]
19. Lee, J.; Cuddihy, M.J.; Kotov, N.A. Three-Dimensional Cell Culture Matrices: State of the Art. *Tissue Eng. Part B Rev.* **2008**, *14*, 61–86. [CrossRef]
20. Wickström, S.A.; Niessen, C.M. Cell adhesion and mechanics as drivers of tissue organization and differentiation: Local cues for large scale organization. *Curr. Opin. Cell Biol.* **2018**, *54*, 89–97. [CrossRef]
21. Rodriguez, M.L.; Beussman, K.M.; Chun, K.S.; Walzer, M.S.; Yang, X.; Murry, C.E.; Sniadecki, N.J. Substrate Stiffness, Cell Anisotropy, and Cell–Cell Contact Contribute to Enhanced Structural and Calcium Handling Properties of Human Embryonic Stem Cell-Derived Cardiomyocytes. *ACS Biomater. Sci. Eng.* **2019**. [CrossRef]
22. Tulloch, N.L.; Muskheli, V.; Razumova, M.V.; Korte, F.S.; Regnier, M.; Hauch, K.D.; Pabon, L.; Reinecke, H.; Murry, C.E. Growth of engineered human myocardium with mechanical loading and vascular coculture. *Circ. Res.* **2011**, *109*, 47–59. [CrossRef] [PubMed]
23. Thavandiran, N.; Dubois, N.; Mikryukov, A.; Massé, S.; Beca, B.; Simmons, C.A.; Deshpande, V.S.; McGarry, J.P.; Chen, C.S.; Nanthakumar, K.; et al. Design and formulation of functional pluripotent stem cell-derived cardiac microtissues. *Proc. Natl. Acad. Sci. USA* **2013**, *110*, E4698–E4707. [CrossRef]
24. Conant, G.; Lai, B.F.L.; Lu, R.X.Z.; Korolj, A.; Wang, E.Y.; Radisic, M. High-Content Assessment of Cardiac Function Using Heart-on-a-Chip Devices as Drug Screening Model. *Stem Cell Rev. Rep.* **2017**, *13*, 335–346. [CrossRef]
25. Vuorenpää, H.; Penttinen, K.; Heinonen, T.; Pekkanen-Mattila, M.; Sarkanen, J.-R.; Ylikomi, T.; Aalto-Setälä, K. Maturation of human pluripotent stem cell derived cardiomyocytes is improved in cardiovascular construct. *Cytotechnology* **2017**, *69*, 785–800. [CrossRef] [PubMed]
26. Vuorenpää, H.; Ikonen, L.; Kujala, K.; Huttala, O.; Sarkanen, J.-R.; Ylikomi, T.; Aalto-Setälä, K.; Heinonen, T. Novel in vitro cardiovascular constructs composed of vascular-like networks and cardiomyocytes. *Vitr. Cell. Dev. Biol. Anim.* **2014**, *50*, 275–286. [CrossRef] [PubMed]
27. Giol, E.D.; Schaubroeck, D.; Kersemans, K.; De Vos, F.; Van Vlierberghe, S.; Dubruel, P. Bio-inspired surface modification of PET for cardiovascular applications: Case study of gelatin. *Colloids Surf. B Biointerfaces* **2015**, *134*, 113–121. [CrossRef] [PubMed]
28. Ribeiro, V.P.; Silva-Correia, J.; Nascimento, A.I.; da Silva Morais, A.; Marques, A.P.; Ribeiro, A.S.; Silva, C.J.; Bonifácio, G.; Sousa, R.A.; Oliveira, J.M.; et al. Silk-based anisotropical 3D biotextiles for bone regeneration. *Biomaterials* **2017**, *123*, 92–106. [CrossRef]
29. Ribeiro, V.P.; Almeida, L.R.; Martins, A.R.; Pashkuleva, I.; Marques, A.P.; Ribeiro, A.S.; Silva, C.J.; Bonifácio, G.; Sousa, R.A.; Oliveira, A.L.; et al. Modulating cell adhesion to polybutylene succinate biotextile constructs for tissue engineering applications. *J. Tissue Eng. Regen. Med.* **2017**, *11*, 2853–2863. [CrossRef]

30. Lahti, A.L.; Kujala, V.J.; Chapman, H.; Koivisto, A.-P.; Pekkanen-Mattila, M.; Kerkelä, E.; Hyttinen, J.; Kontula, K.; Swan, H.; Conklin, B.R.; et al. Model for long QT syndrome type 2 using human iPS cells demonstrates arrhythmogenic characteristics in cell culture. *Dis. Model. Mech.* **2012**, *5*, 220–230. [CrossRef]
31. Lian, X.; Hsiao, C.; Wilson, G.; Zhu, K.; Hazeltine, L.B.; Azarin, S.M.; Raval, K.K.; Zhang, J.; Kamp, T.J.; Palecek, S.P. Robust cardiomyocyte differentiation from human pluripotent stem cells via temporal modulation of canonical Wnt signaling. *Proc. Natl. Acad. Sci. USA* **2012**, *109*, E1848–E1857. [CrossRef] [PubMed]
32. Mummery, C.; Ward-van Oostwaard, D.; Doevendans, P.; Spijker, R.; van den Brink, S.; Hassink, R.; van der Heyden, M.; Opthof, T.; Pera, M.; de la Riviere, A.B.; et al. Differentiation of human embryonic stem cells to cardiomyocytes: Role of coculture with visceral endoderm-like cells. *Circulation* **2003**, *107*, 2733–2740. [CrossRef]
33. Pölönen, R.P.; Penttinen, K.; Swan, H.; Aalto-Setälä, K. Antiarrhythmic Effects of Carvedilol and Flecainide in Cardiomyocytes Derived from Catecholaminergic Polymorphic Ventricular Tachycardia Patients. *Stem Cells Int.* **2018**, *2018*, 1–11. [CrossRef] [PubMed]
34. Kartasalo, K.; Pölönen, R.-P.; Ojala, M.; Rasku, J.; Lekkala, J.; Aalto-Setälä, K.; Kallio, P. CytoSpectre: A tool for spectral analysis of oriented structures on cellular and subcellular levels. *BMC Bioinform.* **2015**, *16*, 344. [CrossRef] [PubMed]
35. Ojala, M.; Prajapati, C.; Pölönen, R.-P.; Rajala, K.; Pekkanen-Mattila, M.; Rasku, J.; Larsson, K.; Aalto-Setälä, K. Mutation-Specific Phenotypes in hiPSC-Derived Cardiomyocytes Carrying Either Myosin-Binding Protein C Or α -Tropomyosin Mutation for Hypertrophic Cardiomyopathy. *Stem Cells Int.* **2016**, *2016*, 1–16. [CrossRef] [PubMed]
36. Livak, K.J.; Schmittgen, T.D. Analysis of relative gene expression data using real-time quantitative PCR and the 2(-Delta Delta C(T)) Method. *Methods* **2001**, *25*, 402–408. [CrossRef]
37. Carson, D.; Hnilova, M.; Yang, X.; Nemeth, C.L.; Tsui, J.H.; Smith, A.S.T.; Jiao, A.; Regnier, M.; Murry, C.E.; Tamerler, C.; et al. Nanotopography-Induced Structural Anisotropy and Sarcomere Development in Human Cardiomyocytes Derived from Induced Pluripotent Stem Cells. *ACS Appl. Mater. Interfaces* **2016**, *8*, 21923–21932. [CrossRef]
38. Xu, C.; Wang, L.; Yu, Y.; Yin, F.; Zhang, X.; Jiang, L.; Qin, J. Bioinspired onion epithelium-like structure promotes the maturation of cardiomyocytes derived from human pluripotent stem cells. *Biomater. Sci.* **2017**, *5*, 1810–1819. [CrossRef]
39. Huethorst, E.; Hortigon, M.; Zamora-Rodriguez, V.; Reynolds, P.M.; Burton, F.; Smith, G.; Gadegaard, N. Enhanced Human-Induced Pluripotent Stem Cell Derived Cardiomyocyte Maturation Using a Dual Microgradient Substrate. *ACS Biomater. Sci. Eng.* **2016**, *2*, 2231–2239. [CrossRef]
40. Maitz, M.F. Applications of synthetic polymers in clinical medicine. *Biosurf. Biotribol.* **2015**, *1*, 161–176. [CrossRef]
41. Wang, P.Y.; Yu, J.; Lin, J.H.; Tsai, W.B. Modulation of alignment, elongation and contraction of cardiomyocytes through a combination of nanotopography and rigidity of substrates. *Acta Biomater.* **2011**, *7*, 3285–3293. [CrossRef]
42. Sarantitis, I.; Papanastasopoulos, P.; Manousi, M.; Baikoussis, N.G.; Apostolakis, E. The cytoskeleton of the cardiac muscle cell. *Hell. J. Cardiol.* **2012**, *53*, 367–379.
43. Khan, M.; Xu, Y.; Hua, S.; Johnson, J.; Belevych, A.; Janssen, P.M.L.; Gyorke, S.; Guan, J.; Angelos, M.G. Evaluation of changes in morphology and function of human induced pluripotent stem cell derived cardiomyocytes (hiPSC-CMs) cultured on an aligned-nanofiber cardiac patch. *PLoS ONE* **2015**, *10*, e0126338. [CrossRef] [PubMed]
44. Feinberg, A.W.; Ripplinger, C.M.; Van Der Meer, P.; Sheehy, S.P.; Domian, I.; Chien, K.R.; Parker, K.K. Stem Cell Reports Repor t Functional Differences in Engineered Myocardium from Embryonic Stem Cell-Derived versus Neonatal Cardiomyocytes. *Stem Cell Rep.* **2013**, *1*, 387–396. [CrossRef] [PubMed]

© 2019 by the authors. Licensee MDPI, Basel, Switzerland. This article is an open access article distributed under the terms and conditions of the Creative Commons Attribution (CC BY) license (http://creativecommons.org/licenses/by/4.0/).

Article

The Preparation and Characterization of Polyacrylonitrile-Polyaniline (PAN/PANI) Fibers

Iwona Karbownik [1], Olga Rac-Rumijowska [2,*], Marta Fiedot-Toboła [3], Tomasz Rybicki [1] and Helena Teterycz [2]

1. Faculty of Electrical, Electronic, Computer and Control Engineering, Technical University of Łódź, Żeromskiego 116, 90-924 Łódź, Poland; ivakabari@gmail.com (I.K.); trybicki@p.lodz.pl (T.R.)
2. Faculty of Microsystem Electronics and Photonics, Wrocław University of Science and Technology, Janiszewskiego 11/17, 50-372 Wrocław, Poland; helena.teterycz@pwr.wroc.pl
3. Polish Centre for Technology Development PORT, Stabłowicka 147, 54-066 Wrocław, Poland; marta.fiedot-tobola@eitplus.pl
* Correspondence: olga.rac@pwr.edu.pl

Received: 21 January 2019; Accepted: 18 February 2019; Published: 22 February 2019

Abstract: The paper presents a method of modifying polyacrylonitrile (PAN) fibers using polyaniline (PANI). The PAN fibers were doped with polyaniline that was obtained in two different ways. The first consisted of doping a spinning solution with polyaniline that was synthesized in an aqueous solution (PAN/PANI blended), and the second involved the synthesis of polyaniline directly in the spinning solution (PAN/PANI in situ). The obtained fibers were characterized by the methods: X-ray powder diffraction (XRD), scanning electron microscope (SEM), fourier-transform infrared spectroscopy (FTIR), thermogravimetry (TG) and differential scanning calorimetry (DSC). Analysis of the results showed strong interactions between the nitrile groups of polyacrylonitrile and polyaniline in the PAN/PANI in situ fibers. The results of mechanical strength tests indicated that the performance of the PAN/PANI mixture significantly improved the mechanical parameters of polyaniline, although these fibers had a weaker strength than the unmodified PAN fibers. The fibers obtained as a result of the addition of PANI to PAN were dielectric, whereas the PANI-synthesized in situ were characterized by a mass-specific resistance of 5.47 kΩg/cm^2.

Keywords: polyacrylonitrile; polyaniline; conductive fibers

1. Introduction

Fibers, regardless of whether they are a natural or synthetic product, are used primarily for the production of clothing and technical fabrics. Recently, more and more attention has been paid to giving completely new properties to fabrics, such as antibacterial or photocatalytic properties, as well as properties of electrical conductivity. Electrically conductive fibers can be used in textronic products, which are multifunctional textile materials that have the features of standard textile products, but at the same time have the functions of electronic components. Electroconductive fibers, depending on their resistance, can be used as electrical connections, sensors, or electromagnetic field shielding materials. However, the mechanical properties of the electrically conductive fibers must allow their further processing in order to obtain a textile product.

Polyaniline, apart from polypyrrole (PPy) or poly(3,4-ethylenedioxythiophene) (PEDOT), is one of the most popular conductive polymers because of its reversible redox, pH-switching, and sensing properties, and its simple synthesis [1]. However, like most of these types of materials, it has poor mechanical properties and thermal stability. Due to the fact that it is difficult to use melt techniques like extrusion, polyaniline fibers are formed only from spinning solution in a spinning process [2]. The polyaniline fibers were initially obtained from pure polyaniline, and in order to increase its electrical conductivity it was doped with concentrated sulfuric acid during the preparation of the spinning solution. This improves the electrical conductivity, but weakens the mechanical properties of the fibers [3]. One of the solutions is to use high molecular weight polyaniline [4,5]. However, the most common solution is to combine polyaniline with other polymers to increase its mechanical strength and to facilitate its processing. For this purpose, polymers known for their high strength parameters, poly–ω–aminoundecanoyle (nylon-11) and poly-phenylene terephthalamide (Kevlar), were used in a wet-spinning process [6–9]. These fibers were characterized by a much higher mechanical strength in comparison to pure polyaniline fibers; however, they had several orders of magnitude lower electrical conductivity [2].

One of the polymers used for this purpose is also a polyacrylonitrile, which is characterized by high mechanical strength, as well as thermal and chemical resistance, and also low price. In addition, it is a fiber-forming polymer popularly used for the production of carbon fibers.

Zhai et al. obtained flexible and conductive layers as a result of direct polymerization of aniline on the surface of microporous polyacrylonitrile. The use of PAN improved the mechanical properties and thermal stabilities of the polyaniline layers [10]. The combination of PAN and PANI polymers is used in medicine, where copolymer is used as scaffolds for muscle cells [11]. Polymer membranes were made from the PANI-doped polyacrylonitrile, and their color depended on the pH of the solution [12]. Moreover, fibers from the PAN/PANI composition were made by electrospinning, where an increase in electrical conductivity towards polyacrylonitrile was observed [13]. Toptas et al. obtained conductive polyaniline/polyacrylonitrile composite fibers by sorption of aniline and polyacrylonitrile fiber [14]. These fibers had satisfactory electrical conductivities, mechanical strengths, and resistances for laundering. Xia and Lu obtained conductive fibers in which polyacrylonitrile fibers were coated with polypyrrole, polyaniline, and poly (3,4-ethylenedioxythiophene). The polymerization occurred directly on the surface of the PAN fibers, by way of a chemical connection between polyacrylonitrile chains and conjugated polymers that was created [15]. There are also studies in which polyaniline fibers are doped with carbon nanotubes (CNT) with high electrical conductivity and high mechanical strength [16].

This article presents the results of research on PAN/PANI fibers, obtained using the wet method, from solutions in which the polymer matrix was polyacrylonitrile (PAN) and admixed polyaniline (PANI). PANI was introduced to PAN fibers using two methods:

- the blending method, where PANI was mixed with the spinning solution,
- in situ synthesis of polyaniline in a spinning solution.

The present method, where polyaniline is synthesized in situ during the preparation of a spinning solution for forming PAN/PANI composite fibers by wet spinning, is new and has not been described in literature.

2. Materials and Methods

2.1. Materials

As a fiber-forming polymer, polyacrylonitrile (PAN) (Good-Fellow Co., Huntingdon, UK) was used, which contained 99.5 wt.%. polyacrylonitrile and 0.5 wt.%. methyl polyacrylate. Polyaniline was prepared from aniline chloride $C_6H_7N \cdot HCl$ (Sigma-Aldrich Co., St. Louis, MO, USA) and ammonium persulphate $(NH_4)_2S_2O_8$ (POCH, Gliwice, Poland). Dimethylformamide (DMF) (Chempur, Piekary Śląskie, Poland) was used as a solvent for the polyacrylonitrile.

2.2. Blended Polyaniline (PANI) with Spinning Dope

Polyaniline (PANI) was obtained by mixing an aqueous solution of 0.5 M aniline hydrochloride $(C_6H_7N \cdot HCl)$ and 0.6 M ammonium persulfate $((NH_4)_2S_2O_8)$. Ammonia was added to the mixture after 4 h and the resulting precipitate was filtered using a Schott funnel. The obtained precipitate was washed with methanol, dried at room temperature, and then ground to a fine powder. The polyaniline powder was dissolved in dimethylformamide under the influence of ultrasound. Polyacrylonitrile (PAN) was added to the obtained solution. The weight concentration of polyacrylonitrile in the spinning solution was 13.5%, and polyaniline was approximately 1 wt.% to the weight of polyacrylonitrile. The fibers obtained using this method were marked as PAN/PANI blended.

2.3. Synthesis of PANI in Spinning Dope

The literature data shows that the best convertible polyaniline is obtained in the solution polymerization process [17]. For this reason, in the second method, polyaniline was synthesized in situ during the preparation of the polyacrylonitrile spinning solution in dimethylformamide (DMF). Two solutions in dimethylformamide (containing 1% of dissolved PAN) were prepared with 0.5 M aniline hydrochloride and 0.6 M ammonium persulfate. Both solutions were mixed together for 4 h. The reaction was carried out at a temperature not exceeding 4 °C. After 20 h, the obtained reaction mixture was filtered because colorless crystals were precipitated during the reaction (Figure 1).

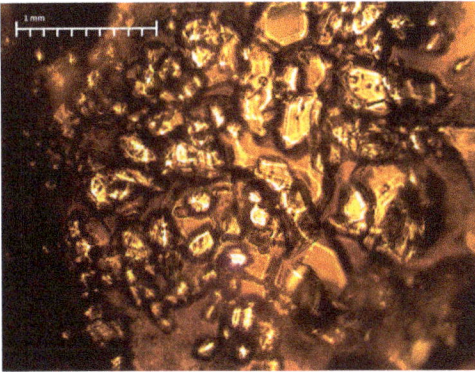

Figure 1. Crystals precipitated during the reaction of polyaniline in dimethylformamide (DMF) after filtration.

Polyacrylonitrile was added to the filtered solution. The solution was heated to 40 °C and stirred for 4 h, then cooled for 2 h. In the synthesis of the polyaniline, the same amount of substrates as in the blended method was used. The fibers obtained using this method were marked as PAN/PANI in situ.

2.4. Spinning of Fibers

Polyacrylonitrile (PAN) fibers and both types of polyacrylonitrile-polyaniline fibers (PAN/PANI) were obtained using the wet method. A reference fiber spinning solution (PAN) was prepared by dissolving polyacrylonitrile (13.5%) in dimethylformamide (DMF) at 40 °C. After passing through the nozzle, the spinning solution was introduced into a solidifying bath, in which PAN coagulated in the form of fibers (Figure 2). In a solidifying bath (60% aqueous DMF solution) at 20 °C, and in a plasticizing bath (50% aqueous solution of DMF) at a temperature of 70 °C, intensive removal of the solvent and impurities from the formed fibers took place. The fibers underwent a two-stage stretching process—in the plasticizing bath and in water vapor at 135 °C. During these processes, the orientation of macromolecules took place and the fiber structure was ordered, which improved the mechanical properties of the fibers. The PAN/PANI in situ fibers were not subjected to stretching in water vapor because of tearing during the tensile test. The formed and stretched fibers were wound on reels.

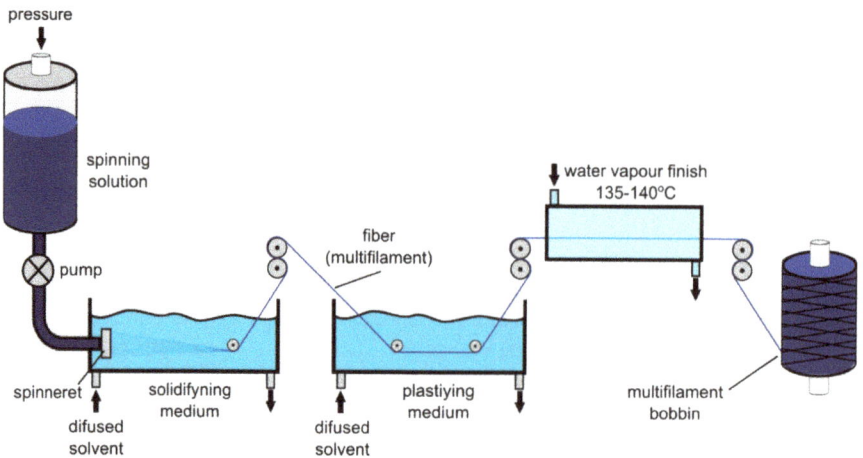

Figure 2. Scheme of fiber spinning using the wet method.

2.5. Characterization

The molecular weights of polymers were determined by gel permeation chromatography (GPS) using dimethylacetamide (DMAc) as a solvent with the addition of 0.5 wt.% lithium chloride (LiCl). Rheological measurements of the spinning solutions were obtained by a ReolabQC Anton Paar rheoviscometer (Anton Paar, Ostfildern, Germany). The kinetics of polyaniline formation in the DMF were investigated using UV-Vis spectroscopy with a UV-Vis-NIR JASCO V-570PC spectrophotometer (Jasco, Easton, MD, USA). Photographs of the fibers were made using a Leica DM 4000M Led microscope (Leica, Wetzlar, Niemcy). The chemical composition was determined on the basis of infrared spectroscopy with a Nicolet 380 FTIR spectrophotometer (Thermo Fisher Scientific, Waltham, MA, USA) with an ATR Smart Performer (Thermo Fisher Scientific, Waltham, MA, USA). A PC-controlled low-load machine was used for testing the mechanical properties of the fibers formed (Zwick Z2.5/TN1S, Ulm, Germany), according to the PN-EN ISO 2062:2010 standard [18]. The fiber samples' morphology was analyzed with a TESCAN VEGA3–EasyProbe (TESCAN, Brno, Czech Republic), and a scanning electron microscope (TESCAN, Brno, Czech Republic) equipped with VEGA TG software (TESCAN, Brno, Czech Republic) was used for morphological analysis of the doped polyacrylonitrile fibers (high vacuum mode (SE); accelerating voltage 7–20 kV). Before the measurements, the samples were sputtered with Au-Pd (SC7620 Mini, Quorum Technologies Ltd., Lewes, UK) for 120 s. The cross-section of the fibers was observed and obtained using a high-resolution SEM Xe-PFIB FEI Helios (Brucker, Billerica, MA, USA). A single fiber was covered with a platinum layer, and then a cross-section was made with a focus ion beam (FIB). Thermal gravimetric analysis (TGA) was carried out using a thermobalance TGA Mettler Toledo (Columbus, OH, USA) in the temperature range from 25 to 600 °C and a heating speed of 10 °C/min. A corundum crucible with a hole in its lid of 70 µL was used. Measurements were carried out under nitrogen, and the flow rate was 30 mL/min. Differential scanning calorimetry (DSC) measurements were performed using a DSC Mettler Toledo (Columbus, OH, USA) in the temperature range from 25 to 200 °C with a heating/cooling rate of 10 °C/min. Measurements were carried out under nitrogen, and the flow rate was 30 mL/min in a corundum crucible with one hole in its lid of 40 µL. The study of electrical conductance was investigated using the direct current (DC) method with the use of a Keithley 2000 multimeter (Cleveland, OH, USA). Measurements were made on 1 cm lengths between two silver electrodes.

3. Results and Discussion

3.1. Molecular Weight of Polymers

Based on the results of exclusion chromatography (GPC) analysis, the molecular weights of the polymers were determined. The average molecular weight of polyacrylonitrile was Mw 644,600 and Mn 177,400. Its polydispersity was 3.6, which indicates a very large weight distribution of molecules in the polymer. On the other hand, the average molecular weight of the PANI obtained in the aqueous solution, which was Mw 12,920 and Mn 4263, respectively, and the degree of polydispersity of PANI were also large, amounting to 3.0.

3.2. Rheological Properties of Spinning Solutions

The basic properties of each spinning solution, which demonstrated its quality, were rheological properties. To modify the composition of these solutions, it was necessary to determine the effect of modifications on these properties. When analyzing the rheological properties of spinning solutions, the dependence of tangential stress on shear rate (A power-law fluid—the Ostwald–de Waele relationship) is analyzed (1):

$$\tau = k\gamma^n \tag{1}$$

where n is the flow behavior index (-), and k is the flow consistency index (Pa·s^n). These two basic rheological parameters of each spinning solution were determined experimentally (Table 1).

Table 1. The value of flow behavior index n and the flow consistency index k of the spinning solutions.

Spinning Solution	n	k (Pa·s^n)
PAN	0.85	32.84
PAN/PANI blended	0.89	19.97
PAN/PANI in situ	0.87	24.03

The value of the flow behavior index depended on the way in which polymer macromolecules were placed in motion, and the way they formed a system of mutually sliding layers. The linear polymer chains in the spinning solution (with a sufficiently high concentration) can be highly tangled. The determined value of n (Table 1) indicated that all the obtained spinning solutions were shear thinned liquids. This is desirable for technological reasons (the ease of pressing the solution through the nozzles). The flow rates of solutions containing polyaniline were slightly higher. This means that polyaniline particles minimally affected the orientation of polyacrylonitrile macromolecules. This gentle effect on the *n*-value may be due to the difference in the molecular structure of both polymers. The flow behavior index depended on several factors, e.g., the average molecular weight of the polymer and its degree of polydispersity, the degree of branching of macromolecules, and the content of additional components (especially fillers) and plasticizers. In the studied system, both polymers were linear, but the PAN macromolecules formed entangled structures, and the PANI macromolecules were largely rigid due to the presence of quinone rings in them. This parameter, in addition to temperature, pressure, and concentration, had a significant impact on the value of the flow consistency index k. Taking into account technological requirements, the k value of spinning solutions should be greater than 20 Pa·s^n. In comparison to changes in the n value, the presence of polyaniline molecules in the spinning solution caused a significant change in the k value. This change was mainly due to the addition of a polymer with a much lower average molecular weight relative to the spinning solution. The average molecular weight of PANI was 50 times less than PAN. Moreover, a significantly lower value of the consistency coefficient of spinning solutions containing PANI may be related to the occurrence of interactions between the PAN and PANI chains [19,20].

3.3. Fiber Morphology

It is well known that polyaniline obtained by aniline polymerization can occur in three oxidation states [21]:

- leucoemeraldine—$(C_6H_4NH)_n$—completely reduced, white or colorless;
- emeraldine—$([C_6H_4NH]_2[C_6H_4N]_2)_n$—blue for the emeraldine base, green for the emeraldine salt;
- (per)nigraniline—$(C_6H_4N)_n$—completely oxidized, blue or violet.

Of all forms of polyaniline, only emeraldine salt (green) has an electrical conductivity at the level of semiconductors, i.e., around 1 S/cm. Emeraldine salt is obtained as a result of emeraldine base modification by HCl or HBr. Each type of polyaniline can be transformed in a reversible process of oxidation or reduction, during which changes in color and electrical conductivity occur [22].

The PAN/PANI blended fibers obtained by adding polyaniline powder to the spinning solution had a dark green color (Figure 3a), which, after immersing them in a 3% aqueous solution of hydrochloric acid, turned to dark blue (Figure 3b). After rinsing again in water, their color returned to the previous color. The fibers obtained as a result of the synthesis of polyaniline in situ in the spinning solution were dark navy (Figure 3c).

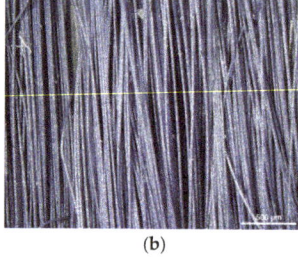

Figure 3. Fiber images made with an optical microscope. Fibers obtained as a result of the addition of polyaniline (PANI): (**a**) after rinsing in water; (**b**) after rinsing in HCl; and (**c**) fibers obtained as a result of in situ synthesis of PANI.

Based on the color of the obtained fibers, it can be concluded that polyaniline in the PAN/PANI blended fibers was in an emeraldine base form, and in an HCl solution it transformed into emeraldine salt (Figure 3b). Moreover, in the PAN/PANI in situ fibers, polyaniline was in the emeraldine base form.

(Per)nigraniline—$(C_6H_4N)_n$—a completely oxidized form of PANI—also has a blue or violet color. However, taking into account that the synthesis of polyaniline was carried out in a solvent (DMF), which is a strong reducer, it cannot be assumed that a fully oxidized form of PANI was formed. In addition, during the synthesis of PANI in DMF, colorless crystals were formed, which indicated that a part of polyaniline precipitated in the form of leucoemeraldine—a completely reduced form. A significant part of the leucoemeraldine was filtered during the synthesis, and, therefore, the fibers were mostly polyaniline fractions with shorter chains.

3.4. UV–Vis Spectroscopy

On the UV-Vis spectrum of PANI solution in DMF, two peaks were visible (Figure 4). The peak at 323 nm was connected with π-π* excitation, and the peak at 618 nm was connected with the exciton excitation of the quinone rings in the PANI (Figure 4a). The presence of these peaks and the dark blue coloration of the solution were characteristic for the emeraldine base [23]. On the spectrum of the spinning solution containing 1 wt.% of polyacrylonitrile and polyaniline synthesized in DMF, the peak near 320 nm was more visible and shifted in towards the shorter wave (Figure 4b). However, in this spectrum there was no clear peak around 618 nm, which corresponded to the exciton excitation of the quinone rings in the PANI.

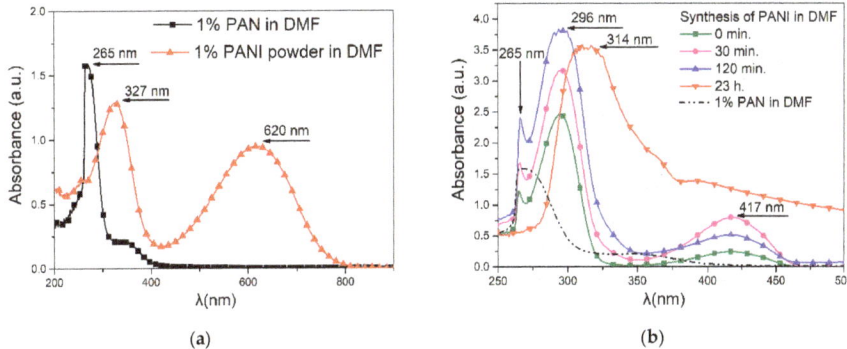

Figure 4. UV-Vis spectrum: (**a**) polyacrylonitrile (PAN) and PANI powder dissolved in Dimethylformamide (DMF) (blended); (**b**) change in absorbance during the PANI synthesis in DMF (in situ).

The PANI synthesis in the DMF solution containing 1 wt.% of PAN occurred very slowly. In this time, the peak characteristic for PAN disappeared. In contrast, the characteristic peak, corresponding to the π-π* excitation during synthesis, moved towards the 327 nm band. This can indicate the interaction between PAN and PANI molecules [24].

3.5. FTIR Spectroscopy

The chemical structure of the polyaniline, polyacrylonitrile, and obtained fibers was determined on the basis of infrared spectroscopy (Figure 5). A description of the characteristic vibrations is shown in Table 2 [15,25].

On the fibers' FTIR spectra, the bands characteristic of polyaniline and polyacrylonitrile were visible. In the case of PAN/PANI blended fibers, all absorption bands characteristic of polyacrylonitrile were clearly visible.

In the case of PAN/PANI blended fibers, all absorption bands characteristic of polyacrylonitrile were clearly visible at 3445, 2926, 2243, 1731, 1615, 1453 and 1250 cm^{-1}. The peaks corresponded to N–H stretching and C–H stretching in the polymer structure, stretching in C≡N, stretching in C=O, stretching in C=C, bending C–H in CH_2, and stretching in C–N. The gentle absorption bands characteristic of PANI were also visible at 1588, 1160 and 804 cm^{-1}, which were characteristic of C=C stretching in the quinoid rings, C–N bond stretching in quinoid and benzenoid rings, and δ C–H in 1,2,4 trisubstituted benzene rings. FTIR analysis indicated that these fibers were composed of polyacrylonitrile and had a low concentration of polyaniline incorporated in them, as evidenced by the low intensity of the bands characteristic of this polymer.

On the other hand, in the PAN/PANI in situ fiber spectrum, the absorption bands associated with the presence of PANI-specific bonds were much more intense. A fuzzy peak above 3000 cm^{-1} was also visible, which was associated with the presence of water. This is not present in the PAN/PANI blended spectrum. The bands occurring at 1414, 1194 and 879 cm^{-1} were characteristic of stretching in C–N between benzene and quinoid rings, and bending in C–H and 1,2,4 trisubstituted benzene rings. The absorption bands characteristic for polyacrylonitrile were also visible at 2940, 2244, 1726, 1629, 1445 and 1260 cm^{-1}, which was characteristic of C–H stretching in polymer structures, stretching in C≡N, stretching in C=O, stretching in C=C, bending C–H in CH$_2$, and stretching in C–N.

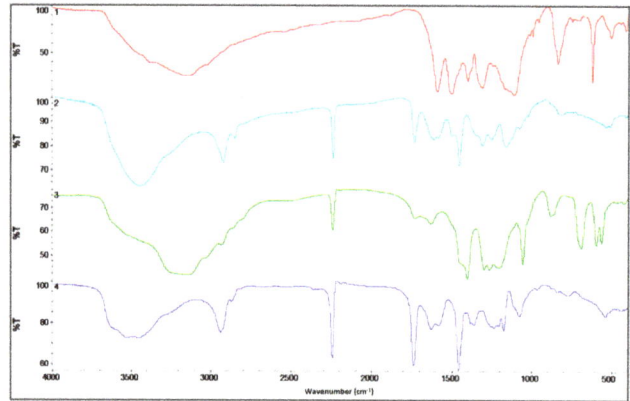

Figure 5. FTIR Spectra from the top: PANI powder (red); fibers: PAN/PANI blended (cyan); PAN/PANI in situ (green); and PAN (blue).

Table 2. Description of characteristic vibrations observed in the FTIR spectra of the tested substance.

Band Origin [1]	PAN	PANI	PAN/PANI Blended	PAN/PANI In Situ
υ N–H	3480	-	3445	3440
υ C–H in CH$_2$	2940	-	2926	2940
υ C≡N	2243	-	2243	2244
υ C=O	1733	-	1731	1726
υ C=C	1629	-	1615	1629
υ C=N in Q	-	1576	1588	-
δ C–H in CH$_2$	1454	-	1453	1445
υ C–N$^+$	-	1414	-	1414
υ C–N in Q and B	-	1302	1306	1300
δ C–H in CH	1227	-	1250	1260
υ Q=N–B and B–NH–B (deprotonated PANI)	-	-	1160	1194
υ Q=N$^+$H–B and B–N$^+$H–B (protonated PANI)	-	1144	-	-
δ C–H in 1,2,4 trisubstituted B and Q	-	832	804	879

[1] Where: υ—stretching, δ—bending, B—benzene ring, and Q—quinoid ring.

In addition, a band corresponding to the vibrations of the deprotonated polyaniline form (1160 cm^{-1}) was present in both fibers containing polyaniline. Moreover, pure polyaniline was in a protonated form.

From the shifts in absorption bands in the PAN/PANI in situ fiber spectra, relative to unmodified polyacrylonitrile and polyaniline, it can be concluded that electrostatic interactions arose between the chains of these polymers between the polyaniline and nitrile groups. The confirmation of the interaction between polyaniline and the nitrile group of polyacrylonitrile was in the reduction of the intensity of the absorption band around 2244 cm^{-1} associated with stretching vibrations in C–N, and also the band around 3480 cm^{-1} associated with stretching N–H vibrations. In both cases, this reduction was more visible in the PAN/PANI in situ fibers. Such interactions between different conductive polymers and polyacrylonitrile are known in the literature [15]. These results confirmed observations drawn from the research of the rheological properties of the obtained solutions and UV-Vis spectra.

3.6. SEM Images

The microstructure of the fibers was observed using a scanning electron microscope (Figure 6). Cross-sections of individual fibers were also made using a focus ion beam (FIB) microscope (Figure 7). The surface morphology and fiber cross-sections were different for every fiber. The standard PAN fibers had a visible fibrous structure with characteristic furrows on their surface (Figure 6a). The cross-sections of these fibers had a characteristic bean-like shape (Figure 6b). The surface of the PAN/PANI blended fibers treated with hydrochloric acid was very smooth, and their cross-section was oval (Figures 6c,d and 7b). Deep furrows were clearly visible on the surface of the PAN/PANI in situ fibers. The cross-sections of these fibers showed pores and cracks (Figure 7c). The cross-section shape was also bean-like, but the surface structure was clearly different from the inside of the fibers (Figure 6e,f). The structure of the PAN/PANI in situ fibers was discontinuous and contained many defects. It was clearly more heterogeneous than in the PAN/PANI blended fibers (Figure 6e,f).

Figure 6. SEM images of fibers: (**a**,**b**) PAN; (**c**,**d**) PAN/PANI blended; and (**e**,**f**) PAN/PANI in situ.

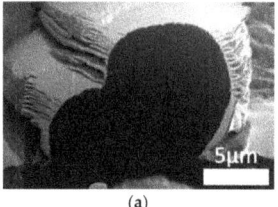 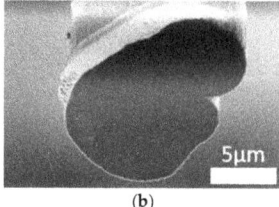 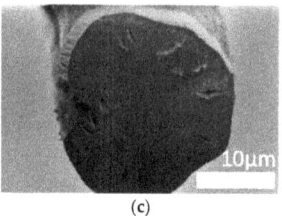

(a)   (b)   (c)

Figure 7. SEM images of the cross-section of: (**a**) PAN; (**b**) PAN/PANI blended; and (**c**) PAN/PANI in situ.

The reason for such differences in the microstructure of fibers can be, above all, the difference in the size of the macromolecules that have been in the solution (see Section 3.1), and also the amount of tension during the spinning process (see Section 2.4). As previously mentioned, during the synthesis of the PANI in the DMF and PAN solution, transparent crystals were formed (Figure 1), which were then filtered off. Thus, only low molecular weight molecules were left in the solution, which hindered the fiber formation process. Moreover, according to FTIR results (Figure 5) (Table 2), during the synthesis of PANI in the DMF and PAN solution, new interactions appeared between the functional groups of both polymers. This could weaken the interaction between the polyacrylonitrile chains themselves.

3.7. Mechanical Strength

Polyaniline is a highly used material, primarily due to its high electrical conductivity, low price, and its easy method of obtaining. However, the biggest difficulty associated with its use is its low mechanical strength and processing problems. For this reason, the fibers were made from a mixture of polyaniline with fiber-forming polyacrylonitrile, which, in this system, was a polymer responsible for mechanical strength.

The fiber linear mass was determined using the following Equation (2):

$$m_l = \frac{m}{l} \times 1000 \left[\frac{g}{m} \times 1000 = \text{tex} \right], \tag{2}$$

where m is the average sample weight and l is the length of the sample.

The linear mass of the PAN/PANI blended fibers was slightly higher than the PAN fibers (Table 3). However, the linear mass of the PAN/PANI in situ fibers was more than twice as high. This was because the fibers were thick as a result of them both being subjected to stretching in steam.

Table 3. Mechanical parameters of the fibers.

Sample	m_l (tex)	E (cN/tex)	F_{brk} (cN)	F_{max} (cN)	WtP (cN/tex)	ε (%)	λ (mm)
PAN	95.59	583.49	1983.35	3680.89	38.51	18.59	9.30
PAN/PANI blended	100.03	472.59	847.48	2011.36	18.39	8.73	4.37
PAN/PANI in situ	233.7	164.85	275.89	1465.97	6.27	11.27	5.64

where: m_l—linear mass; E—Young's modulus; Fbrk—average breaking force of fiber multifilament; Fmax—maximum breaking force of fiber multifilament; WtP—specific strength of fiber multifilament; ε—average relative elongation; and λ—average absolute elongation.

Based on the breaking curve, the most important parameters of the mechanical strength of the fibers were determined (Table 3). The Young's modulus and the specific strength of both polyaniline-containing fibers was lower than that of the pure polyacrylonitrile fibers, with the reduction being more pronounced in the case of the fibers modified in situ (Table 3).

During the stretching of fiber-forming polymers, their crystalline structure is shaped, which increases the mechanical strength of the fibers. However, as previously mentioned, the PAN/PANI in situ fibers were not subjected to stretching in steam. While stretching, the fibers were easier to delaminate and less likely to break. This effect may occur when adding an incompatible polymer or when adding a polymer that changes the interactions between the fiber-forming polymer chains. Both the PANI/PANI blended and PAN/PANI in situ fibers contained the same polymer, and, therefore, the delamination effect was difficult to explain by the presence of an incompatible polymer.

However, on the basis of the FTIR results (Figure 5), interactions between these polymers were found, which probably caused a reduction in the interactions between the PAN macromolecules, as well as a significant reduction in the consistency coefficient (Table 1). In addition, if it was assumed that a small molecule polyaniline remained in the polyacrylonitrile solution during the fiber coagulation process, it would pass into the coagulation bath, causing pores to be visible on SEM images (Figure 6f,c).

Typically, thermoplastic polymers such as Nylon or Kevlar [7,8] are used as PANI strength-enhancing materials. However, the presence of these polymers causes a decrease in the electrical conductivity of the PANI. This inversely proportional relationship is associated with the presence of micropores in the fibers, which on the one hand reduces the mechanical strength, but on the other hand facilitates the penetration of an admixture, such as HCl, into the structure of a fiber. This penetration, in turn, increases the conductivity of the fibers [26,27]. For this reason, the generation of electrically conductive fibers based on electronically conductive polymers has many denominations, of which, the greatest seems to be the achieving of both the desired mechanical properties and a satisfactory electrical conductivity.

3.8. XRD

The fiber crystal structure was determined based on XRD studies (Figure 8). On all diffractograms, peaks characteristic of polyacrylonitrile were visible. Particularly clear was the peak at 2θ~18°, which was characteristic of the hexagonal structure of PAN. In the PAN/PANI blended fiber spectrum there were no peaks characteristic of PANI, most probably due to its small amount in the fibers. However, they were clearly visible in the PAN/PANI in situ fibers, and corresponded to the pseudo-orthorhombic phase of PANI [28].

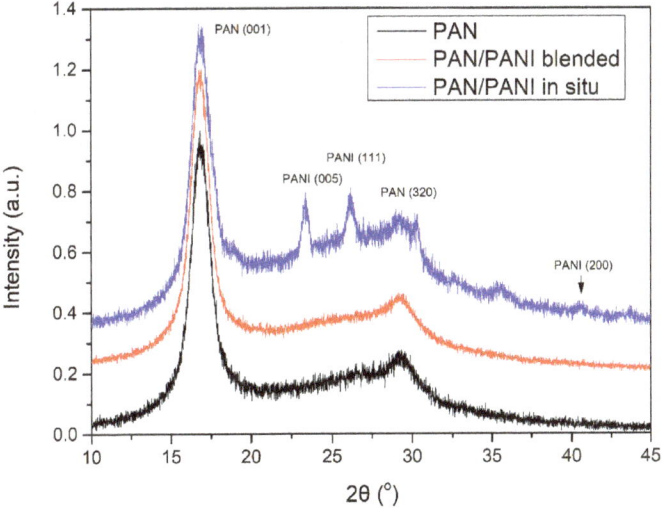

Figure 8. Diffractogram of the fibers.

The degree of fiber crystallinity (X_c) was calculated using the Hinrichen method (3):

$$X_c = \frac{I_c}{I_c + I_a},\qquad(3)$$

where I_c is the integral under the peaks corresponding to the crystalline phase of the polymer, and I_a is the integral under the peaks corresponding to the amorphous phase of the polymer.

Crystal and amorphic peaks were determined based on the optimization carried out in the WAXFIT program. Curve fitting was performed using the Rosenbrock method. The degree of crystallinity for the PAN and PAN/PANI blended fibers was the same at 0.53, while for the PAN/PANI in situ fibers it was significantly lower at 0.32 (Table 4). The significantly lower crystallinity degree of the PAN/PANI in situ fibers was caused by a lack of stretching, and was correlated with their mechanical strength.

Table 4. The degree of fiber crystallinity.

Sample	The Degree of Fiber Crystallinity
PAN	0.53
PAN/PANI blended	0.53
PAN/PANI in situ	0.32

The higher degree of crystallinity, in the case of fibers, was associated primarily with a larger ordering of polymer chains along the fiber axis [29]. This parameter depended on the type and degree of interactions between macromolecules.

The polyacrylonitrile macromolecules contained positively charged CN groups (Figure 9a), and caused the chains to twist into a helix, from which the crystallites are made [30]. In the PANI, due to the presence of amine groups, hydrogen bonds and dipole-dipole interactions were formed between the macromolecules (Figure 9b). H-bonding interactions between adjacent chains caused the stiffness of the PANI chain [31], and an easier orientation of the chains in the material containing only polyaniline.

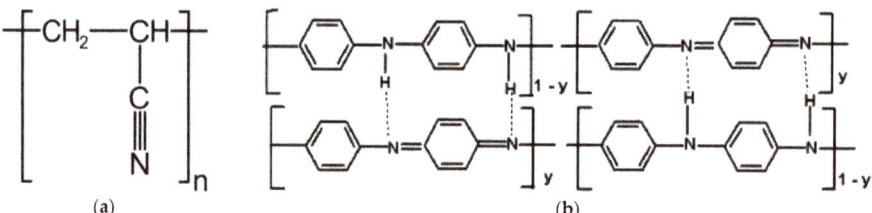

Figure 9. Chemical structure of: (a) PAN; and (b) PANI.

However, in the material containing both of these polymers, the crystallization process was difficult due to the interactions between nitrile (PAN) and amine (PANI) groups (Figure 10). In addition, numerous cracks and voids appeared in the structure of the fiber (Figure 6c). Analysis of the literature data [32] indicated that the nitrile group, in some cases, may have interfered with or acted as a barrier when ordering polymer chains. This can take place in the PAN/PANI in situ fibers, where the strong interaction between the polymer chains blocks their movement, thereby limiting the crystallization process.

The average size of PAN crystallites was determined based on the half-width of the peak $2\theta \sim 17°$, characteristic of the plane (001) in a hexagonal structured PAN. The size of the PANI crystallites was determined based on the peaks $2\theta \sim 23°$ and $26°$, characteristic of the planes (005) and (111). The calculations were made using the Debye–Scherrer Equation (4):

$$L_{hlk} = \frac{K \times \lambda}{\beta \times \cos \theta_{max}}, \quad (4)$$

where $L_{(hlk)}$ is the average size of crystallites, K is the Scherrer constants (0.89), θ_{max} is the angle for the maximum peak (rad), λ is the length of the radiation beam (Å), and β is the half-width of the peak (rad).

The inter-chain separation length (R) was determined based on the analysis of the most intense crystalline peak from the following Equation (5) [31]:

$$R = \frac{5\lambda}{8 \sin \theta_{max}}. \tag{5}$$

The distance between planes (the d-spacing) was determined from the Bragg Equation (6):

$$n\lambda = 2d \sin \theta, \tag{6}$$

where n is the deflection integer.

Lattice strain ε was determined from Formula (7):

$$\beta = 4\varepsilon \tan \theta \tag{7}$$

Data on the crystal structure of the obtained fibers was collected in Table 5. The crystal phase of the polyacrylonitrile in all the fibers had a similar structure. Crystallite size, d-spacing, inter-chain separation, and lattice strain all had very similar values. The size of the polyaniline crystallites was larger and amounted to 15.8 (111) and 17.9 (005) nm. Differences in the structure of both chemical and crystalline polyacrylonitrile and polyaniline present in fibers were reflected in their macroscopic structure and mechanical properties.

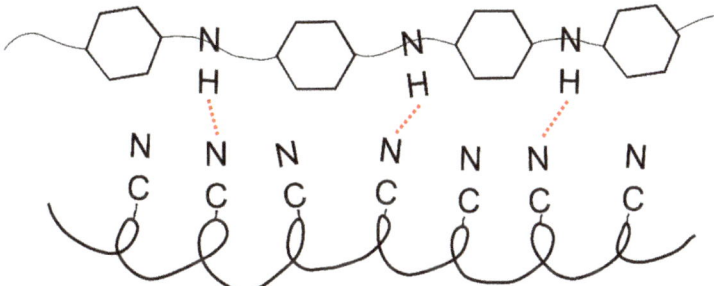

Figure 10. Diagram of interaction between the PAN and PANI chains.

Table 5. Crystallinity properties of fibers.

Sample	hkl	Polymer	Diffraction Peak (°)	L (nm)	d-Spacing (Å)	R (Å)	ε (-)
PAN reference	(001)	PAN	16.85	5.7	3.95	6.58	0.00092
PAN/PANI blended		PAN	16.80	6.0	3.96	6.60	0.00086
PAN/PANI in situ		PAN	16.77	5.7	3.97	6.61	0.00091
	(111)	PANI	26.19	15.8	3.40	4.26	0.00053
	(005)	PANI	23.42	17.9	3.80	4.75	0.00041

Inter-chain separation length is a parameter affecting the electrical conductivity of fibers. This value represents the distance between the electron jump between the chains. The smaller it is, the higher the probability of charge hopping. The distance of the order of ~4 Å corresponded to the literature concerning polyaniline, and allowed hopping to be charged between the chains [24,33].

3.9. Thermal Properties

Thermogravimetric analysis of the pure PAN fiber and PANI-containing fibers showed some similarity between the samples (Figure 11). Three mass losses were observed for all materials. The first between 25 and 100 °C (Table 6—T_1), the second between 200 and 300 °C (Table 6—T_2), and the third at around 420 °C (Table 6—T_3).

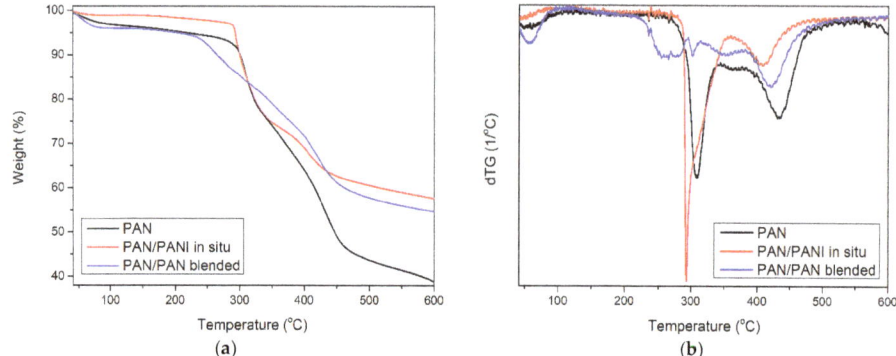

Figure 11. TG (**a**) and DTG (**b**) curves of polymer fibers.

Table 6. Data obtained from the thermogravimetric curve, $T_{1,2,3}$—temperature of subsequent losses in mass.

Sample	Onset (°C)	T_1 (°C)	T_2 (°C)	T_3 (°C)	Δm_{tot} (%)	m_{PANI} (%)
PAN reference	296	-	308	434	61.22	-
PAN/PANI blended	290	-	294	409	42.32	18.9
PAN/PANI in situ	231	265	303	421	45.26	16.0

The first mass loss was caused by the loss of water bound in the material, which was due to polyaniline adsorbing large amounts of water. Water can be associated with polyaniline macromolecules in two ways: strong and weak. Poorly bound water molecules are connected to the PANI chain with one hydrogen bond (Figure 12a). In contrast, strongly bound water is connected by two hydrogen bonds (Figure 12b), most often between two adjacent PANI chains [25].

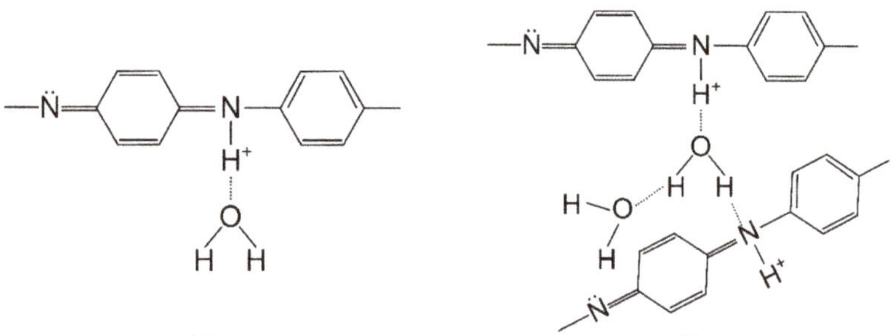

Figure 12. Polyaniline with weak (**a**) and strong (**b**) bound water [30].

Poorly adsorbed water can be removed while drying with nitrogen at room temperature, while the strongly bound one is mostly removed at a temperature in the range of 60–140 °C. Removal of this water is reversible and it can be re-absorbed. However, further removal of water from the polymer at 150–220 °C is an irreversible process and leads to polymer degradation [34,35].

The first derivatives of the TG curve showed that the most moisture was adsorbed in the PAN/PANI in situ fiber. Less water was present in the pure PAN fiber, and the smallest amount was found in the PAN/PANI blended fiber. As pure polyaniline can bind up to 40% by weight of water [36], it is possible to determine its presence in fibers based on the analysis of the water-related peak. An intense peak, much larger than in the reference fiber, was associated with the presence of water in the PAN/PANI in situ fiber. This demonstrated the presence of polyaniline in fibers, which was also confirmed by the FTIR results (Table 5) and XRD (Table 6). A very small amount of adsorbed water in the PAN/PANI blended fiber resulted from the fact that the polyaniline, after the synthesis process, was dried before its dissolution in dimethylformamide.

Further fiber degradation was associated primarily with the phenomenon of polyacrylonitrile cyclization occurring under nitrogen [37]. Analysis of the first derivative of TG curve (i.e., DTG curve) showed that in the case of the pure PAN and PAN/PANI blended samples, the degradation can be described as one-step and starts at around 280 °C. The PAN/PANI blended fiber was characterized by a very rapid decomposition in this temperature range, which persisted up to 350 °C.

The PAN/PANI in situ fiber was decomposed in two stages. It had a wide weight loss of about 10% in the temperature range of 200–290 °C. The initial weight loss in this range was associated with the removal of the rest of the strongly bound water. On the other hand, according to the obtained results and literature data, the weight loss was due to the presence of low-molecular polyaniline fractions in the material, which underwent gradual decomposition [38]. This process is characteristic for PANI degradation, which is why it indicated its presence in the fiber. Another loss beginning at 290 °C and ending at 380 °C was associated with PAN cyclization. Comparing the results obtained in all the fibers showed that the in situ synthesis of polyaniline was a difficult cyclization. This was closely correlated with the FTIR conclusions (Table 5), which indicated electrostatic interactions between the macromolecules of polyacrylonitrile and polyaniline in the PAN/PANI in situ fibers.

The third loss of fiber mass corresponded to the carbonation process of the PAN, which consisted of dehydrogenation and denitrogenation processes [39]. As a result of these processes, a cyclic carbon structure was created. However, when combining the maximum transformation temperature, it can be seen that the lowest value of this parameter was for the PAN/PANI blended fiber. This clearly indicated that the addition of polyaniline introduced by the blended method facilitated the carbonization process.

The total weight loss (Δm_{tot}) after measurement was about 15% higher for the pure PAN fiber than for the modified fibers (Table 6). Considering the high thermal resistance of PANI, this proved the effectiveness of both methods of introducing a polymer additive to the fiber.

To verify the phase composition of the fibers, DSC measurements were conducted (Figure 13). The obtained results indicated that the pure PAN fiber showed a glass transition at a temperature of about 100 °C both for heating and cooling runs, which was characteristic for polyacrylonitrile [40,41].

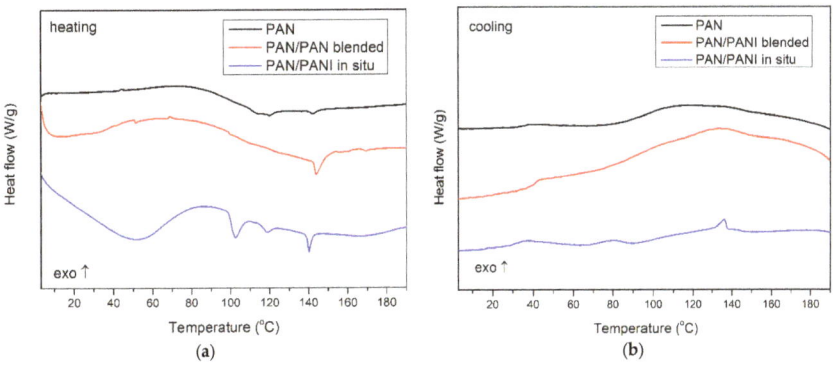

Figure 13. Differential scanning calorimetry (DSC) curve of polymer fibers: (**a**) heating curves; (**b**) cooling curves.

In the case of the PAN/PANI in situ fibers' heating curve, a wide endothermic peak was seen with a maximum at 50 °C, associated with the removal of poorly bound water. An endothermic peak was seen at 102 °C, associated with the desorption of strongly bound water to the PANI [38]. Subsequently, the glass transition in the PAN was observed at about 110 °C, but its exact value was difficult to accurately determine because of the overlap of both processes. At 140 °C, one more endothermic peak was visible, which may be the result of the presence of low molecular weight fractions in the material remaining after the PANI polymerization process (Figure 13a). In the cooling curve, two exothermic peaks were observed, which could be connected with crystallization of previous melted low molecular fractions during heating. Also, a glass transition of PAN at about 100 °C was present (Figure 13b). The obtained results indicated a large heterogeneity of the system, which was also confirmed by previous SEM, mechanical strength, and TG measurements.

In the case of the PAN/PANI blended fibers, only the PAN glass transition (100 °C) was visible in the heating curve, with an additional transition at around 150 °C. Because this temperature was not identical to the glass transition of polyaniline, and was not observed in the front samples (Figure 13a), the authors believe that it was the result of the presence of small amounts of low molecular weight PANI fractions in the material structure, which was well mixed with pure polyacrylonitrile; therefore, the obtained fibers had good mechanical parameters (Table 3) and a smooth structure (Figure 6b). This fraction was homogenized with the rest of the material because during the cooling run, no crystallization peaks were present. Only T_g of PAN was observed (Figure 13b). This phenomenon also explains the problem in observing peaks from PANI to XRD, and the presence of only partial peaks in the FTIR.

3.10. Electrical Measurements

In order to determine the effect of polyaniline on the electrical properties of the fibers, resistance tests were carried out. The resistance of fibers on the 0.3 cm section was measured using the constant current method, and the results were given as mass-specific resistance Rs ($\Omega g/cm^2$) (8). This is a parameter used to describe fiber resistance [42]. This measure allows the electrical properties of fibers to be shown in relation to their construction, and not geometric parameters (such as diameter), which are difficult to determine in yarns composed of many fibrils.

$$Rs = \frac{RNm_l}{1 \times 10^5}, \qquad (8)$$

where R is the resistance of the sample, N is the number of fibers in the yarn, m_l is the linear mass (tex) = (g/km), and l is the length of the sample (cm).

Standard polyacrylonitrile (reference) fibers are dielectrics with a resistance of 10^{11} Ω (Table 7). Both types of fibers doped with polyaniline were tested for resistivity. In the case of the PAN/PANI blended fibers, no change in the fiber resistance was observed, whereas the PAN/PANI in situ fibers were characterized by mass-specific resistances of 5.47 $k\Omega g/cm^2$. These fibers, as shown by the FTIR, XRD, and TG studies, consisted of two phases of different polymers, which were additionally visible in the SEM images. In these fibers, the polyacrylonitrile matrix was responsible for the mechanical properties and thermal stability of the fibers. Polyaniline was responsible for their electrical conductivity.

In order to compare the obtained results with literature data, it can be assumed that the conductance of the obtained PAN/PANI in situ fibers is of the order 10^{-4} Scm^{-1}. This conductance value is lower than that of pure PANI fibers. According to literature data, polyaniline fibers have conductance on the order of 10–100 Scm^{-1} [43,44], but mechanical properties of these fibers are weak. Better mechanical properties, but lower conductance, usually is obtained in composite fibers. Toptaş et al. obtained composite PAN/PANI fibers in a different method that was presented in this paper, but mechanical and electrical properties are comparable [14].

Table 7. Electrical properties of obtained fiber.

Sample	Mass-Specific Resistance	Units
PAN reference	~10^{11}	$\Omega g/cm^2$
PAN/PANI blended	~10^{11}	$\Omega g/cm^2$
PAN/PANI in situ	5470	$\Omega g/cm^2$

4. Conclusions

The article presents two methods of obtaining polyacrylonitrile-polyaniline fibers (PAN/PANI), as well as their physicochemical properties. Polyaniline was used to dope the polyacrylonitrile, which was obtained by synthesis in aqueous solution or dimethylformamide during the preparation of the spinning solution.

It should be noted that in the PAN fiber production process that was doped with PANI, an identical polyacrylonitrile amount of 13.5% was assumed in both the non-doped and PANI-doped fibers, which, as a consequence, influenced the spinning process and physical properties of the obtained fibers.

During spinning, an attempt was made to maintain identical solution concentrations, spinning bath temperature and water vapor, as well as the pickup and feed speed at the pickup and delivery points. In turn, this was cause for the PAN/PANI in situ fibers to resist stretching in steam, which, in turn, translated into a lowering of their specific strength. During the spinning of the identical solution concentrations, the spinning bath temperature and water vapor, as well as the pickup and feed speed at the pickup and delivery points were maintained. This was cause for the PAN/PANI in situ fibers to resist stretching in steam, which resulted in a lowering of their specific strength.

The fibers obtained by mixing the previously synthesized polyaniline with the polyacrylonitrile fiber spinning solution (PAN/PANI blended) were characterized by:

- blue or green color (depending on pH);
- good mechanical strength;
- smooth and homogeneous surface and cross-section;
- lack of visible peaks from the PANs on the diffraction patterns;
- poor visibility of the peaks—characteristic of PANI on the FTIR spectra;
- one-stage thermal decomposition;
- being dielectric.

Analysis of the above results led to the conclusion that the PANI could be introduced into the fiber in amount and form, which blended well with polyacrylonitrile and formed a homogeneous structure. This additive did not disturb the fiber crystal structure and morphology, but only slightly changed the thermal parameters of the fibers. However, this degree of combination of polymers prevented the transport of electric charges in the fibers; therefore, they remained dielectric. Further research on this modification method will focus on the effective increase of polyaniline content in fibers.

The fibers obtained as a result of the direct synthesis of polyaniline in the spinning solution (PAN/PANI in situ) were characterized by:

- black color;
- poor mechanical strength;
- a heterogeneous and porous surface and cross-section;
- visible peaks from PANI on the diffraction patterns;
- visible peaks characteristic of PANI on the FTIR spectra;
- interactions between the chains of functional groups of the PAN and PANI;
- high water content;
- a multi-stage thermal decomposition;
- good electrical conductivity.

The low mechanical strength of the fibers was mainly due to the lack of fiber stretching, their heterogeneous structure, and low degree of crystallinity. The reduced crystal crystallinity in comparison to the PAN fibers resulted from the stiffness of the structure caused by the presence of interactions between the PAN and PANI chains visible on the FTIR spectra. These fibers exhibited electro-conductive properties at the level of 5.47 kΩg/cm^2, which can be used to construct electrical connections in textronic clothing. Moreover, thanks to the reversible color change, which depends on the pH of the environment in which they are found, they can also be used as pH sensors in protective clothing. In further work, the authors will develop research to obtain electro-conducting PAN/PANI fibers with the highest possible mechanical strength. For this purpose, the study will focus primarily on the modification of parameters for obtaining a spinning solution, as well as the spinning itself.

Author Contributions: Conceptualization, I.K.; Investigation, O.R.-R. and M.F.-T.; Methodology, I.K. and O.R.-R.; Resources, T.R.; Supervision, H.T.; Visualization, O.R.-R.; Writing—Original Draft, O.R.-R.; Writing—Review & Editing, H.T.

Funding: This work was supported by the National Science Centre [decision number DEC-2013/09/B/ST8/03679]; a statutory activity subsidy from the Polish Ministry of Science and Higher Education for the Faculty of Microsystem Electronics and Photonics of Wroclaw University of Science Technology.

Conflicts of Interest: The authors declare no conflict of interest. The funders had no role in the design of the study; in the collection, analyses, or interpretation of data; in the writing of the manuscript, or in the decision to publish the results.

References

1. Wallace, G.G.; Teasdale, P.R.; Spinks, G.M.; Kane-Maguire, L.A. *Conductive Electroactive Polymers: Intelligent Polymer Systems*, 3rd ed.; CRC Press: Boca Raton, FL, USA, 2008.
2. Foroughi, J.; Spinks, G.M.; Wallace, G.G. *Nanotechnology and Conducting Polymer Fibres: Towards the Development of Nano-structured Conducting Polymers and Nano-composite Fibres*; LAP Lambert Academic Publishing: Germany, 2010.
3. Andreatta, A.; Cao, Y.; Chiang, J.C.; Heeger, A.J.; Smith, P. Electrically-conductive fibers of polyaniline spun from solutions in concentrated sulfuric acid. *Synth. Met.* **1988**, *26*, 383–389. [CrossRef]
4. Wang, H.L.; Romero, R.J.; Mattes, B.R.; Zhu, Y.; Winokur, M.J. Effect of processing conditions on the properties of high molecular weight conductive polyaniline fiber. *J. Polym. Sci. Part B Polym. Phys.* **2000**, *38*, 194–204. [CrossRef]
5. Bowman, D.; Mattes, B.R. Conductive fibre prepared from ultra-high molecular weight polyaniline for smart fabric and interactive textile applications. *Synth. Met.* **2005**, *154*, 29–32. [CrossRef]
6. Hsu, C.H.; Cohen, J.D.; Tietz, R.F. Polyaniline spinning solutions and fibers. *Synth. Met.* **1993**, *59*, 37–41. [CrossRef]
7. Hsu, C.H.; Shih, H.; Subramoney, S.; Epstein, A.J. High tenacity, high modulus conducting polyaniline composite fibers. *Synth. Met.* **1999**, *101*, 677–680.
8. Zhang, Q.; Jin, H.; Wang, X.; Jing, X. Morphology of conductive blend fibers of polyaniline and polyamide-11. *Synth. Met.* **2001**, *123*, 481–485. [CrossRef]
9. Zhang, Q.; Wang, X.; Chen, D.; Jing, X. Preparation and properties of conductive polyaniline/poly-ω-aminoundecanoyle fibers. *J. Appl. Polym. Sci.* **2002**, *85*, 1458–1464.
10. Zhai, G.; Fan, Q.; Tang, Y.; Zhang, Y.; Pan, D.; Qin, Z. Conductive composite films composed of polyaniline thin layers on microporous polyacrylonitrile surfaces. *Thin Solid Films* **2010**, *519*, 169–173. [CrossRef]
11. Mohamadali, M.; Irani, S.; Soleimani, M.; Hosseinzadeh, S. PANi/PAN copolymer as scaffolds for the muscle cell-like differentiation of mesenchymal stem cells. *Polym. Adv. Technol.* **2017**, *28*, 1078–1087. [CrossRef]
12. Fryczkowska, B.; Piprek, Z.; Sieradzka, M.; Fryczkowski, R.; Janicki, J. Preparation and Properties of Composite PAN/PANI Membranes. *Hindawi Int. J. Polym. Sci.* **2017**. [CrossRef]
13. Bertea, A.; Manea, L.R.; Hristian, L. Associated Polymers, Solvents and Doping Agents to Make Polyaniline Electrospinnable. *IOP Conf. Ser. Mater. Sci. Eng.* **2017**, *209*, 012073. [CrossRef]
14. Toptaş, N.; Karakışla, M.; Saçak, M. Conductive polyaniline/polyacrylonitrile composite fibers: Effect of synthesis parameters on polyaniline content and electrical surface resistivity. *Polym. Compos.* **2009**, *30*, 1618–1624. [CrossRef]

15. Xia, Y.; Lu, Y. Conductive polymers/polyacrylonitrile composite fibers: Fabrication and properties. *Polym. Compos.* **2010**, *31*, 340–346. [CrossRef]
16. Mottaghitalab, V.; Spinks, G.M.; Wallace, G.G. The influence of carbon nanotubes on mechanical and electrical properties of polyaniline fibers. *Synth. Met.* **2010**, *152*, 77–80. [CrossRef]
17. Kuramoto, N.; Tomita, A. Chemical oxidative polymerization of dodecylbenzenesulfonic acid aniline salt in chloroform. *Synth. Met.* **1997**, *88*, 147–151. [CrossRef]
18. PN-EN ISO 2062:2009 *Textiles-Yarns from Packages—Determination of Single-End Breaking Force and Elongation at Break Using Constant Rate of Extension (CRE) Tester*; International Organization for Standardization: Geneva, Switzerland, 2009.
19. Karbownik, I.; Rac, O.; Fiedot, M.; Suchorska-Woźniak, P.; Teterycz, H. In situ preparation of silver–polyacrylonitrile nanocomposite fibres. *Eur. Polym. J.* **2015**, *69*, 385–395. [CrossRef]
20. Feast, W.J.; Tsibouklis, J.; Pouwer, K.L.; Groenendaal, L.; Meijer, E.W. Synthesis, processing and material properties of conjugated polymers. *Polymer* **1996**, *37*, 5017. [CrossRef]
21. Ćirić-Marjanović, G. Recent advances in polyaniline research: Polymerization mechanisms, structural aspects, properties and applications. *Synth. Met.* **2013**, *177*, 1–47. [CrossRef]
22. Stejskal, J.; Sapurina, I.; Trchová, M. Polyaniline nanostructures and the role of aniline oligomers in their formation. *Prog. Polym. Sci.* **2010**, *35*, 1420–1481. [CrossRef]
23. Mazzeu, M.A.C.; Faria, L.K.; Cardoso, A.D.M.; Gama, A.M.; Baldan, M.R.; Gonçalves, E.S. Structural and morphological characteristics of polyaniline synthesized in pilot scale. *J. Aerosp. Technol. Manag.* **2017**, *9*, 39–47. [CrossRef]
24. Du, X.; Xu, Y.; Xiong, L.; Bai, Y.; Zhu, J.; Mao, S. Polyaniline with high crystallinity degree: Synthesis, structure, and electrochemical properties. *J. Appl. Polym. Sci.* **2014**, *131*. [CrossRef]
25. Yin, C.; Gao, L.; Zhou, F.; Duan, G. Facile Synthesis of Polyaniline Nanotubes Using Self-Assembly Method Based on the Hydrogen Bonding: Mechanism and Application in Gas Sensing. *Polymers* **2017**, *9*, 544. [CrossRef]
26. Mottaghitalab, V. Development and Characterisation of Polyaniline-Carbon Nanotube Conducting Composite Fibres. Ph.D. Thesis, University of Wollongong, Wollongong, Australia, 2006.
27. Butoi, B.; Groza, A.; Dinca, P.; Balan, A.; Barna, V. Morphological and Structural Analysis of Polyaniline and Poly (O-anisidine) Layers Generated in a DC Glow Discharge Plasma by Using an Oblique Angle Electrode Deposition Configuration. *Polymers* **2017**, *9*, 732. [CrossRef]
28. Popescu, N.; Despa, V.; Ungureanu, D. Structural analysis of PAN fiber by X-ray diffraction. *J. Sci. Arts* **2010**, *1*, 89–94.
29. Urbańczyk, G.W. *Fizyka włókna*; Wydawnictwo Politechniki Łódzkiej: Łódź, Poland, 2002.
30. Zeghioud, H.; Lamouri, S.; Safidine, Z.; Belbachir, M. Chemical synthesis and characterization of highly soluble conducting polyaniline in mixtures of common solvents. *J. Serb. Chem. Soc.* **2015**, *80*, 917–931. [CrossRef]
31. Bhadra, S.; Khastgir, D. Determination of crystal structure of polyaniline and substituted polyanilines through powder X-ray diffraction analysis. *Polym. Test.* **2008**, *27*, 851–857. [CrossRef]
32. Maron, J.; Winokur, M.J.; Mattes, B.R. Processing-induced changes in the local structure of amorphous polyaniline by radial distribution function analysis of X-ray scattering data. *Macromolecules* **1995**, *28*, 4475–4486. [CrossRef]
33. Alix, A.; Lemoine, V.; Nechtschein, M.; Travers, J.P.; Menerdo, C. Water absorption study in polyaniline. *Synth. Met.* **1989**, *29*, 457–562. [CrossRef]
34. Lubentsov, B.; Timofeeva, O.; Saratovskikh, S.; Krinichnyi, V.; Pelekh, A.; Dmitrenko, V.; Khidekel, M. The study of conducting polymer interaction with gaseous substances. IV. The water content influence on polyaniline crystal structure and conductivity. *Synth. Met.* **1992**, *47*, 187–192. [CrossRef]
35. Lubentsov, B.Z.; Timofeeva, O.N. Conducting polymer interaction with gaseous substances II. PANI-H_2O, PANI-NH_3. *Synth. Met.* **1991**, *45*, 235–240. [CrossRef]
36. Geniès, E.M.; Lapkowski, M.; Tsintavis, C. Preparation, properties and applications of polyaniline. *New J. Chem.* **1988**, *12*, 181.
37. Karbownik, I.; Fiedot, M.; Rac, O.; Suchorska-Woźniak, P.; Rybicki, T.; Teterycz, H. Effect of doping polyacrylonitrile fibers on their structural and mechanical properties. *Polymer* **2015**, *75*, 97–108. [CrossRef]

38. Zeghioud, H.; Lamouri, S.; Mahmoud, Y.; Hadj-Ali, T. Preparation and characterization of a new polyaniline salt with good conductivity and great solubility in dimethyl sulfoxyde. *J. Serb. Chem. Soc.* **2015**, *80*, 1435–1448. [CrossRef]
39. Zhang, L.; Luo, J.; Menkhaus, T.J.; Varadaraju, H.; Sun, Y.; Fong, H.J. Antimicrobial nano-fibrous membranes developed from electrospun polyacrylonitrile nanofibers. *Membr. Sci.* **2011**, *369*, 499–505. [CrossRef]
40. Alarifi, I.M.; Alharbi, A.; Khan, W.S.; Swindle, A.; Asmatulu, R. Thermal, electrical and surface hydrophobic properties of electrospun polyacrylonitrile nanofibers for structural health monitoring. *Materials* **2015**, *8*, 7017–7031. [CrossRef]
41. Saade, W.; Dalmolin, C.; Becker, D.; Neto, A.R.S.; Pachekoski, W.M.; Mattoso, L.H.C.; Marconcini, J.M. Effect of the pre-treatment of banana pseudostem fiber for polyaniline adhesion. In Proceedings of the 2nd Brazilian Conference on Composite Materials—BCCM2, São José dos Campos-SP, Brazil, 15–18 September 2014.
42. Lewin, M.; Pearce, E.M. (Eds.) *Handbook of Fiber Chemistry, Revised and Expanded*; CRC Press: Boca Raton, FL, USA, 1998.
43. Chacko, A.P.; Hardaker, S.S.; Gregory, R.V.; Samuels, R.J. Viscoelastic characterization of concentrated polyaniline solutions: New insights into conductive polymer processing. *Synth. Met.* **1997**, *84*, 41–44. [CrossRef]
44. Gregory, R.V.; Mallya, A.N.; Joseph, A.; Harrell, W.R.; Gregory, R.V. Synthesis and characterization of high molecular weight polyaniline for organic electronic applications. *Polym. Eng. Sci.* **2012**, *52*, 1811.

© 2019 by the authors. Licensee MDPI, Basel, Switzerland. This article is an open access article distributed under the terms and conditions of the Creative Commons Attribution (CC BY) license (http://creativecommons.org/licenses/by/4.0/).

Article

Enhancing the Dyeability of Polyimide Fibers with the Assistance of Swelling Agents

Dongyan Shao, Changhai Xu, Hongbo Wang * and Jinmei Du *

Key Laboratory of Eco-textiles, Ministry of Education, College of Textiles and Clothing, Jiangnan University, Jiangsu 214122, China; dongyan-shao@foxmail.com (D.S.); changhai_xu@jiangnan.edu.cn (C.X.)
* Correspondence: wxwanghb@163.com (H.W.); jinmei_du@jiangnan.edu.cn (J.D.)
  Tel.: +86-130-5730-7010 (H.W.); +86-138-1205-0292 (J.D.)

Received: 27 December 2018; Accepted: 21 January 2019; Published: 22 January 2019

Abstract: Polyimide (PI) fibers have outstanding thermal stability and mechanical properties, but are difficult to dye with disperse and basic dyes. In this work, it was proposed to use N-methylformanilide (MFA), phenoxyisopropanol (PIP), and acetophenone (AP) as swelling agents to enhance the dyeability of PI fibers. The PI fibers treated with swelling agents were characterized by thermal gravimetric analysis, scanning electronic microscopy, tensile testing, and crystalline analysis. It was found that the swelling agents penetrated into the PI fibers in amounts greater than 10% (pertaining to the weight of PI fibers). The swelling agents did not really swell the PI fibers, but broke the interaction forces between the PI macromolecules. With the assistance of swelling agent, the PI fibers could be dyed with disperse and basic dyes in strong color strengths. AP exhibited the best performance for enhancing the dyeability of PI fibers, followed by MFA and PIP. The dyed PI fibers were found to have good colorfastness to washing.

Keywords: Polyimide fiber; thermal stability; swelling agent; dyeability

---

1. Introduction

Aromatic Polyimides (PIs) are a series of heterocyclic polymers synthesized by polycondensation of dianhydrides with diamines, and exhibit excellent thermal stability because of their cyclic and rigid molecular chain backbones [1–3]. So far PIs have been applied as high-performance thermally stable materials for films, membranes, coatings, and fibers [4–9]. As spinning technologies have advanced in enhancing the mechanical properties of PI fibers [10–12], PI fibers have been proposed for use in textiles such as thermal protective clothing for firefighters [13,14]. However, PI fibers are difficult to dye with common dyestuffs due to the fact that they contain few functional groups for dyeing, and also the highly compact arrangement of the PI macromolecular backbones limits their interactions with dye molecules. A facile approach to coloration of PI fibers is dope-dyeing which is carried out by mixing pigments or dyestuffs with PI before spinning. However, the dope-dyeing method has drawbacks of reduction in stability of the spinning solution and spinneret clogging. It has been recently reported that the dyeability of PI fibers for disperse dyes can be improved by pretreatment with alkali [15,16], but PI fibers undergo apparent strength loss due to alkali hydrolysis. The swelling agents are so called because they can expand the amorphous region in the fibers for easier penetration of dye molecules, so that the fiber can be dyed at a lower temperature. Coloration with swelling agents has been well studied and practiced on polyester, aramid, and other synthetic fibers [17–20]. Anilide, phenoxy, and phenone types of compounds are often used as swelling agents because these organic compounds are hydrophobic with strong polarity which provides them with intense interactions with fibers. In this work, it was proposed that PI fibers be dyed with disperse dyes and basic dyes by using N-methylformanilide (MFA), phenoxyisopropanol (PIP), and acetophenone (AP) as swelling agents.

The properties of the dyed PI fibers were evaluated by color measurement, thermal analysis, crystalline analysis, and surface morphology.

2. Materials and Methods

2.1. Materials

PI fibers in double yarns (29.2 tex) were provided by Aoshen New Material Inc. (Jiangsu, China). C.I. Disperse Red 153 (DR 153), C.I. Disperse Blue 60 (DB 60), C.I. Basic Red 46 (BR 46) and C.I. Basic Blue 41 (BB 41) were used for dyeing of PI fibers, and provided by DyStar (Shanghai, China). MFA, PIP, and AP were used as swelling agents, and provided by Sigma-Aldrich (Shanghai, China). Figure 1 shows the chemical structures of PI fibers, dyes and swelling agents. Soaping agent, hydrosulphite ($Na_2S_2O_4$) and sodium hydroxide (NaOH) were used as auxiliaries for soaping and reduction clearing of the dyed PI fibers.

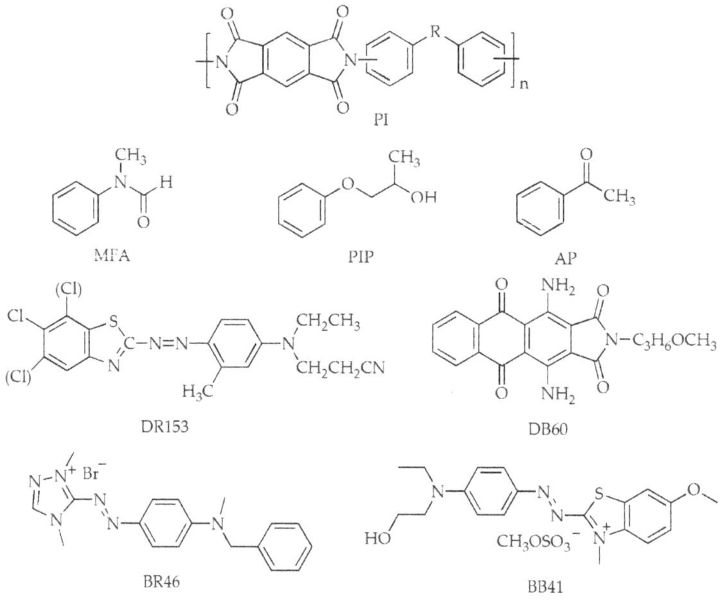

Figure 1. Chemical structures of Polyimide (PI) fibers, dyes and swelling agents.

2.2. Dyeing of PI Fibers

Dyeing experiments were carried out on an Ahiba IR dyeing machine (Datacolor, Lawrenceville, NJ, USA). PI fibers were dyed at a liquor-to-material ratio of 20:1 with 5% of disperse and basic dyes (owf) by adding 50 g/L of swelling agents. The temperature of the dyebath was raised to 130 °C at a rate of 2 °C/min, and held for 60 min. When the dyeing was complete, the temperature was lowered to 85 °C, and the dyed PI fibers were taken out for washing. The PI fibers dyed with basic dyes were washed by soaping with 2 g/L of soaping agent at 85 °C for 20 min, and the PI fibers dyed with disperse dyes were washed by reduction cleaning with 2 g/L of hydrosulphite and 2 g/L of sodium hydroxide at 85 °C for 20 min. All the dyed PI fibers were rinsed thoroughly with fresh water and dried under ambient conditions.

2.3. Testing of PI Fibers

2.3.1. Thermal Analysis

Thermal degradation behavior of PI fibers was tested on a TA Q500 thermal analysis machine (TA Instruments, New Castle, DE, USA) in a nitrogen atmosphere, and the heating rate was 20 °C/min with the temperature ranging from 50 °C to 900 °C.

2.3.2. Surface Morphology

PI fibers were sputter-coated with gold and scanned for surface morphology with a magnification of 3000 through a SU1510 scanning electron microscope (Hitachi, Tokyo, Japan).

2.3.3. Tensile Properties

Tensile strength and breaking elongation of PI fibers were measured in terms of the ISO 2062-2009 on the YG020B Single Yarn Tensile Tester (Futai Machinery Co., Ltd, Changzhou, China) which was set with a drawing speed of 250 mm/min and a test length of 250 mm. All samples were measured 20 times to give an average value.

2.3.4. Crystalline Analysis

The crystallinity of PI fibers before and after treating with swelling agents was analyzed through an AXS D8 wide angle X-ray diffractometer (Bruker, Karlsruher, German) (WAXD) with a scan angel 2θ of 10° to 40° at a scanning speed of 5°/min.

2.3.5. Color Strength

The dyed PI yarns were knitted into single jersey circular fabric (121 g/m^2). The reflectance of the dyed PI fabric was measured on a Datacolor 650 Spectrophotometers (Datacolor, Lawrenceville, NJ, USA) under the CIE Standard Illuminant D65 and the CIE 1964 10° Standard Observer. The color strength(K/S) of the dyed fibers was calculated from the reflectance by the Kubleka–Munk equation as shown in Equation (1),

$$K/S = \frac{(1.0 - R_\lambda)^2}{2R_\lambda} \quad (1)$$

where K is the absorption coefficient, S is the scattering coefficient, and R_λ is the reflectance at a specific wavelength (λ) ranging from 400 to 700 nm.

2.3.6. Colorfastness to Washing of the Dyed PI Fibers

The colorfastness to washing of the dyed PI fibers was tested in terms of ISO 105-C10: 2006 (https://www.iso.org/standard/31775.html, accessed on 10 September 2017). The dyed yarns were stitched to a multifiber adjacent fabric containing wool, acrylic, polyester, polyamide, cotton, and acetate, and washed in a solution containing 5 g/L of a standard detergent (ECE) and 2 g/L of sodium carbonate at 60 °C for 30 min. Colorfastness rating was estimated by using the AATCC grey scale.

3. Results and Discussion

3.1. Effect of Swelling Agents on PI Fibers

It was assumed that the swelling agents play the role of enhancing the dyeability of PI fibers by penetrating into the PI fibers and expanding the pore apertures in the PI fibers. This assumption was confirmed by characterizing the physical and mechanical properties of PI fibers treated with swelling agents.

Figure 2 shows the curves of thermal gravimetric analysis (TGA) and differential thermal gravimetric analysis (DTG) for PI fibers treated with swelling agents. The DTG curve of PI fibers shows only one peak at 595.9 °C that can be ascribed to the pyrolysis of PI fibers. The PI fibers exhibited extremely high stability within the temperature range from 50 °C to 350 °C, and only had a weight loss at 350 °C of less than 0.24%. However, two peaks were observed on the DTG curves of PI fibers treated with swelling agents, respectively, being ascribed to the vaporization of swelling agents from the PI fibers and the pyrolysis of PI fibers.

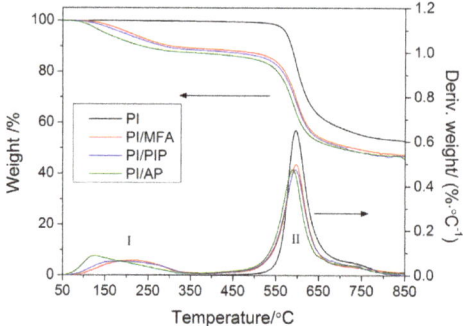

Figure 2. Curves of thermal gravimetric analysis and differential thermal analysis for PI fibers treated with swelling agents (I: the first peak at DTG curve, II: the second peak at DTG curve).

Table 1 summarizes the temperatures of the maximum degradation rate of the PI fibers treated with swelling agents, in which the weight losses at 350 °C of the PI fibers treated with swelling agents are also given for quantifying the amounts of the swelling agents penetrating into the PI fibers. As can be seen in Table 1, the PI fibers treated with swelling agents had a temperature of maximum degradation rate (II) which had changed little. This indicates that the appearance of swelling agents on PI fibers would not significantly impact the thermal properties. The weight loss at 350 °C indicates that swelling agents could penetrate into PI fibers in amounts of 10% or greater pertaining to the amount of PI fibers.

Table 1. Temperatures of maximum degradation rate of the PI fibers treated with swelling agents and the weight loss at 350 °C.

Sample	Temperature of Maximum Degradation Rate (°C)		Weight Loss at 350 °C (%)
	I	II	
Polyimide (PI)	None	595.9	0.24
PI/N-methylformanilide (MFA)	213.1	596.9	10.72
PI/Phenoxyisopropanol (PIP)	209.6	595.2	11.36
PI/Acetophenone (AP)	128.3	588.8	12.55

Figure 3 shows the scanning electron microscope (SEM) images of PI fibers. The treatment with the swelling agents caused no apparent effect on the surface morphology of PI fibers in spite of an amount greater than 10%. This is most likely due to the fact that the swelling agents having smaller molecular sizes than dyes preferentially penetrated into the PI fibers under high temperature conditions, and were embedded into the pore apertures of the PI fibers when cooled down to room temperature. The diameters of PI fibers were slightly increased from 11.7 ± 0.14 μm to 12.06 ± 0.34 μm for treatment with MFA, 12.17 ± 0.10 μm for treatment with PIP, and 12.18 ± 0.25 μm for treatment with AP, respectively. Therefore, the swelling agents did not really swell the PI fibers.

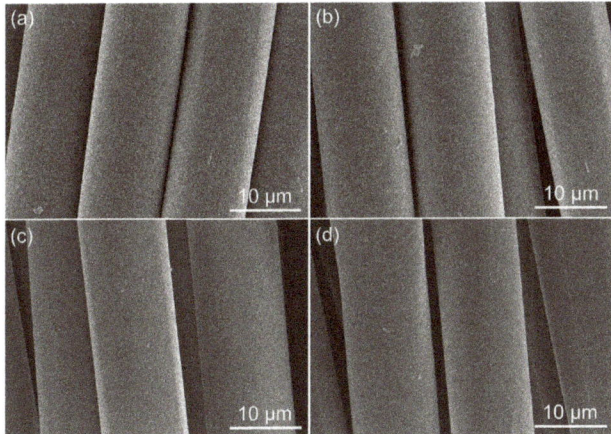

Figure 3. SEM images of PI fibers (**a**) and the PI fibers treated with N-methylformanilide (MFA) (**b**), phenoxyisopropanol (PIP) (**c**), and acetophenone (AP) (**d**).

PI fibers were spun into yarns for measurement of the tensile properties as shown in Table 2. It was found that the breaking strength and elongation of the PI yarns decreased more or less when the PI fibers were treated with swelling agents. The decrease of the breaking strength and elongation of PI yarns could be ascribed to the fact that, as the swelling agents penetrated into PI fibers, the interaction forces between the PI macromolecules were replaced by the interaction forces between the polyimide and swelling agent so that the polyimide chains readily slid at a tensile strength.

Table 2. Tensile properties of PI yarns treated with swelling agents.

Yarn	Breaking Strength (cN/tex)	Breaking Elongation (%)
PI	25.32	9.78
PI/MFA	20.53	8.04
PI/PIP	20.43	7.84
PI/AP	21.16	9.41

The results of WAXD analysis are presented in Figure 4. It can be seen that there is a wide diffraction peak accompanied by several peaks on the PI fiber pattern at 14.7°, 22.3°, and 26.5°, corresponding to the crystals of PI fibers. After treated with swelling agents, these three accompanying peaks become sharper, indicating that the crystalliniy of PI fibers increased slightly. This could be ascribed to the solvent-introduced crystallization between fibers and swelling agents. It has been reproted that the interactions between fibers and swelling agents could result in the movements of fiber molecular chains under high-temperature and high-pressure conditions, and crystallinity of the fibers would occur as the temperature increased [21,22]. From Figure 4, however, it can be seen that there is no evident shift of the peaks on the WAXD patterns when PI fibers were treated with swelling agents. This indicates that the crystal form of PI fibers remained almost the same when the PI fibers were treated with the swellling agents.

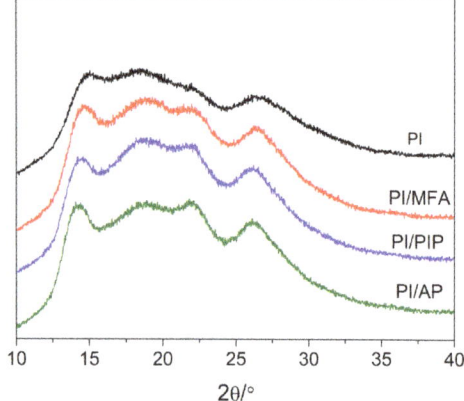

Figure 4. WAXD patterns of PI fibers and PI fibers treated by MFA, PIP and AP.

3.2. Effect of Swelling Agents on the Dyeability of PI Fibers

Considering the inherent brilliant golden yellow shade of the PI fibers, red and blue dyes were used for sensitively examining dyeability. Figure 5 shows the color strength of the dyed PI fibers. As can been seen, the PI fibers without dyeing present a strong color strength in the wavelength from 400 nm to 500 nm, which matches their inherent brilliant golden yellow shade. For successful dyeing, the red dyes (DR 153 and BR 46) and blue dyes (DB 60 and BB 41) would need to provide PI fibers with strong color strengths in the wavelength range from 500 nm to 600 nm and the wavelength range from 600 nm to 700 nm, respectively. However, the dyed PI fibers failed to present such strong color strengths as anticipated. This indicates that the PI fibers were hardly dyed with disperse and basic dyes without assistance under high-temperature dyeing conditions (i.e., 130 °C). The poor dyeability of PI fibers is mainly ascribed to the highly compact arrangement of PI macromolecular backbones and the lack of functional groups for dyeing.

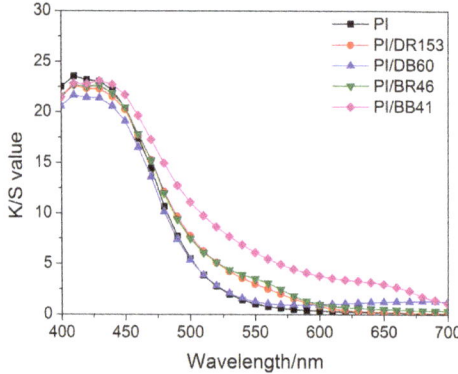

Figure 5. Effect of disperse dyes and basic dyes on the color strengths of PI fibers.

Figure 6 shows the effect of swelling agents on the color strengths of the dyed PI fibers. It can been seen that the color strengths of the dyed PI fibers were enhanced more or less with the addition of swelling agents. It is thought that the swelling agents enhance the dyeability by interacting with fibers as well as dyes. On the one hand the swelling agents preferentially penetrated into the PI fibers from the dyeing bath to expand the pore apertures of PI fibers through which the dye molecules diffused. On the other hand the swelling agents had a strong solubilization effect on the hydrophobic dyes by which the dye molecules could readily diffuse from the dyeing bath into the PI fibers. Therefore, the dyeability of PI fibers depends on the types of swelling agents as well as the dyes to a great extent. Among the three swelling agents, AP has the highest hydrophobicity, followed by MFA and PIP. According to the similarity-intermiscibility principle, AP could be preferentially adsorbed onto the PI fibers and adequately expand the pore apertures of PI fibers. Additionally, the disperse dyes are more hydrophobic than basic dyes, and more readily adsorbed into the PI fibers than basic dyes with the assistance of the swelling agents. Therefore, it is seen in Figure 6 that, with the addition of swelling agents, the dyeability of the PI fibers was improved more effectively for disperse dyes than basic dyes, and AP was the most effective swelling agent for enhancing the dyeablity of PI fibers, followed by MFA and PIP, with the exception that PIP was used in the dyeing of PI fibers with BB 41. In Figure 6d, it is not to be expected that, when using PIP as the swelling agent, the maximal color strength of the PI fibers dyed with BB 41 would shift to a wavelength of 500 nm. It indicates that PIP most likely interacted with BB 41 so as to result in a change in the chromophore of BB 41.

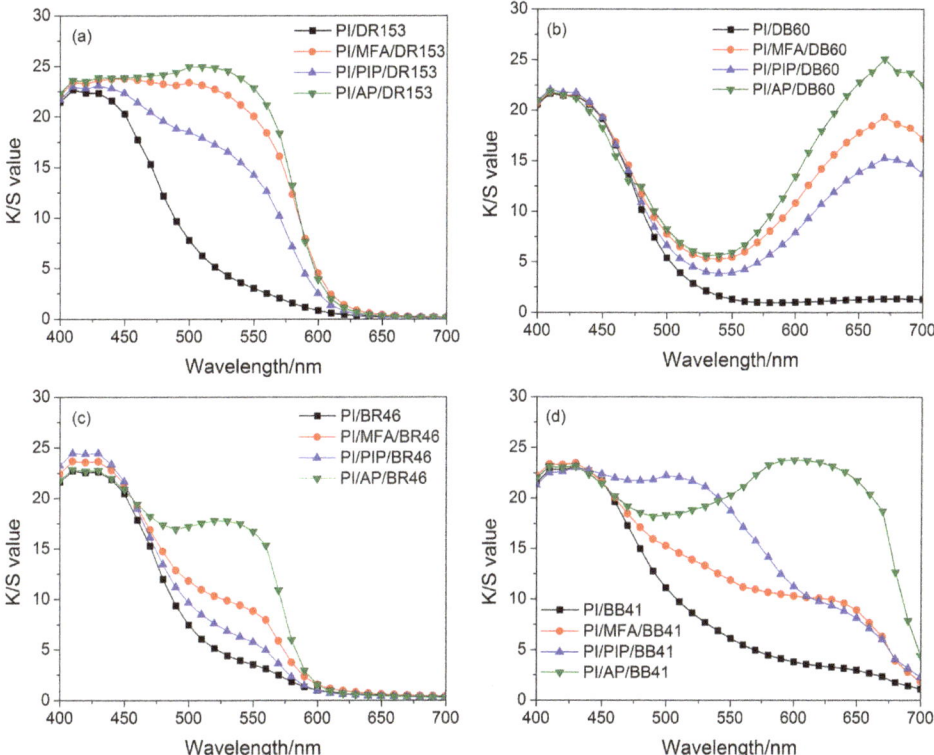

Figure 6. Effect of swelling agents on the color strengths of the PI fibers dyeing with Disperse Red 153 (DR 153) (**a**), Disperse Blue 60 (DB 60) (**b**), Basic Red 46 (BR 46) (**c**), and Basic Blue 41 (BB 41) (**d**).

The PI yarns dyed with the assistance of swelling agents were tested for colorfastness to washing. As shown in Table 3, all the dyed PI yarns exhibited good colorfastness to washing tested by color change. The colorfastness to washing tested by color stain was dependent on dyes to some extent, but did not correlate with swelling agents. This indicates that the PI yarns could be dyed with various color shades with good colorfastness by using swelling agents.

Table 3. Colorfastness to washing of PI fabrics dyed with the assistance of swelling agents.

Dye	Swelling Agents	Color Change	Color Stain					
			Wool	Acrylic	Polyester	Polyamide	Cotton	Acetate
DR 153	MFA	4–5	3	4–5	3–4	3	4	3
	PIP	4–5	3	3–4	2–3	2	3	2–3
	AP	4–5	3–4	4	3	2–3	3–4	3
DB 60	MFA	5	5	5	5	5	5	5
	PIP	5	5	5	5	5	5	5
	AP	5	5	5	5	4	5	5
BR 46	MFA	5	5	5	5	5	3–4	5
	PIP	5	4–5	5	5	4–5	3	4–5
	AP	4–5	4	4–5	4	4–5	3	4–5
BB 41	MFA	5	5	5	5	4	5	4–5
	PIP	4–5	4	5	4	3	3	3
	AP	5	4–5	5	4–5	3–4	3	3–4

4. Conclusions

PI fibers were shown to be difficult to dye with disperse and basic dyes. When treated with swelling agents such as MFA, PIP, and AP, the dyeability of the PI fibers could be significantly enhanced. It was found that swelling agents could penetrate into the PI fibers in amounts greater than 10% (pertaining to the weight of PI fibers). The swelling agent penetrating into the PI fibers did not really swell the PI fibers, but broke the interaction forces between the PI macromolecules. When the swelling agents were used for dyeing of PI fibers, they preferentially penetrated into PI fibers and expanded the pore apertures of PI fibers so that the dye molecules readily diffused into the PI fibers. The swelling agents exhibited different performances in enhancing the dyeablity of PI fibers, and AP was most effective, followed by MFA and PIP. The PI fibers dyed with the assistance of the swelling agents exhibited good colorfastness to washing, tested by color change. The dyed PI fibers' colorfastness to washing tested by color stain was moderate, which was mainly dependent on the dyes but not the swelling agents. Therefore, using a swelling agent would be an applicable approach for enhancing the dyeability of PI fibers.

Author Contributions: Conceptualization, D.S.; validation, J.D. and H.W.; formal analysis, D.S. and C.X.; investigation, D.S.; resources, J.D. and H.W.; writing—original draft preparation, D.S.; writing—review and editing, D.S. and C.X.; funding acquisition, C.X., J.D. and H.W.

Funding: This work was supported by the National First-class Discipline Program of Light Industry Technology and Engineering under Grant LITE2018-21 (China) and the 111 Project under Grant B17021 (China).

Conflicts of Interest: The authors declare no conflict of interest.

References

1. Liaw, D.J.; Wang, K.L.; Huang, Y.C.; Lee, K.R.; Lai, J.Y.; Ha, C.S. Advanced polyimide materials: Syntheses, physical properties and applications. *Prog. Polym. Sci.* **2012**, *37*, 907–974. [CrossRef]
2. Kotek, R. Recent Advances in Polymer Fibers. *Polym. Rev.* **2008**, *48*, 221–229. [CrossRef]
3. Zhang, M.Y.; Niu, H.Q.; Wu, D.Z. Polyimide Fibers with High Strength and High Modulus: Preparation, Structures, Properties, and Applications. *Macromol. Rapid Commun.* **2018**, *39*, 1800141. [CrossRef] [PubMed]
4. Clausi, D.T.; Koros, W.J. Formation of defect-free polyimide hollow fiber membranes for gas separations. *J. Membr. Sci.* **2000**, *167*, 79–89. [CrossRef]
5. Heo, C.; Chang, J.H. Polyimide nanocomposites based on functionalized graphene sheets: Morphologies, thermal properties, and electrical and thermal conductivities. *Solid State Sci.* **2013**, *24*, 6–14. [CrossRef]
6. Karataş, S.; Kayaman-Apohan, N.; Demirer, H.; Güngör, A. Polyimide–silica hybrid coatings: Morphological, mechanical, and thermal investigations. *Polym. Adv. Technol.* **2007**, *18*, 490–496. [CrossRef]

7. Vanherck, K.; Koeckelberghs, G.; Vankelecom, I.F.J. Crosslinking polyimides for membrane applications: A review. *Prog. Polym. Sci.* **2013**, *38*, 874–896. [CrossRef]
8. Dong, J.; Yin, C.Q.; Zhao, X.; Li, Y.Z.; Zhang, Q.H. High strength polyimide fibers with functionalized graphene. *Polymer* **2013**, *54*, 6415–6424. [CrossRef]
9. Qu, C.Z.; Hu, J.S.; Liu, X.; Li, Z. Morphology and Mechanical Properties of Polyimide Films: The Effects of UV Irradiation on Microscale Surface. *Materials* **2017**, *10*, 1329. [CrossRef]
10. Kaneda, T.; Katsura, T.; Nakagawa, K.; Makino, H.; Horio, M. High-strength-high-modulus polyimide fibers II. Spinning and properties of fibers. *J. Appl. Polym. Sci.* **1986**, *32*, 3151–3176. [CrossRef]
11. Xu, Y.; Wang, S.H.; Li, Z.T.; Xu, Q.; Zhang, Q.H. Polyimide fibers prepared by dry-spinning process: Imidization degree and mechanical properties. *J. Mater. Sci.* **2013**, *48*, 7863–7868. [CrossRef]
12. Wang, S.H.; Dong, J.; Li, Z.T.; Xu, Y.; Tan, W.J.; Zhao, X.; Zhang, Q.H. Polyimide fibers prepared by a dry-spinning process: Enhanced mechanical properties of fibers containing biphenyl units. *J. Appl. Polym. Sci.* **2016**, *133*, 43727. [CrossRef]
13. Bourbigot, S.; Flambard, X. Heat resistance and flammability of high performance fibres: A review. *Fire Mater.* **2002**, *26*, 155–168. [CrossRef]
14. Morgan, A.B.; Putthanarat, S. Use of inorganic materials to enhance thermal stability and flammability behavior of a polyimide. *Polym. Degrad. Stab.* **2011**, *96*, 23–32. [CrossRef]
15. Wang, Z.H.; Rao, Z.J.; Zhan, Y.Z.; Hao, T.Q.; Wang, W.; Yu, D. Improving the dyeability of polyimide by pretreatment with alkali. *Color. Technol.* **2016**, *132*, 481–487. [CrossRef]
16. Bai, G.; Liu, Y.C.; Qian, H.F. Surface characteristics and dyeing properties of polyimide fabric treated with sodium hydroxide. *Color. Technol.* **2018**, *135*, 53–59. [CrossRef]
17. Cao, G.Y.; Sheng, D.; Xu, W.L.; Wang, X. Structural and dyeing properties of aramid treated with 2-phenoxyethanol. *Color. Technol.* **2015**, *131*, 384–388. [CrossRef]
18. Islam, M.T.; Aimone, F.; Ferri, A.; Rovero, G. Use of N-methylformanilide as swelling agent for meta-aramid fibers dyeing: Kinetics and equilibrium adsorption of Basic Blue 41. *Dyes Pigm.* **2015**, *113*, 554–561. [CrossRef]
19. Kinashi, K.; Suzuki, T.; Yasunaga, H.; Tsuchida, H.; Sakai, W.; Tsutsumi, N.; Yamane, H. Carrier-assisted dyeing of poly (L-lactic acid) fibers with dispersed photochromic spiropyran dyes. *Dyes Pigm.* **2017**, *145*, 444–450. [CrossRef]
20. Kim, I.S.; Cho, H.M.; Koh, J.; Kim, J.P. Low-temperature carrier dyeing of poly(vinyl chloride) fibers with disperse dyes. *J. Appl. Polym. Sci.* **2003**, *90*, 3896–3904. [CrossRef]
21. Kulshreshtha, A.K.; Khan, A.H.; Madan, G.L. X-ray diffraction study of solvent-induced crystallization in polyester filaments. *Polymer* **1978**, *19*, 819–823. [CrossRef]
22. Chidambaram, D.; Venkatraj, R.; Manisankar, P. Solvent-induced modifications in polyester yarns. II. Structural and thermal behavior. *J. Appl. Polym. Sci.* **2003**, *89*, 1555–1566. [CrossRef]

© 2019 by the authors. Licensee MDPI, Basel, Switzerland. This article is an open access article distributed under the terms and conditions of the Creative Commons Attribution (CC BY) license (http://creativecommons.org/licenses/by/4.0/).

MDPI
St. Alban-Anlage 66
4052 Basel
Switzerland
Tel. +41 61 683 77 34
Fax +41 61 302 89 18
www.mdpi.com

Materials Editorial Office
E-mail: materials@mdpi.com
www.mdpi.com/journal/materials

www.ingramcontent.com/pod-product-compliance
Lightning Source LLC
LaVergne TN
LVHW070440100526
838202LV00014B/1632